Vibration Control Engineering

Vibration Control Engineering

Passive and Feedback Systems

Ernesto Novillo

CRC Press
Taylor & Francis Group
Boca Raton London New York

CRC Press is an imprint of the
Taylor & Francis Group, an **informa** business

First edition published 2022
by CRC Press
6000 Broken Sound Parkway NW, Suite 300, Boca Raton, FL 33487-2742

and by CRC Press
2 Park Square, Milton Park, Abingdon, Oxon, OX14 4RN

CRC Press is an imprint of Taylor & Francis Group, LLC

ISBN: 978-1-032-00699-4 (hbk)
ISBN: 978-1-032-00702-1 (pbk)
ISBN: 978-1-003-17523-0 (ebk)

DOI: 10.1201/9781003175230

Typeset in Times
by SPi Technologies India Pvt Ltd (Straive)

To Marité, my dear wife, who supported and encouraged me to write this book.

Contents

Part II Turbo Machines And Ship Vibrations

Part III Vibration Control Systems

Preface

This book is about vibration engineering aimed at undergraduates, graduates and field engineers involved in installing, commissioning, maintenance, and operation as applied to turbomachinery, ships, high voltage overhead lines and feedback control systems to control vibration. It is based on vibration and automatic control theories and substantial experience collected in the field since the first steam machines suffered vibration problems in essential mechanical parts. In this text, problems and solutions are discussed using theoretical principles for discrete and continuous systems.

This is an accessible text to acquire a practical insight into the usual problems in machinery vibration and control. Therefore, the book has a theoretical and practical nature, with easy-to-use formulas, figures, and charts. It thoroughly discusses the mathematical basis of vibration problems to understand the solutions to be implemented. There are extensive practical examples, using calculation procedures explained in the text, that the undergraduate or engineer may use as a guide to creating other models.

Many mathematical models in this text can be readily implemented using standard commercial spreadsheets, which can be used as a first approach to the problem at hand. It is a generally recommended procedure to create an initial single or multi-degree of freedom (SDOF or MDOF) model as the first approach to a vibration problem. This simple model provides a fast understanding of the problem and guidelines for the potential solution. If this model does not correctly explain the observed vibration; then, the engineer should resort to more sophisticated models, like finite elements based on specialized software, to design a more accurate solution. In general, spreadsheets are useful for creating a simple model and assessing it with the "what if" analysis tool. Another advantage of this procedure is that spreadsheets are always available, and the same cannot be said about more sophisticated software. Nonetheless, this book uses Maple and other specialized mathematical tools, like Wolfram Alpha.

Metal parts of rotating and reciprocating machinery may suffer from vibration fatigue. This phenomenon has special attention in this book because it inevitably harms any metal piece that vibrates as time evolves. Therefore, in mechanical installations, vibration is a problem that requires immediate and professional attention. Vibration and fatigue are currently the most common causes of turbomachine failures. Given enough time, vibration produces internal micro-cracks, which finally causes the collapse of the affected piece. These failures are costly because most machines fulfill essential functions in industries, power plants, ships' propulsion systems, oil and gas fields, gas pipelines and other production installations. Therefore, machine stoppages inevitably produce significant economic damage. For these reasons, the book presents examples using a multidisciplinary procedure by including Vibration Engineering concepts and Mechanics of Materials and Feedback Control Systems.

As an engineer with field experience, I have authored this book focusing on using vibration theory to understand and solve practical vibration problems that may affect a machine's mechanical integrity. Why is the integrity of mechanical parts affected by vibration? Because vibration produces damage such as bearing deterioration, foundation settlement, chipping of gear teeth, blade erosion, loose bolts and fasteners and many other serious issues that may result in a catastrophic collapse. Fortunately, it is possible to anticipate such failures by methodically monitoring vibration amplitude and frequency to detect changes and create trends. Chapter 10 describes vibration control techniques and introduces the organization of monitoring procedures and the standards used to diagnose and prevent vibration.

The book covers the following main subjects, using the coherent International System of Units.

- Part I: Vibration Theory of SDOF, MDOF and Continuous Dynamic Systems. This part describes the theoretical basis of discrete and continuous systems applied to linear and torsional vibration. Concepts, formulas, and figures of this chapter are intensively used in the rest of the book. Multi-degree of freedom and continuous systems vibration are discussed based on matrixial formulas and the Euler-Bernoulli and wave equations.
- Part II: Turbomachines and Ship Vibrations. In this part, the turbomachines' critical velocity, rotor balancing, lateral vibration, vibratory forces, and ships' vibration are modelled from a theoretical and practical perspective. Readers interested in rotordynamics may consider this part as an introduction to this field of turbomachine technology. The last chapter of this part is dedicated to ship rolling and pitching in navigation and its hull and propulsion machinery vibration with a special explanation about the thrust-bearing longitudinal vibration and its rigidity calculation. The most successful anti-roll systems are described using formulas of the vibration theory.
- Part III: Vibration Control Systems. It describes the machine foundation isolation and the absorbers' design frequently used to overcome machines' vibration problems. An example of overhead line vibration due to Karman vortices is discussed, and the Stockbridge vibration absorber is explained. The last chapter presents the application of dynamic systems theory to feedback control of vibration using the Laplace transform to determine the system transfer functions and PID controllers' optimum settings. Examples of a structure's vibration and a ship's roll controlled with a feedback system are explained.

There exists an extensive bibliography about vibration and mitigating techniques. Therefore, it is almost impossible to mention all authors, books, papers, and manufacturer's information that contribute to vibration engineering and this book. The footnotes in the text mention many of them, including some specialized Internet sites that provide valuable information.

I am grateful to Christine Soostmeyer Novillo, who patiently read and efficiently edited the manuscript. I wish to express my special thanks to engineers Rodrigo Azcueta and Ramiro Barbosa, Dr. Roberto Pascual and navy commanders Fernando Azcueta and Hugo Alvarez, who contributed with their good ideas and excellent professional knowledge in various parts of the book. Finally, I would like to pay homage to my professors at the Military Naval Academy and the National University of Córdoba in Argentina, and many other technicians and engineers with whom I carried out projects of installation, maintenance, and commissioning of power generation, high voltage overhead lines, and naval propulsion in several countries of three continents.

Ernesto Novillo

Acknowledgements

I wish to express my gratitude to the following professionals whose expert and wise comments have contributed greatly to making this book a reality.

Ramiro Barbosa, Electromechanical Engineer
Roberto Pascual, PhD
Fernando Azcueta, Navy Warship Commander
Rodrigo Azcueta Repetto, Naval Doctor Engineer
Hugo Fernando Alvarez, Navy Warship Commander

About the Author

Ernesto Novillo has a strong academic background and field experience in power plants, cement industry, oil fields and naval ship propulsion, where vibration problems are significant. A former Naval officer, he earned two university engineering degrees: Electrical-Electronics (Summa Cum Laude) and Mechanical. Working for General Electric, he delivered large energy engineering projects to many countries in Asia, Europe, and America throughout his long professional career. He was a university professor of Automatic Control Systems and Oil Engineering. His experience encompasses living in several countries, where he assumed important management positions. He has published books about relativity science and turbine engineering. In this book, Ernesto combines vibration theory with his valuable experience in land and marine machinery and automatic control systems.

Abbreviations

2-DOF	2 Degrees of Freedom
ABS	American Bureau of Shipping
BOP	Best Operating Point
CAD	Computer Aided Design
CFD	Computational Fluid Dynamics
DVA	Dynamic Vibration Absorber
FDTR	Frequency Difference to Resonance
FEM	Finite Element Method
HF	High Frequency (disturbance)
HRD	Hull Resonance Diagram
LF	Low Frequency
MDOF	Multi-degree of Freedom
ODE	Ordinary Differential Equation
P	Proportional (controller)
PD	Proportional plus Derivative (controller)
PDD	Proportional Derivative second Derivative (controller)
PI	Proportional and Integral (controller)
PI-D	PI with Restricted Derivative (controller)
PID	Proportional Integral Derivative (controller)
PLC	Programmed Logic Controller
PRU	Power Related Unbalance
PZ	piezoelectric
RLC	Resistor, Inductor, Capacitor Circuit
ROI	Return on Investment
RPM	Revolutions Per Minute
SCF	Stress Concentration Factor
SDOF	Single Degree of Freedom
SF	Safety Factor
SYS	Shear Yield Strength
TMD	Tuned Mass Damper
TMM	Transfer Matrices Method
TVN	Tuned Vibration Neutralizer
UTS	Ultimate Tensile Strength
YS	Yield Strength

Part I

Vibration Theory of SDOF, MDOF and Continuous Dynamic Systems

1 Dynamics of Linear SDOF Systems

1.1 INTRODUCTION TO MACHINE'S VIBRATION

Mechanical vibration is a quick and small alternating motion experienced by a machine or mechanism. Technically speaking, it is a harmonic displacement of minimal amplitude produced by unbalanced rotating parts or transmitted by vibrating nearby machines. Under some circumstances, wave amplitudes rise to such an elevated level that dynamic forces may harm the rotating machine. This phenomenon is known as resonance. The ability to predict and mitigate or prevent vibration is of utmost importance for the machine's life span and surrounding constructions.

Rotating machines are usually impacted by vibration because of the imbalance produced by shafts, wheels, and other rotating parts revolving at high speed. A centrifugal force deflects the axle as the shaft, wheels and other revolving parts are never perfectly straight and balanced. The axle reacts with a rigid force in opposition to the centrifugal force. It is trying to recover its initial straightness, which of course, is impossible in practice because the shaft is not entirely rigid. Thus, the axle is bent by its weight or mechanically attached unbalanced loads. Although rigid and centrifugal forces are in equilibrium, vibration and axle bending are inevitable aftermaths.

Turbomachine shafts and wheels behave like mass-spring sets known as the mechanical system, where the shaft is equivalent to a spring, and wheels are equivalent to rotating masses. Their combination has a characteristic frequency at which the system oscillates even if no external force is applied. This frequency is known as natural frequency. There is more than one natural frequency in machines as they are complex configurations of mass and springs. A harmonic disturbance with a frequency close to any natural frequency produces an abnormal amplification of the vibration amplitude.

The content of this chapter is to prepare for the following problems that usually affect rotating machines' performance or are harmful to their mechanical integrity, such as:

- Vibration at a critical velocity
- Lateral vibration
- Torsional vibration
- Axial vibration
- Vibration transmission to foundations and other installations

DOI: 10.1201/9781003175230-2

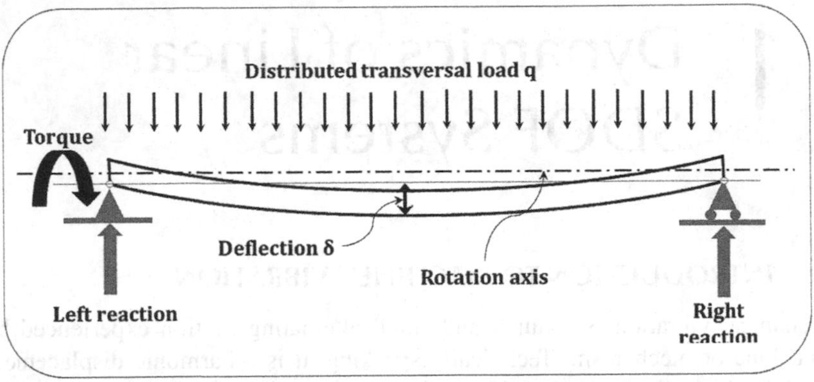

FIGURE 1.1 Deflected shaft predicts lateral vibration.

Lateral or transverse vibration is due to shaft bending produced by transversal loads (see Figure 1.1). It is also known as bending or whirling vibration. This type of vibration is less harmful than torsional vibration but, in any case, requires strict monitoring due to associated risks of fatigue and eventual breakage. Another common cause of lateral vibration is when the shaft's center of gravity does not lie on the shaft's geometric axis of symmetry. This issue produces a centrifugal force that creates a cyclic shaking perpendicular to the axle, like a shaft bending.

Vibration puts many parts of rotating machines (i.e., bearings, blades and other moving parts) in danger of catastrophic damage. Therefore, it is essential to identify their origin and mitigate vibrations as soon as possible during the machine operation. Current state-of-the-art instrumentation is very sophisticated; it facilitates vibration measurements to help diagnose problems and guide design solutions. These measurements help to understand where the problems originate and to make decisions to prevent future damages.

1.2 THE BASICS OF VIBRATING SYSTEMS

- A mechanical system is a machine or mechanism conceived for a specific technical application. The components are characterized by the type of reaction force they produce when triggered by an external stimulus. These reaction forces are inertia, rigidity, and friction.
- A system's order is physically the quantity of energy storing elements, like mass or springs, being the system's reaction to external excitations, such as a force or a torque.
- The system's degrees of freedom are the minimum number of coordinates needed to determine a body's position at any point in time. The degrees of freedom are equal to the masses or the quantity of wheels times the number of possible motions on one coordinate of each mass or wheel.
- The output of a mechanical system, as shown in Figure 1.8, is the mass motion produced by an excitation input like a force. The mass does not

exactly follow the input force shape because of inertial, rigid and friction reaction forces. The model to study the mass motion is known as the system's equation of motion. It is based on the equilibrium between the applied external force and the inertial, rigid, and friction reactions opposed by the mass, springs, and dampers that form the system.

- Systems without friction are called conservative or frictionless because they do not dissipate heat to the environment. Systems with friction are called non-conservative because they dissipate heat to the environment.
- It is said that the system has initial conditions if the spring is stretched or compressed or the mass is in motion at the instant the system is excited by an external force.
- A frictionless system with initial conditions oscillates when activated by the spring stretching with a frequency called natural frequency. Oscillations are due to the energy interchange between the mass and the spring. This oscillation is undamped.
- In mechanical vibration engineering, a dynamic system is formed by vibrating parts; therefore, its forces and displacements are a function of time and frequency.

1.2.1 Time Response

A system's time response is the displacement of some significant part of a mechanical system as a function of time. This response depends on the system configuration and the type of input received. In this section, typical systems responses to an excitation input are described. See a graphic summary of the most common input and output response in Figure 1.2.

There are two types of system motion: linear or angular. Vibration theory applies to any of these systems with similar mathematical models to describe their time response. Figures, equations, and formulas of linear mechanical systems and

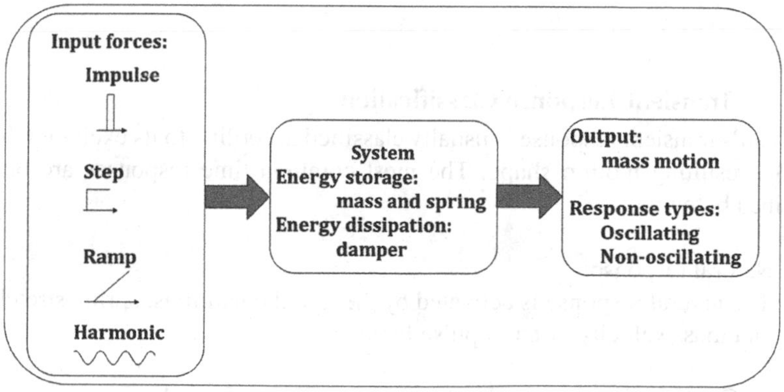

FIGURE 1.2 System inputs and time responses.

torsional systems are analogous. Only some mechanical equivalent properties need to be replaced to convert linear vibration equations to torsional equations. For example, mass is replaced by a polar moment of inertia, linear rigidity by torsional rigidity, and linear displacement by angular displacement.

The most used input forces in vibration study are:

- Impulse:
  a force of negligible duration, for example, a hammer blow. The impulse force only generates a transient response. This transient is known as a natural response because no external force remains acting after the impulse instant.
- Step:
  a force rising suddenly and remaining acting on the system with a constant value; for example, an abrupt sustained load increase in a power plant when the electrical demand produces a constant step torque in the generator driving machine. The step force generates a natural response superimposed to the forced response. In machines, the first force disappears due to friction forces after a few cycles.
- Ramp:
  an input force that linearly increases. Mainly used to assess servomechanism performance.
- Harmonic:
  an external disturbance cyclically acting on the system. Shaft eccentricity, flexural deflection, or rotor imbalance of a rotating machine generate a cyclical centrifugal force transmitted to the bearings, foundations, and other machine parts. The frequency of this force is equal to the machine's angular velocity. However, rotor gyration is not the only vibration source because close rotating machines can induce forced vibration at a frequency other than the shaft angular velocity. The forced response initially shows a natural regime and a transient harmonic regime. After these regimes are damped, a harmonic force of constant amplitude and frequency remains.

1.2.1.1 Transient Response Classification

A system's transient response is usually classified according to its excitation input and the resulting motion shape. The most common time responses are briefly described below:

A. Natural response.
   The natural response is activated by the initial conditions: spring stretching and mass velocity or an impulse force.

   1. Conservative system. No friction exists; therefore, the system response is oscillatory and undamped.

2. Non-conservative system. Friction exists; then, the system response is:
   a. Oscillatory
   b. Non-oscillatory

B. Forced response.
External forces act on the system, following the same pattern as the excitation, regardless of the existing friction forces. It means that external forces overcome friction forces.

C. Transient response.
The natural and forced responses form the transient response. Both responses act superimposed. They are:
   a. Oscillatory
   b. Non-oscillatory

The non-oscillatory regime rarely occurs in rotating machines. In any case, as this type of motion is not a vibration, fatigue and subsequent aftermaths do not happen. As this is not a vibration, it is not discussed in this text.

1.2.2 Frequency Response

The frequency response is a function that describes a system's behavior under different frequencies of the input excitation. In practice, the frequency response may be obtained by introducing a harmonic signal at the system input, as shown in Figure 1.3.

Procedures to obtain the frequency response are based on mathematical models or empirical tests. The frequency response measures two system's physical properties, namely the output amplitude and the phase at distinct input frequencies. A graphic representation of amplitude and phase curves help to interpret the system behavior. The output amplitude may be higher, equal, or lower than the input amplitude, and the system output is a harmonic wave not in phase with the input.

The frequency response is a powerful tool to predict or design the system performance according to technical specifications. In mechanical machines, the amplitude curve always presents a peak, as shown in Figure 1.3, because their design avoids friction parts as much as possible.

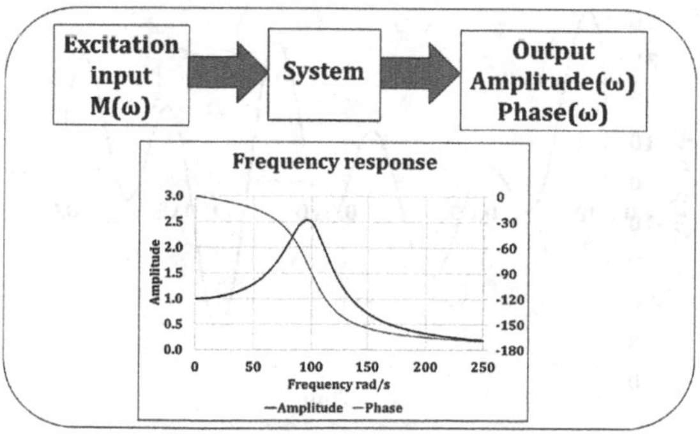

FIGURE 1.3 Frequency response test.

1.2.3 Vibration Graphical Representation

There are three types of vibration representation, shortly described in this section: a) vibration amplitude against time, b) amplitude spectrum, and c) frequency response. These three help understand the complexity of vibration and interpret formulas used to assess an equipment vibration. Field engineers responsible for vibration problems must consult the extensive bibliography about vibration theory and what modern instrumentation can do. Modern vibrometers readily produce displays showing the mechanical systems' time reactions when they are vibrating. These plots are easy to interpret and help reach primary conclusions about machine vibration. An example of vibration amplitudes versus time shown by modern instruments is shown in Figures 1.4 and 1.5.

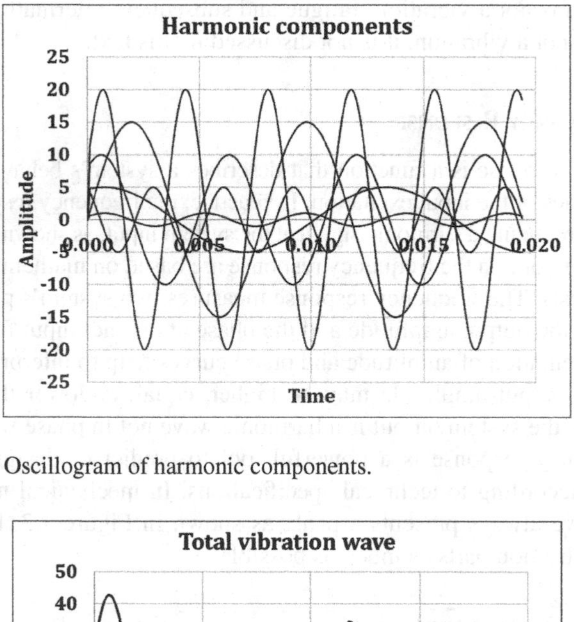

FIGURE 1.4 Oscillogram of harmonic components.

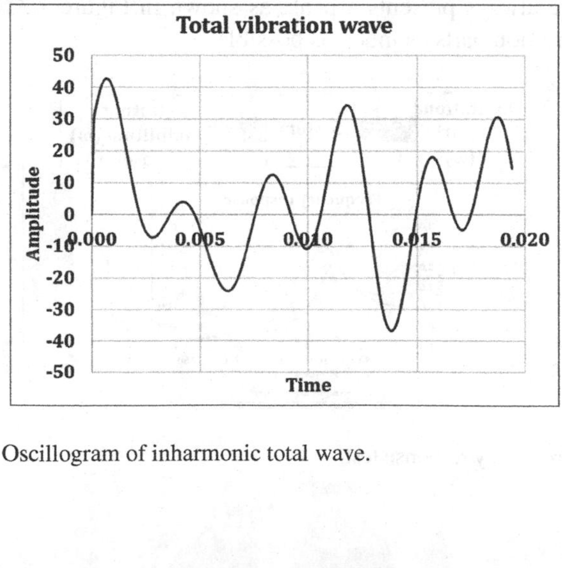

FIGURE 1.5 Oscillogram of inharmonic total wave.

Due to their complex mechanism, machines have several vibration sources with different superimposed amplitudes and frequencies. The whole wave, formed by the sum of all harmonic components, has no perfect harmonic shape because of several vibration sources with different amplitude and frequency, as shown in Figure 1.5. Fortunately, modern instrumentation allows visualizing a noisy wave by each wave component based on the Fourier analysis. This analysis is done by hand or using specialized software. With current technology, vibration instrumentation can complete Fourier analysis using software incorporated into the instruments. It is also possible to check if critical parts, like machine blades, have cracks produced by fatigue using a laser sensor[2].

The frequency spectrum is a figure representing the amplitude of each component wave versus its frequency. For example, the frequency spectrum of the oscillogram represented in Figure 1.4 is displayed in Figure 1.6. In practice, the frequency spectrum is a powerful tool to identify the harmonic disturbance sources because there are standard patterns of frequency spectrums that identify the vibration source. This technique helps to diagnose the problem and conceive its mitigation. This task requires a technician trained in vibration assessment and solutions implementation to prevent or mitigate their impact.

1.2.3.1 Resonance

Resonance is produced when a disturbance frequency is close to the structure or the machine's natural frequencies. As machines have low friction during rotation, their resonant frequency is of the same order as the natural frequency. At the ideal case of zero friction, both resonance and natural frequencies are the same.

In resonance, the wave amplitude suddenly rises to exceedingly high values. In an ideal frictionless machine, amplitudes reach infinite. Of course, an infinite amplitude is a practical impossibility, but in any case, resonance creates dangerous situations to the machine's mechanical integrity.

At startup, rotating machines pass over a wide range of velocities before meeting the rated speed. If the system has a resonance at a velocity slower than its rated

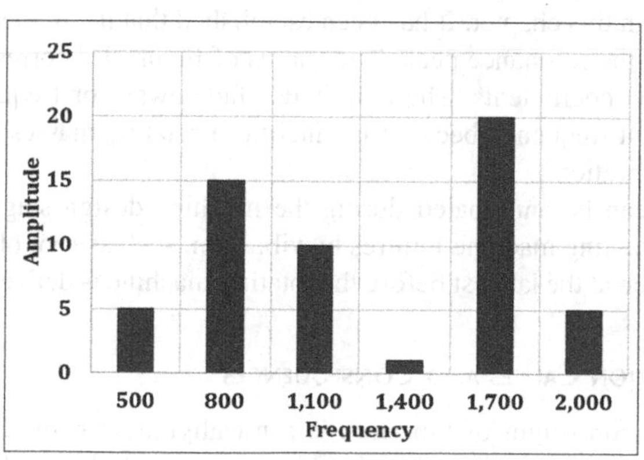

FIGURE 1.6 Frequency spectrum of Figures 1.4 and 1.5.

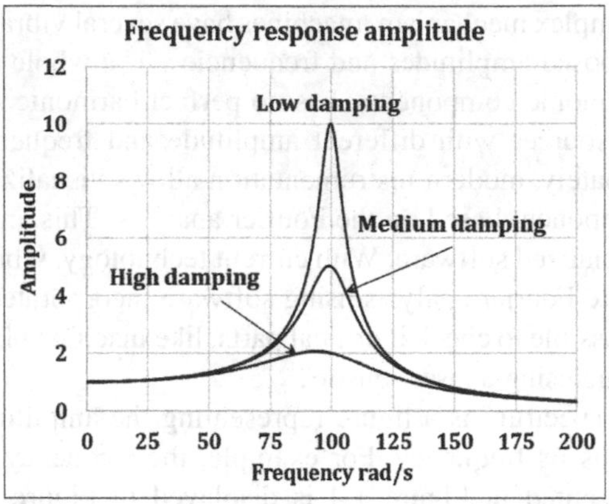

FIGURE 1.7 Frequency response for different damping ratios.

speed, the machine undergoes an abnormal vibration when passing through that resonant frequency.

Resonance is easy to interpret in a plot of amplitudes versus the input frequency. The typical curves of this type of graph are in Figure 1.7. An abrupt amplitude rise indicates the existence of resonance. However, as shown in this figure, friction inside the rotating machine mechanism damps the resonance peak.

1.2.4 Friction Damping

Resonance peaks are damped because of heat dissipation in bearings surfaces, labyrinthic packages, and other rotating machine parts. This heat dissipation is responsible for dampening the waves' amplitude, preventing them from reaching an infinite value. Nonetheless, peak duration is short. For example, in shaft torsional oscillations, the transient regime might be lower than 0.1 seconds.

In vibration theory, friction intensity is characterized by the damping ratio ζ, as discussed later in this chapter. It has been established that the higher the damping ratio, the lower the resonance peak. The curves of Figure 1.7 correspond to three different friction coefficients. The amplitudes fade away for frequencies higher than the resonant frequency because the inertia of rotating masses cannot follow a high-speed vibration.

Resonance can be anticipated during the machine design stage. Experience indicates that rotating machine failures by vibration, such as root blade breakage, may happen even at the lab test before the rotating machine is delivered to the site.

1.2.5 Vibration Causes and Consequences

The most common origin of vibration is a misalignment between the rotating machine shaft and the driven load shaft. Other causes include unbalanced rotating parts, electromagnetic issues in the generator due to oscillating demand and short circuits, loose rotating machine housing, bearings in poor condition, deficient lubricating oil, cracked or bent rotors, excessive clearances, and many others.

Any machine can resonate at different angular velocities due to static and dynamic imbalance. As discussed previously, bending is a common cause of vibration produced by flexural moments on the rotor and shaft assembly[3] produced by its weight. When the shaft is bent, centrifugal forces appear that increase the shaft deflection and steel stress. If the rotor-shaft resonates, the high vibration amplitudes may break the shaft or other parts. That is why long heavy axles need to be rotated when stored to prevent creep and consequent bending.

A common cause of shaft bending in machines is the uneven cooling of wheels and stationary parts after shutdown. The machine shaft might bend during out of service hours because of thermal stress and the shaft and wheel weights. This deflection is risky because it might be permanent and produce a contact between the rotor and stationary parts. This rubbing contact creates an uneven rotor heating and increases the shaft bending and vibration during the operation.

Rotor and shaft uneven cooling may produce deformation and rubbing between the tip blades and the rotating machine stator and cause significant damages. This rubbing raises the temperature and provokes further shaft bending. If the rotating machine continues running under this condition, the steel stress could reach limits higher than its yield strength and produce severe damages.

In turbomachines, wheel disks can also resonate due to excitations imposed by the gas or steam flow through nozzles at different angular locations. Another vibration source is a variable electric load in turbogenerator sets, which may produce alternating resisting torques and induce vibrations in the shaft and other mechanical parts. In any case, the result is the material's fatigue. Famous accidents have happened due to fatigue fracture of blades, for example, the nuclear power plant of Narora in Northern India and the RMS Queen Elizabeth 2. Even vibrations of small amplitude may cause cracks by the steel fatigue, which grow in time and finally produce a catastrophic rupture of the affected piece. Even significant structures are subject to vibration induced by agents like the wind. For example, the Tacoma Narrows bridge fell due to torsional resonance.

1.3 LINEAR MECHANICAL SYSTEM DESCRIPTION

This chapter's vibration models are called discrete or concentrated because they assume that each physical property is concentrated in only one system component (mass, spring, or damper). Therefore, each part has only one reaction type to an applied force. For instance, a concentrated mass only has an inertial response. Therefore, no friction or rigid properties exist in the mass. Properties discretization is an ideal assumption because mass, friction and rigid properties are not concentrated in only one machine's point but are uniformly distributed throughout the machine parts. Systems with uniform distribution are called continuous.

Figure 1.8 shows one of the most known lumped vibrating systems formed by one mass, one spring and one damper. Formulas and conclusions of this model

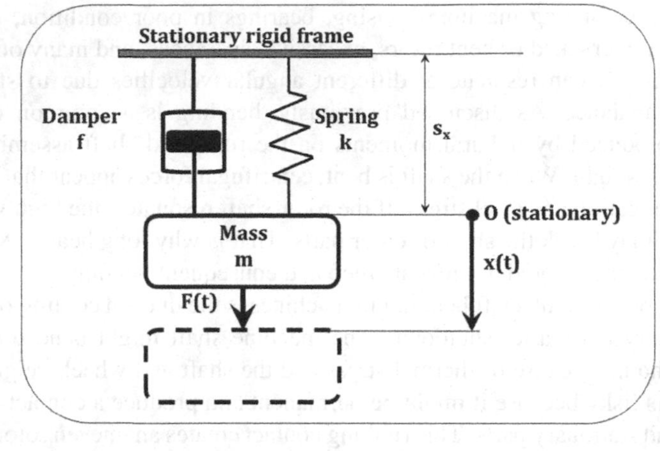

FIGURE 1.8 Schematic of a linear mass-spring-damper system.

are extendable to more complex systems. This example has two energy-storing elements (mass and spring); thus, it is a second-order system.

The x-axis indicates the mass position. The x(t) function is a harmonic function or a mix of harmonic functions. The x-axis center O is in a fixed position, just at the node where mass, damper and spring are linked. In a machine at a standstill, this node is at a constant distance s_x from the frame. Distance s_x is the deflection produced by the mass weight, done against the rigid spring reaction. Its value is equal to the mass weight over the spring stiffness coefficient ($s_x = W_{mass}/k$).

As only one number is needed to determine the mass position (abscissa x), it is said that the system has a single degree of freedom (SDOF). This system is essential because it is used as the most common "basic cell" to explain more complex systems vibration. The mass and spring store potential energy, but they do not do it simultaneously because it is alternately stored in one element or the other. Therefore, the stored energy is transferred from the mass to the spring and vice versa during the mass movement. If the mass is forced to be at a distance higher than s_x and then released, the system starts to oscillate freely after that instant; this is the system's natural response.

1.4 EQUATION OF MOTION OF DYNAMIC SYSTEMS

Newton's second law of motion explains the equilibrium of acting and resisting forces on the system shown in Figure 1.8. The expression that represents this equilibrium is known as the equation of motion of dynamic systems. For rotating systems, the equation of motion arises from the equilibrium of acting and resisting torques. For a system with mass, friction and rigid reactions, the equations of motion for linear and rotating systems formed by several second-order systems are written below.

Formula 1.1 Equations of motion of linear and rotating systems

Linear systems Rotating systems

$$\sum_{i=1}^{n} F_i(t) = \sum_{i=1}^{n} m_i \cdot \ddot{x}_i(t) + f_i \cdot \dot{x}_i(t) + k_i \cdot x_i(t) \sum_{i=1}^{n} T_i(t)$$

$$= \sum_{i=1}^{n} J_{pi} \cdot \ddot{\vartheta}_i(t) + f_{ti} \cdot \dot{\vartheta}_i(t) + k_{ti} \cdot \vartheta_i(t)$$

F_i and T_i are the applied forces and moments to the system.
x and ϑ are the linear and angular displacements.
m and J_p are the mass and mass moment of inertia.
k and k_t are the linear and torsional rigidity coefficients.
f and f_t are the linear and torsional friction coefficients.
$\ddot{x}(t)$ is the mass acceleration, then, the product $m \cdot \ddot{x}(t)$ is the inertial reaction force.
$\dot{x}(t)$ is the mass velocity; then, $f \cdot \dot{x}(t)$ is the friction reaction force. This force is directly proportional to the velocity, which corresponds to a viscous force usually produced by a dashpot with air or oil.
x(t) is the mass position; then, $k \cdot x(t)$ is the rigidity reaction force. The mass displacements are referred to as the x-axis. The inverse of k is the flexibility influence coefficient or elasticity coefficient[4].

The analogy between linear and rotating systems defines the rotating reaction forces similar to linear systems. In a linear system, the solution to the equation of motion is the oscillating mass position x as a function of time. In rotating systems, it is the shaft torsional angle ϑ versus time. In linear systems, the equation of motion is an ordinary differential equation (ODE) of nth order that stands for the equilibrium of all applied and resisting forces. Correspondingly, in rotating systems, the equation of torsional motion is also an ODE of nth order that stands for the equilibrium of applied and resisting moments.

For a mechanical linear second-order system, like the configuration shown in Figure 1.8, the equation of motion is:

Formula 1.2 Equation of motion of a mechanical second-order system

$$F_a(t) = m \cdot \ddot{x}(t) + f \cdot \dot{x}(t) + k_w \cdot x(t)$$

This equation of motion represents a second-order system composed of one mass, one damper and one spring, excited by the applied force $F_a(t)$. However, a machine is a more complex mechanism than a second-order system, but in any case, this system is a useful tool for vibration modelling. This text adopts viscous friction because it has a more straightforward mathematical expression than the other friction types. This simplicity is due to the direct proportionality between the damping force and the piston speed inside an oil or air dashpot. Therefore, the damper only acts when the mass is moving. Unlike a spring, at a standstill, it does not produce any force. Many other friction types of complicated nature exist in mechanical systems, but they are beyond this book's scope. The solution to Formula 1.2 is formed by the sum of the homogeneous equation and a particular solution.

The homogeneous formula is:

Formula 1.3 Homogeneous equation of a mechanical second-order system

$$0 = m \cdot \ddot{x}(t) + f \cdot \dot{x}(t) + k_w \cdot x(t)$$

The classical solution to this equation is $x_a \cdot e^{-j \cdot \omega t}$. This formula is the harmonic output that describes the system vibration when it has no input signal. This solution's vibration frequency is ω, and the amplitude x_a is obtained by replacing this solution in Formula 1.2. As no force is applied to the system, the solution to the homogeneous equation is the system's natural response.

This particular solution depends on the type of input force. For example, if it is a harmonic force, the system input is $F(t) = F \cdot \sin \omega \cdot t$, and the solution may be assumed as $x(t) = x_a \cdot \sin \omega \cdot t$. This formula is introduced in Formula 1.2 to calculate the amplitude x_a.

The solutions to the equation of motion must consider two initial conditions: mass position and mass velocity that are arbitrarily set to study the system's transient performance. For example, $x_i \neq 0$ means that the spring is stretched or compressed at the instant $t = 0$; therefore, the mass starts moving upward or downward.

Formula 1.4 Non-zero initial conditions

$$xi \neq 0 \text{ and } \dot{x}_i \neq 0$$

It is assumed that initial conditions are known for they are arbitrarily imposed. If any of these conditions is not zero and there is no applied force, the system's reaction is its natural response.

The equation of motion is a vector expression because it represents forces or moments in equilibrium. The solution to the equation of motion 1.2 is the sum of two components: the natural response (homogeneous equation) and the forced response (particular solution). The natural response to the equation of motion is the solution $F(t) = 0$. The forced response equation is the solution $F(t) \neq 0$. The forced response always replicates the mathematical form of the input force F(t). For example, if F(t) is a step force, the transient response is a constant displacement after damping the natural response. If F(t) is a harmonic force, the mass displacement is also harmonic. In real life, the natural regime is a damped harmonic curve that disappears after a definite period. After the natural regime ends, only the forced regime remains. This regime disappears when external forces are removed.

In summary:

- The solution to the equation of motion reveals the mass motion amplitude and frequency as a function of time or frequency. The first solution is known as the time response, and the second solution is the frequency response.
- The system's natural response is due to initial conditions which differ from zero.
- The forced response replicates the input force shape and only vanishes when the input force disappears.
- The transient response is the sum of the natural reaction plus the forced reaction.

1.4.1 Vector Interpretation of the Equation of Motion

Vectors that form the equation of motion are gyrating at an angular velocity equal to the system vibration frequency. The law of physics for harmonic motion states that a gyrating vector at velocity ω may represent harmonic oscillations, whose modulus is the oscillation amplitude. In the equation of motion 1.2, the oscillation is described by $\overline{x}(t) = \overline{x}_a \cdot \sin\omega{\cdot}t$. The vector of modulus $\overline{x}_a$ is the gyrating vector that represents x(t). The velocity of the friction term is expressed as $\dot{x}(t) = \overline{x}_a{\cdot}\omega{\cdot}\cos\omega{\cdot}t$. As the derivative of a rotating vector is another vector multiplied by the imaginary unit j (it is reminded that $j = \sqrt[2]{-1}$), the velocity vector is $j{\cdot}\omega{\cdot}\overline{x}_a$. Due to the imaginary unit j, this vector is at 90° ahead with respect to vector $\overline{x}_a$. The inertial term contains the acceleration of x(t), then, its trigonometric formula is $x(t) = -\overline{x}_a{\cdot}\omega^2{\cdot}\sin\omega{\cdot}t$. In the vector form, the acceleration is $j^2 - \overline{x}_a \cdot \omega^2$, a vector 90° ahead of the velocity vector, in the same direction as the vector $\overline{x}_a$ but heading oppositely. The product of these vectors, multiplied by the mechanical properties of rigidity, friction, and inertia, returns the equation of motion's reaction forces.

Formula 1.5 Reaction forces of a mechanical SDOF system

$$\overline{F}_{rigidity} = k{\cdot}\overline{x}_a \; \overline{F}_{friction} = j{\cdot}f{\cdot}\omega{\cdot}\overline{x}_a \; \overline{F}_{inertia} = -m{\cdot}\omega^2{\cdot}\overline{x}_a$$

Therefore, the equation of motion 1.2 can be written as follows:

Formula 1.6 Vector form of the equation of motion

$$\overline{F}_a = \left(k - m{\cdot}\omega^2 + j{\cdot}f{\cdot}\omega\right){\cdot}\overline{x}_a$$

The vector sum of reaction forces equals the applied force $\overline{F}_a$, shown in the vector diagram of Figure 1.9. As all vectors are rotating at the same speed ω, their relative angular position is constant. The left diagram shows the angular relations between a gyrating vector V and its first and second derivatives. The right diagram shows that the dashed vector represents the inertial reaction's net force minus the rigidity reactions.

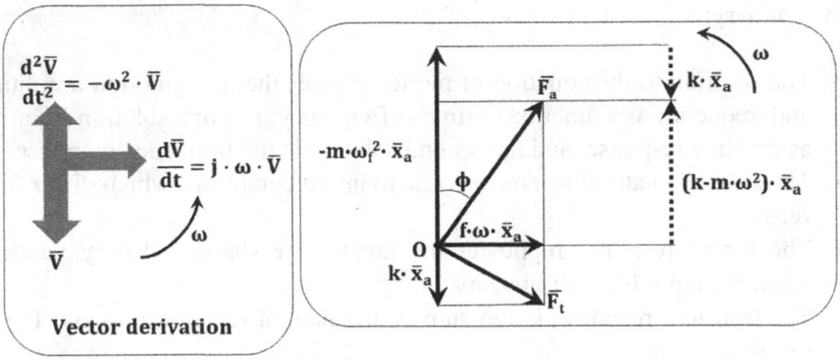

FIGURE 1.9 Vector equilibrium of the harmonic force $F \cdot \cos(\omega \cdot t)$.

Sign (-) of the inertial force is opposed to the rigidity force, which by convention is positive. The applied force $\bar{F}$ is in equilibrium with the reaction force $-\bar{F}$. Force $\bar{F}$ is applied by an external source or by a centrifugal force due to the shaft eccentricity or shaft deflection. Mass displacement x(t) is the vibration amplitude represented by the dotted vector $\bar{x}_a$. The vector $\bar{F}$ is ahead of the inertial force, an angle φ, known as the phase of vibration motion.

1.5 NATURAL FREQUENCY

Because of negligible duration applied force or non-zero initial conditions, frictionless free systems vibrate with a characteristic frequency called natural frequency. Natural frequency only depends on the mass and the spring rigidity. The natural frequency is always constant; it does not change if initial conditions change. Therefore, it is an inherent physical property of the system.

The sudden application of a force on a shaft will produce a lateral (or transversal) vibration, a quick oscillating motion in a perpendicular direction to the shaft axis. The shaft has a natural frequency calculated using Formula 1.9; therefore, resonance may happen if the applied force is harmonic with a frequency equal to the shaft's natural frequency. Furthermore: any type of force may produce vibration of different frequencies. If any of them matches the shaft's natural frequency, resonance will happen.

1.5.1 The Natural Frequency of Linear Systems

As the natural frequency is the oscillation frequency of a frictionless system with no external forces or torques acting on the system, the equation of motion under these conditions is:

Formula 1.7 Equation of motion of a frictionless system with no external excitation

$$0 = m \cdot \ddot{x}(t) + k_w \cdot x(t)$$

The solution to this equation is: $x(t) = x_a \cdot \sin(\omega_n \cdot t)$, therefore, the acceleration formula is $\ddot{x}(t) = -x_a \cdot \omega_n^2 \cdot \sin(\omega_n \cdot t)$. These two formulas are introduced in the equation of motion, and the following equation is returned:

Formula 1.8 The solution to the equation of Formula 1.7

$$0 = -m \cdot \omega_n^2 + k_w$$

From this expression, the natural frequency is cleared. The result is rad/s.

Formula 1.9 The natural frequency of a linear second-order system

$$\omega_n = \sqrt[2]{\frac{k_w}{m}}$$

k_w is the stiffness coefficient for concentrated load in kg/m.
m is the mass expressed in kg mass.
ω_n is the natural frequency, in radian/s.

The natural frequency ω_n is not equal to the resonant frequency, but it is close to it in low friction systems like machines. That is why in many technical publications, the resonant frequency is referred to, with little error, as natural frequency.

1.5.2 The Natural Frequency of Rotating Systems

There is an analogy between linear and rotating systems based on their physical properties to store or dissipate energy. In rotating systems, the equation of motion has an analogous form to that of linear systems. For rotating frictionless systems, the equation of motion is:

Formula 1.10 Equation of motion of rotating frictionless systems

$$0 = -J_p \cdot \ddot{\vartheta}(t) + k_t \cdot \vartheta(t)$$

With this equation, a rotating system's natural frequency is derived; the same was done for linear systems. Similarly, with rotating systems, the energy storing elements are the polar mass moment of inertia and the torsional rigidity. Therefore, this simple analogy helps to write the natural frequency formula of rotating systems:

Formula 1.11 Natural frequency of a rotating system

$$\omega_n = \sqrt[2]{\frac{k_t}{J_p}}$$

k_t is the torsional rigidity coefficient, given by:

Formula 1.12 Torsional rigidity coefficient

$$k_t = G \cdot \frac{I_p}{L}$$

J_p is the polar mass moment of inertia, expressed in kgmass · m^2, and I_p is the polar moment of inertia, expressed in m^4. G is the steel shear modulus of rigidity in kg/m^2. The typical value of G for stainless steel is 77 GPa. The formula of J_p for a uniform circular shaft is:

Formula 1.13 Mass moment of inertia of a uniform solid shaft

$$J_p = \frac{m \cdot D^2}{8} = \rho \cdot I_p \cdot L$$

Like a linear system, if an impulse torque is applied to a rotating system, torsional oscillations will be induced. If the system is conservative, these oscillations will never end unless they are mechanically stopped. Amplitudes of torsional vibration are larger the longer the shaft. This behavior is due to the torsional rigidity coefficient that is inversely proportional to the shaft length. See Formula 1.13.

1.6 NATURAL RESPONSE OF SECOND-ORDER SYSTEMS

If the mass of Figure 1.8 is displaced downward a distance x_0 and then released at a standstill position, the system starts the transient response regime, a quick harmonic motion. The release marks the instant zero. After the mass is liberated, the energy stored in the spring impulses the mass upward. In this upward motion, the mass increases its potential energy. When all the spring energy is transferred to the mass, the movement stops, and the mass comes down, returning its potential energy to the spring. As a result, the mass oscillates, free of forces, until vibration fades away because of the damper's friction. As the system remains vibrating and no force is applied, it is said that the system is freely vibrating.

As conservative systems do not exist because friction is unavoidable, frictionless systems have a theoretical interest only. However, formulas of conservative systems form a useful basic model, readily extended to non-conservative systems.

1.7 DERIVATION OF THE TIME NATURAL RESPONSE

The following homogeneous differential equation expresses the equation of motion to derive a second-order system's natural response. This equation only differs from the equation of motion of a conservative system by the presence of the friction reaction force $f \cdot \dot{x}$.

Formula 1.14 Equation of motion of a mechanical non-conservative system

$$0 = m \cdot \ddot{x} + f \cdot \dot{x} + k \cdot x$$

In the next section, the solution to a differential equation of nth order is briefly described and applied to the above equation of motion to obtain the general response formula, which applies to any system type.

Formulas to be derived in this section are based on a mechanical system; however, they are valid for any other system type. An often-used method in solving a differential equation is with the Laplace transform, explained in Chapter 11, because this method converts the differential equation into an algebraic expression. Alternatively, as the equation to solve is linear and homogeneous and has constant coefficients, it can be used the characteristic equation of the ODE[5] that assumes the solution is $x(t) = e^{p \cdot t}$. Substituting this formula in the ODE, its characteristic equation is returned, and the roots of the characteristic equation give the values of p. In a second-order mechanical system, the equation of motion and its characteristic equation are:

Differential equation of motion:

$$0 = m \cdot \ddot{x} + f \cdot \dot{x} + k \cdot x$$

Characteristic equation:

Formula 1.15 Equation of motion and characteristic equation of a mechanical system

$$0 = m \cdot p^2 + f \cdot p + k$$

The roots of the characteristic equation are:

Formula 1.16 Roots of the characteristic equation

$$p_{1,2} = \frac{-f \pm \sqrt[2]{f^2 - 4 \cdot k \cdot m}}{2 \cdot m} = \sigma \pm j \cdot \omega \sigma = -\frac{f}{2 \cdot m} \omega = \frac{\sqrt[2]{f^2 - 4 \cdot k \cdot m}}{2 \cdot m}$$

If $f < 2 \cdot \sqrt[2]{k \cdot m}$ the roots of Formula 1.16 are complex, predicting that the motion is oscillating, as demonstrated below. Replacing p by its roots in the general solution to the ODE 1.15, the mass displacement as a function of time is returned.

Formula 1.17 First solution to ODE 1.15 exponential functions

$$x(t) = B_1 \cdot e^{(\sigma + j \cdot \omega) \cdot t} + B_2 \cdot e^{(\sigma - j \cdot \omega) \cdot t}$$

Clearing $e^{\sigma \cdot t}$ and replacing the imaginary terms by the Euler's formula: $e^{\pm j \cdot \vartheta} = \cos \vartheta \pm j \cdot \sin \vartheta$, it arises that the solution x(t) is a harmonic motion of the form:

Formula 1.18 Natural response of second-order systems

$$x(t) = e^{\sigma \cdot t} \cdot (C_1 \cdot \cos \omega \cdot t + j \cdot C_2 \cdot \sin \omega \cdot t)$$

$$|x(t)| = e^{\sigma \cdot t} \cdot \sqrt[2]{(C_1^2 \cdot \cos^2 \omega \cdot t + C_2^2 \cdot \sin^2 \omega \cdot t)}$$

Where $C_1 = B_1 + B_2$ and $C_2 = B_1 - B_2$.

The absolute value of x(t) is the lower expression of Formula 1.18. This formula anticipates that the natural system response is oscillatory, and that oscillations amplitude are multiplied by $e^{\sigma \cdot t}$. If σ is positive, amplitudes are theoretically enlarged until they reach infinite values. Conversely, if σ is negative, oscillations are damped, and at an infinite time, they will be zero, but much before that, the response is considered extinguished in practical applications. The settling time of a damped time response is discussed next.

1.7.1 Damping Ratio and Damped Frequency

As discussed previously, when $f < 2 \cdot \sqrt[2]{k \cdot m}$, the equation of motion has imaginary roots; therefore, the mass displacement x(t) is an oscillatory or vibratory motion. Hence, for any value of the friction coefficient equal to or higher than $2 \cdot \sqrt[2]{k \cdot m}$, the system does not vibrate. Therefore, the coefficient $2 \cdot \sqrt[2]{k \cdot m}$ is an inherent property of the mechanical system, named critical friction coefficient.

Formula 1.19 Critical friction coefficient

$$f_c = 2 \cdot \sqrt[2]{k \cdot m}$$

The above definition of critical coefficient allows us to derive one of the most important properties of a second-order system: the damping ratio, defined as the ratio of the actual friction coefficient f to the critical coefficient f_c.

Formula 1.20 Damping ratio

$$\zeta = \frac{f}{f_c}$$

The natural response shape depends on the friction force significance compared to rigidity and inertial forces. The damping ratio ζ is the physical property determining if the system's natural response is damped oscillatory or damped non-oscillatory. The critical value of the damping ratio under which the natural response is oscillatory is 1. If the damping ratio is higher than this critical value, the natural reaction is non-oscillatory. According to this definition, any system with a damping ratio lower than 1 vibrates when it is excited by external forces or torques or initial conditions.

As $f = \zeta \cdot f_c$ and $\omega_n = \sqrt[2]{k/m}$, it is easily demonstrated that the roots of the characteristic equation of a system with $\zeta < 1$ are expressed as:

Formula 1.21 Roots of the characteristic equation as a function of ζ and ω_n

$$p_{1,2} = -\zeta \cdot \omega_n \pm j \cdot \omega_n \cdot \sqrt[2]{1-\zeta^2}$$

The term $\omega_n \cdot \sqrt[2]{1-\zeta^2}$ is the imaginary part of the complex frequency, usually known as damped frequency.

Formula 1.22 Vibration's damped frequency

$$\omega_d = \omega_n \cdot \sqrt[2]{1-\zeta^2}$$

The frequency ω_d is the frequency at which the system oscillates.

1.7.2 Natural Transient Response Formula

The above-defined dimensionless parameters are introduced in Formula 1.18 to obtain:

Formula 1.23 Natural transient response formula with unknown constants C_1 and C_2

$$x(t) = e^{-\zeta \cdot \omega_n \cdot t} \cdot (C_1 \cdot \cos \omega_d \cdot t + j \cdot C_2 \cdot \sin \omega_d \cdot t)$$

Constants C_1 and C_2 are calculated with arbitrary initial conditions, which are the mass position x_0 and its velocity $\dot{x}_0$ at t = 0.

Mass position at time t = 0 is:

Formula 1.24 Constant C_1 formula

$$x(0) = x_0 = C_1$$

Replacing C_1 by x_0 in Formula 1.23 and calculating its derivative with respect to time, the mass velocity formula is returned.

Formula 1.25 Mass velocity formula

$$\dot{x}(t) = -\zeta \cdot \omega_n \cdot e^{-\zeta \cdot \omega_n \cdot t} \cdot x(t) + e^{-\zeta \cdot \omega_n \cdot t} \cdot (-\omega_d \cdot x_0 \cdot \sin \omega_d \cdot t + j \cdot \omega_d \cdot C_2 \cdot \cos \omega_d \cdot t)$$

At t = 0, the mass velocity is:

Formula 1.26 Mass velocity at instant zero

$$\dot{x}(0) = \dot{x}_0 = -\zeta \cdot \omega_n \cdot x_0 + j \cdot \omega_d \cdot C_2$$

Therefore, clearing C_2 from this formula:

Formula 1.27 Constant C_2 formula

$$C_2 = \frac{\zeta \cdot \omega_n \cdot x_0 + \dot{x}_0}{j \cdot \omega_d}$$

Finally, replacing C_1 and C_2 in Formula 1.25, a second-order system's natural transient response is returned.

Formula 1.28 Natural transient response formula with dimensionless parameters

$$x(t) = e^{-\zeta \cdot \omega_n \cdot t} \cdot \left(x_0 \cdot \cos \omega_d \cdot t + \frac{\zeta \cdot \omega_n \cdot x_0 + \dot{x}_0}{\omega_d} \cdot \sin \omega_d \cdot t \right)$$

As C_2 is imaginary, it cancels the j factor of sin $\omega_d \cdot t$. All terms of the above formula are real numbers and apply to any second-order system, regardless of their physical nature. The x(t) formula indicates that the system is damped oscillatory. The damping constant is $T_z = 1/(\zeta \cdot \omega_n)$, and the oscillation frequency is ω_d. If the system has no initial conditions, then x(t) = 0. No natural response happens. If one of the two initial conditions, x_0 or $\dot{x}_0$ is not zero, then the system reacts with its natural response.

In summary: the natural response of a non-conservative second-order system depends on the initial conditions (position x_0 and velocity $\dot{x}_0$), the damping coefficient ζ, and the natural frequency ω_n. The last two properties are measured or calculated by formulas based on the system's physical properties. They are instrumental in predicting system transient response behavior. No other parameter is needed to anticipate if the system time response is or not oscillatory.

1.7.3 Vector Interpretation of the Natural Time Response

As the right term of Formula 1.28 is the sum of a sine plus a cosine function, both terms represent two vectors at 90° one from the other, rotating with frequency ω_n. The magnitude of these two vectors is C_1 and C_2. The vector sum of these two vectors is the vibration amplitude x_a, which also rotates with angular velocity ω_n. See Figure 1.10.

Following this vector diagram, Formula 1.28 may be written in a compact form, an easier expression to interpret. The compact expression arises from the

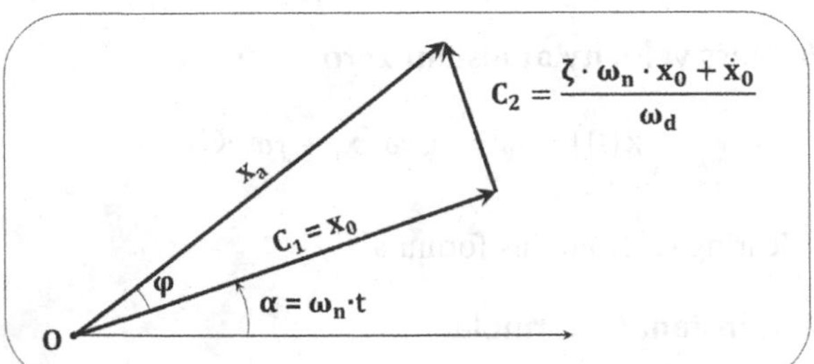

FIGURE 1.10 Vector interpretation of the damped natural response.

amplitude and phase returned by Formula 1.29, and the result is Formula 1.30. The response amplitude x_a and phase φ are calculated with the following formulas, readily derived from Figure 1.10.

Formulas 1.29 Amplitude and phase of a damped second-order system

$$x_a = \sqrt[2]{C_1^2 + C_2^2} = \sqrt[2]{x_0^2 + \left(\frac{\zeta \cdot \omega_n \cdot x_0 + \dot{x}_0}{\omega_d}\right)^2}$$

$$\varphi = \arctan\left(\frac{C_2}{C_1}\right) = \arctan\left(\frac{\zeta \cdot \omega_n \cdot x_0 + \dot{x}_0}{x_0 \cdot \omega_d}\right)$$

The angle φ is formed by the vectors x_a and x_0. This angle represents the phase existing between the curves of $\cos(\omega_n \cdot t)$ and $\cos(\omega_n \cdot t - \varphi)$. This phase is minimum if the system has no initial velocity. In this case, it is equal to $\tan^{-1} \zeta / \sqrt[2]{1-\zeta^2}$. For zero initial velocity $\dot{x}_0$, the amplitude x_a and phase φ only depend on the damping ratio ζ and the initial position x_0.

Therefore, transient response's Formula 1.28 adopts the following compacted form.

Formula 1.30 Damped natural response. Compacted form[6]

$$x_n(t) = x_a \cdot e^{-\zeta \cdot \omega_n \cdot t} \cdot \cos(\omega_d \cdot t + \varphi)$$

Figure 1.10 suggests that the vectors triangle has an angular motion around the center O. During rotation, the three vectors and angle φ remain constant. As the rotation frequency ω_n is constant, the angle α is directly proportional to the time t, elapsed since t = 0.

1.7.4 Concepts to Remember Regarding Second-Order Systems

- Four numbers identify a second-order system's natural response: the damping ratio ζ, the natural frequency ω_n, the initial position x_0 and the initial velocity v_0. These numbers determine the system's natural response.
- Damping ratio and natural frequency depend on the system's physical properties: mass, rigidity coefficient and friction coefficient.
- The friction coefficient is a physical property expressed in units of force over speed. Its value does not predict the transient shape.
- The critical friction coefficient is the limit between the undamped and the damped transient regime.
- The damping ratio is the friction coefficient to the critical friction coefficient ratio.
- The damping ratio value predicts the transient regime shape.

1.7.5 Natural Response and Decay Curves

Figure 1.11 shows two curves: the natural transient response and the decay of the oscillation amplitudes. Its more significant parameters shown in this figure are:

x_a: amplitude of the decay curve at t = 0
x_0: initial mass displacement (arbitrary)
T_z: damping time constant of the decay curve
$t_{s2\%}$: settling time
t_p: first peak time
x_p: first peak or maximum overshoot of the natural response

The decay curve limits the oscillation amplitudes. Therefore, the decay curve is tangent to the oscillatory curve, close to the peak. The decay velocity is determined by the time constant T_z. This constant may be read in the tangent intersection of the decay curve at t =0 and the time axis or calculated with: $T_z = 1/(\zeta \cdot \omega_n)$.

The decay curve formula is directly taken from Formula 1.30. This formula shows that vibration peaks have the amplitude $x_a \cdot e^{-\zeta \cdot \omega_n \cdot t_p}$ and t_p is the time at which peaks are produced. Then, the decay curve is given by the following formula:

Formula 1.31 Decay curve

$$d(t) = x_a \cdot e^{-\zeta \cdot \omega_n \cdot t} = x_a \cdot e^{-t/T_z}$$

Therefore, at t = 0, the decay curve amplitude is x_a. See Figure 1.11. Hence, x_a is not a peak amplitude, but the decay ratio ordinate at t = 0! At t = 0, Formula 1.31 returns: $x_0 = x_a \cdot \cos(\varphi)$. Thus, the cosine of phase φ is equal to the x_0/x_a ratio.

For zero initial velocity, the first peak is x_0 at t = 0. See Figure 1.12. If the system has an initial velocity, the first peak is shifted to the right. Figures 1.12

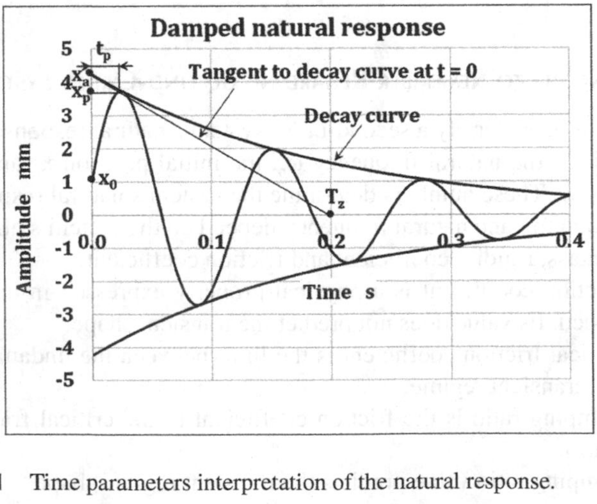

FIGURE 1.11 Time parameters interpretation of the natural response.

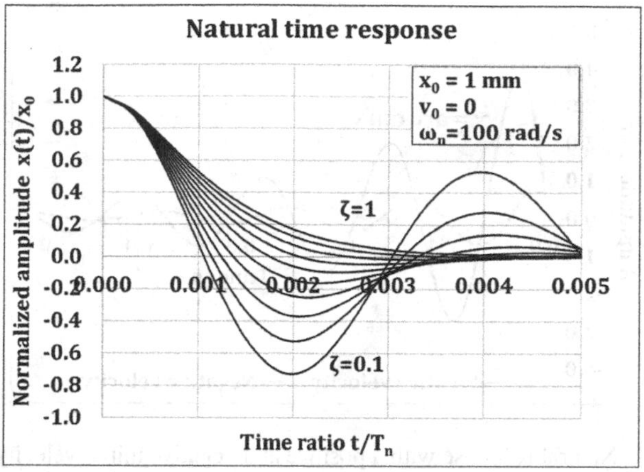

FIGURE 1.12 Natural response with zero initial velocity.

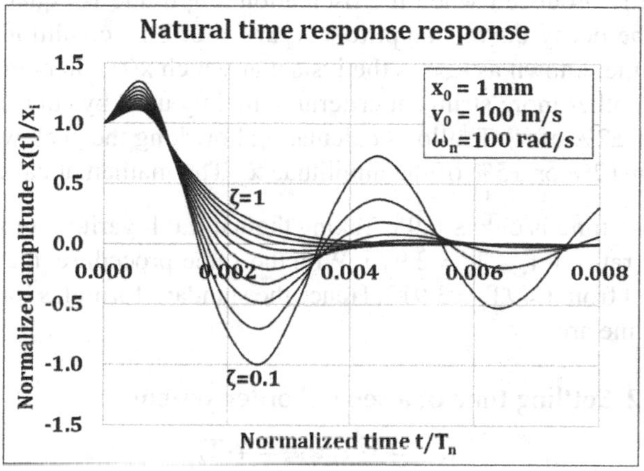

FIGURE 1.13 Natural response with an initial velocity.

and 1.13 show the natural response curves of a second-order system. The left figure shows the transient response with zero initial velocity $\dot{x}_0$ but with an initial position x_0. This peak is the first one of the transient responses. Instead, the right figure shows the transient response with initial position and velocity. Both figures are represented as a function of the dimensionless time t/T_n. Curves of $\zeta = 0.1$ and $\zeta = 1.0$ in Figures 1.12 and 1.13 are tagged to help figure interpretation. Values of ζ differ in 0.1 from one curve to another.

If the mass has an initial velocity at the instant zero, the cosine curve is shifted to the right, as shown in Figure 1.13. In this case, the initial position x_0 is the first peak. Figure 1.14 shows that the initial velocity $\dot{x}_0$ (or v_0) is the tangent of the angle γ formed by the tangent line to the x(t) curve at t = 0 and the time axis. It means that $\dot{x}_0 = \tan \gamma$.

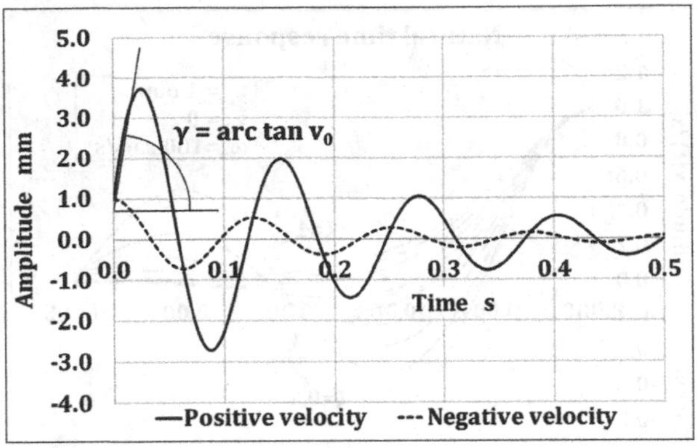

FIGURE 1.14 Natural response with a positive and negative initial velocity.

1.7.5.1 Settling Time and Number of Cycles

Settling time is produced when the oscillation amplitude is equal to or lower than 5% of the decay curve's amplitude x_a at $t = 0$. This condition means that the settling time, known as $t_{s5\%}$, is the instant at which $x(t)$ enters in the band of ±5% of x_a. Another more stringent criterium, mainly used by control engineers, is the band of ±2% of x_a. T_z allows calculating how long the peak waves take to be lesser than ±2% or ±5% of the amplitude x_a. The mathematical condition of the 5% settling time is $e^{-\frac{t}{T_z}} \leq 0.05$. Taking the natural logarithm of this formula results in the ratio of $t_{s5\%}/T_z = 2.996$. With the same procedure, the 2% settling time is cleared from $t_{s2\%}/T_z = 3.912$. Hence the standard formulas for the 5% and 2% settling time are:

Formula 1.32 Settling time of a second-order system

$$t_{s5\%} = 3 \cdot T_z \, t_{r2\%} = 4 \cdot T_z$$

For a better estimation of the 5% settling time, see Figure 1.15. This figure indicates that the minimum settling time is at $\zeta = 0.707$.

Several publications propose different empirical formulas to calculate the 5% settling time. The author proposes using a mix of the following two formulas that agree with the chart of Gille, Decaulne and Pelegrin[7].

- Formula proposed by Benjamin Kuo for $\zeta < 0.7$: $t_{s5\%}/T_n = 0.509/\zeta$
- Formula proposed by the University of Illinois[8] for $\zeta > 0.7$: $t_{s5\%}/T_n = (6.6 \times \zeta - 1.6)/(2 \cdot \pi)$

These formulas were contrasted with others to verify their accuracy. The author used them to construct the chart in Figure 1.15.

The number of cycles in the lapse from 0 to $t_{s5\%}$ or $t_{s2\%}$ only depends on the damping ratio. They are calculated with the following formulas and plotted

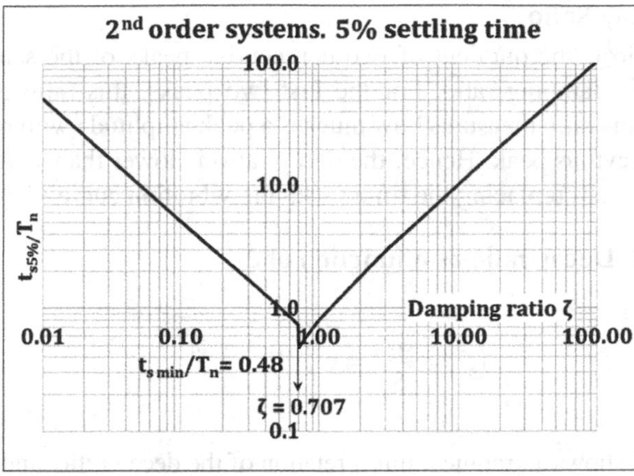

FIGURE 1.15 5% settling time.

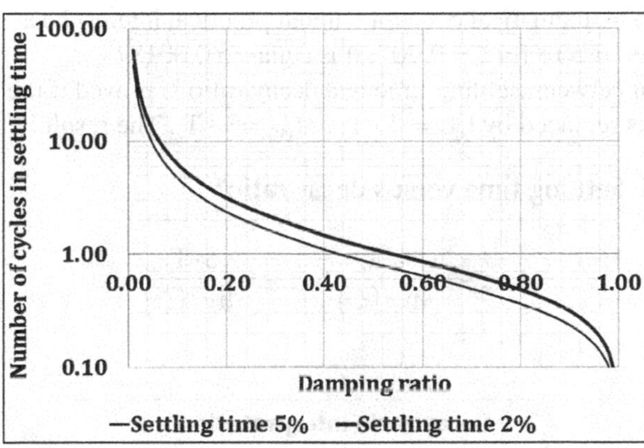

FIGURE 1.16 Cycle quantity within the transient response at the typical machines ζ range.

against ζ in Figure 1.16. The ratio $t_{s5\%}/T_d$ gives the number of cycles in the 5% band. Similar reasoning applies to the 2% band. As $T_z = 1/(\zeta \cdot \omega_n)$ and $T_d = 2 \cdot \pi/\omega_n$, the number of cycles in the 5% and 2% bands are given by the following formulas.

Formula 1.33 Cycle quantity in 2% and 5% settling time

$$C_{y5\%} = \frac{t_{r5\%}}{T_d} = \frac{3}{2 \cdot \pi} \cdot \frac{\sqrt[2]{1-\zeta^2}}{\zeta} \quad C_{y2\%} = \frac{t_{r2\%}}{T_d} = \frac{2}{\pi} \cdot \frac{\sqrt[2]{1-\zeta^2}}{\zeta \cdot \omega_n}$$

1.7.5.2 Decay Ratio

The decay ratio is the quotient of two consecutive peaks of the same sign as a function of the damping ratio. For the first two peaks, this ratio is S_2/S_1. See Figure 1.17. This ratio measures how much is a peak amplitude with respect to the same sign's previous peak. Hence, the decay ratio is lower than 1. It is obtained from Formula 1.33, replacing the time t with the vibration curve's T_d period.

Formula 1.34 Decay ratio as a function of ζ

$$\gamma(\zeta) = \frac{S_{j+1}}{S_j} = e^{-T_d/T_z} = e^{-\frac{2\cdot\pi\cdot\zeta}{\sqrt[2]{1-\zeta^2}}}$$

Figure 1.17 shows a graphical interpretation of the decay ratio, and Figure 1.18 is the decay ratio against the damping ratio curve. This curve is constructed with Formula 1.34. This formula holds for any pair of successive overshoots of the transient response curve. It means that the decay ratio is equal to: $S_2/S_1 = S_3/S_2 = S_4/S_3 \cdots = S_n/S_{n-1}$.

The decay ratio is 1 for $\zeta = 0$ and is zero for an infinite damping ratio. However, the last is just a mathematical concept without practical interest because the decay ratio is considered zero for $\zeta = 0.707$; it is equal to 0.00187.

The relation between settling time and decay ratio is proved if the period T_d of Formula 1.34 is replaced by $t_{s2\%} = 4 \cdot T_z$ or $t_{s5\%} = 3 \cdot T_z$. The result is:

Formula 1.35 Settling time versus decay ratio[9]

$$t_{s2\%} = -\frac{4\cdot T_d}{\ln\gamma(\zeta)} \quad t_{s5\%} = -\frac{3\cdot T_d}{\ln\gamma(\zeta)}$$

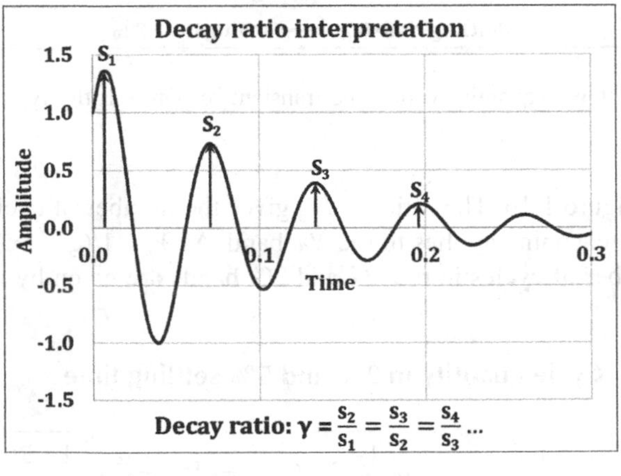

FIGURE 1.17 Graphical interpretation of the decay ratio.

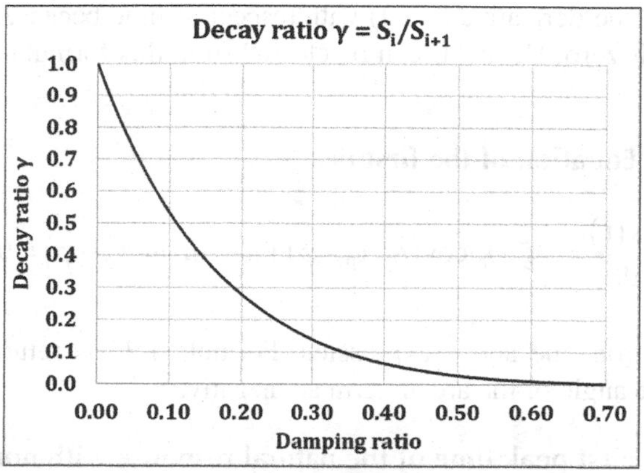

FIGURE 1.18 Decay ratio curve versus damping ratio.

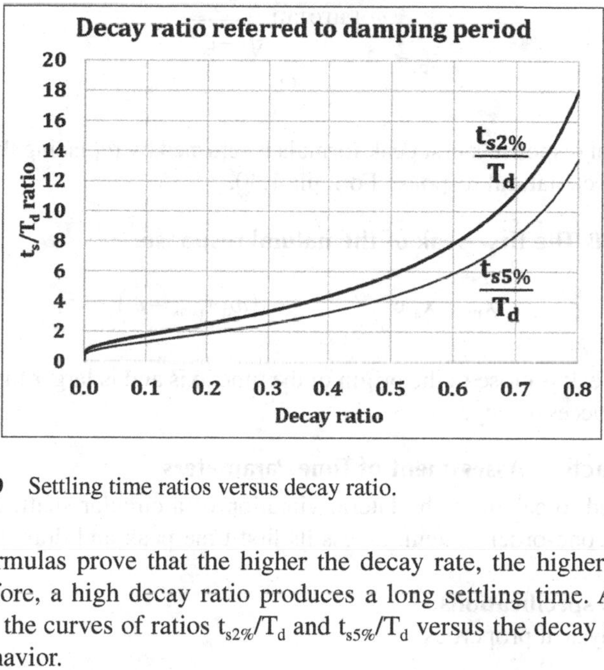

FIGURE 1.19 Settling time ratios versus decay ratio.

These formulas prove that the higher the decay rate, the higher the settling time. Therefore, a high decay ratio produces a long settling time. As shown in Figure 1.19, the curves of ratios $t_{s2\%}/T_d$ and $t_{s5\%}/T_d$ versus the decay ratio γ illustrate this behavior.

In torsional vibration of rotating machines, a typical value of ζ is in the 0.015 and 0.125 range. For $\zeta = 0.015$, the decay ratio is $\gamma = 0.91$ and for $\zeta = 0.125$ the decay ratio is $\gamma = 0.45$. These values anticipate a long settling time because the higher the decay ratio, the longer the settling time.

1.7.5.3 The First Peak Time

As discussed previously, in systems with a non-zero initial velocity, the natural response curve shifts away from the time axis. Therefore, the natural response's first peak is not produced at $t = 0$ but at a peak time t_p. Thus, the peak time formula is derived from the derivative of x(t) with respect to time because, at the peaks, this derivative is zero. Hence, t_p can be cleared from this formula, equating it to zero.

Formula 1.36 Equation of the first peak

$$\frac{dx(t)}{dt} = -\zeta \cdot \omega_n \cdot \cos(\omega_d \cdot t_p - \varphi) + \omega_d \cdot \sin(\omega_d \cdot t_p - \varphi) = 0$$

Clearing t_p from the above expression, Formula 1.37 is returned. Both the phase φ and the angle of the arctan term are negative.

Formula 1.37 First peak time of the natural response with non-zero initial conditions

$$t_p = \frac{\varphi + \arctan \frac{-\zeta}{\sqrt[2]{1-\zeta^2}}}{\omega_d}$$

The natural response's first peak formula is returned by replacing the peak time t_p in the damped natural response Formula 1.30.

Formula 1.38 The first peak of the natural response.

$$x_{pl} = x_a \cdot e^{-\zeta \cdot \omega_n \cdot t_{peak}} \cdot \cos(\omega_d \cdot t_{peak} - \varphi)$$

The first peak is close to the origin of the time axis and is larger than any other peak of the successive cycles.

1.7.5.4 Practical Assessment of Time Parameters

It is requested to calculate the lateral vibration of a circular shaft, modelled as a discrete second-order system, assess its first time peak and draw the transient vibration curve.

Technical specifications:

Shaft's physical properties:

$$m = 386 \text{ kg mass}$$
$$f = 12{,}106 \text{ kg/(m/s)}$$
$$k_w = 5{,}567{,}876 \text{ kg/m}$$

Initial conditions:

$$x_0 = 1 \text{ mm}$$
$$v_0 = 500 \text{ mm/s.}$$

Natural response calculation:
Critical friction coefficient:

$$f_c = 2 \cdot (k_w \cdot m)^{1/2} = 92{,}690 \text{ kg}/(\text{m/s})$$

Damping ratio:

$$\zeta = f/f_c = 0.13$$

Decay ratio:

$$\gamma = \exp\left(-2 \cdot \pi \cdot \zeta / (1-\zeta^2)^{1/2}\right) = 0.439$$

Natural frequency:

$$\omega_n = (k/m)^{1/2} = 120.1 \text{ rad/s}$$

Waves natural period:

$$T_n = 2 \cdot \pi / \omega_n = 0.0523 \text{ s}$$

Damped frequency:

$$\omega_d = \omega_n \cdot (1-\zeta^2)^{1/2} = 119.1 \text{ rad/s}$$

Waves damping period:

$$T_d = 2 \cdot \pi / \omega_d = 0.0528 \text{ s}$$

Transient response constants C_1 and C_2:

$$C_1 = x_0 = 1 \text{ mm}$$

$$C_2 = \zeta / (1-\zeta^2)^{1/2} . x_0 + v_0 / \omega_d = 4.33 \text{ mm}$$

Decay time constant:

$$T_z = 1/(\zeta \cdot \omega_n) = 0.064 \text{ s}$$

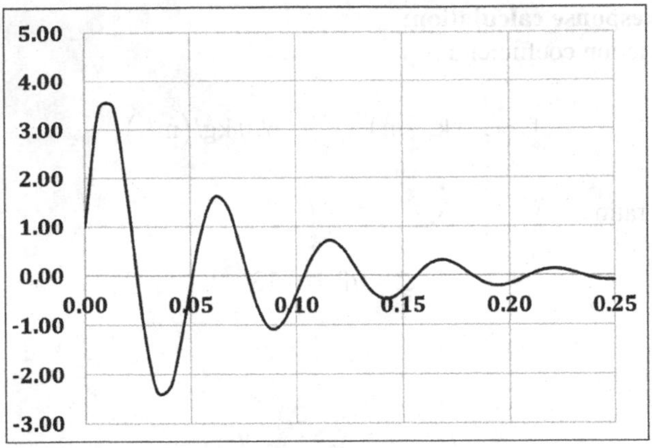

FIGURE 1.20 Natural response of the above example.

2% settling time:

$$ts2\% = 4 \cdot T_z = 0.256 \text{ s}$$

1st peak time:

$$\varphi = \tan^{-1}\left(C_2 / \left(C_1^2 + C_2^2\right)^{1/2}\right) = 1.34 \text{ rad}$$

$$t_p = \left[\varphi + \tan^{-1}\left(-\zeta / \left(1 - \zeta^2\right)^{1/2}\right)\right] / \omega_d = 0.001019 \text{ s}$$

Figure 1.20 shows the transient response of the system calculated in this example

Technical assessment of the results and plot:

The shaft has a damping ratio lower than 1 (0.13), which indicates that the natural response will be damped oscillatory. The settling time shows that the vibration transient will last about ¼ of a second. Suppose a vibrometer controls this vibration. In this case, it is expected that the display will exhibit a first peak time at 0.001019 s, imperceptible if compared to the total duration of the transient regime. The decay ratio indicates that every overshoot is 43.9% of the previous. This decay means that the second overshoot is 43.8% of the first, the third overshoot is 19.2%, and the fourth is 8.4% of the first. See Figure 1.20.

1.8 TRANSIENT RESPONSE TO A STEP FORCE INPUT

A permanent applied force or torque generates a constant displacement of a mass or a constant angular distortion of a body, like a shaft. The result is a transient response formed by the natural response plus a permanent response, which replicates the applied force or torque shape. The transient response modelling requires

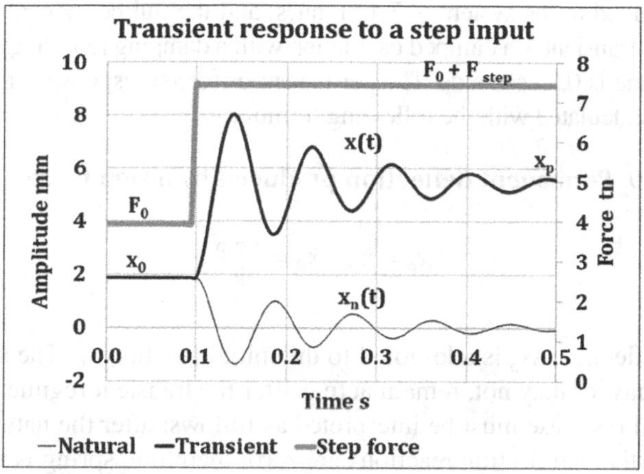

FIGURE 1.21 Natural and transient response to a step input $\zeta = 0.1$.

selecting any of the many force modalities that may excite a system. See the example of Figure 1.21.

1.8.1 Conceptual Description

Figure 1.21 shows the natural and transient response of an SDOF system excited by a step force F. The mass is initially moved to $x_0 = 2$ mm against the spring reaction. This mass displacement is the initial arbitrary stretching imposed on the spring. The mass is then released at time t = 0.1 s, so the natural damped regime disappears at the 2% settling time, in this case, 0.49 s. This regime is the transient natural component (see the fine curve). The other component is the forced deflection $x_p - x_0$ produced by the step force. Both the natural and forced responses are superimposed. The step force produces a permanent deflection $x_p = 6$ mm. As soon as force F is removed, the mass returns to the rest position (x = 2 mm) following a natural response transient.

1.8.2 Transient Response Formula

Refer to the mechanical second-order system of Figure 1.8. The system is initially under static conditions. These conditions mean that a constant force F_0 produces the initial deflection $x_0 = F_0/k$. Under this condition, a step force is applied. See, in Figure 1.21, F_0 and $F_0 + F_{step}$ curves. The step force generates a natural oscillatory response (see the lower curve), and, after this regime is ended, a permanent constant deflection x_p remains.

The most significant figures of the transient of Figure 1.21 are described below:

The mass has an initial deflection x_0= 1.87 mm, produced by its weight of 3.9 tonnes. At time t = 0.1 s, e.g., a step force of 3.5 tonnes is applied, which produces a permanent step deflection $\Delta x_p = x_p - x_0 = 1.67$ mm. Therefore, since t = 0.1 s, the total force applied to the system is 7.42 tonnes, and the full permanent deflection is 3.47 mm. The transient is a damped oscillation with a damping ratio of $\zeta = 0.1$ and the 2% settling time is 0.55 seconds. The permanent or forced response is the difference $x_p - x_0$ and is calculated with the following formula:

Formula 1.39 Permanent deflection produced by a step force

$$\Delta x_p = x_p - x_0 = \frac{F_{step}}{k}$$

So, the deflection Δx_p is additional to the initial position x_0. The step force or step torque may, or may not, remain acting after the transient regime is damped.

The forced response must be interpreted as follows: after the natural response is ended, inertial and friction reactions are zero; therefore, spring is the only element that remains reacting against the applied forces F_0 and F_{step}. This reaction is proportional to the spring stretching $x_0 + \Delta x_p$. Then $F_0 + F_{step} = k \cdot (x_0 + \Delta x_p)$. Once the natural response is damped, the permanent deflection x_p remains if the applied force F_{step} is not removed. If this force is removed, the system returns to the initial position x_0 after experiencing a natural response regime.

1.8.2.1 Equation of Motion for a Step Input Force

According to the above discussion, the transient regime is formed by the permanent response x_p plus the natural response $x_n(t)$.

Formula 1.40 Transient regime produced by a step force

$$x(t) = x_0 + \Delta x_p + x(t) = x_p + x_n(t)$$

As seen in Section 1.7, the natural response x(t) is the solution to the equation of motion given by the following second-order differential equation:

Formula 1.41 Equation of motion for a step input force

Applied step force = inertial + friction + rigidity reactions

$$F_{step} = m \cdot \ddot{x}(t) + f \cdot \dot{x}(t) + k \cdot x(t).$$

1.8.2.2 Natural Response to a Step Input

The solution to this differential equation is given by Formula 1.23, reproduced below:

Formula 1.42 Natural response

$$x_n(t) = e^{-\zeta \cdot \omega_n \cdot t} \cdot (C_1 \cdot \cos \omega_d \cdot t + C_2 \cdot \sin \omega_d \cdot t)$$

Constants C_1 and C_2 formulas are obtained using the same procedures used to derive the natural response constants in Section 1.7.1. Therefore, the general solution is:

Formula 1.43 General solution to a step force excitation

$$C_1 = -\Delta x_p \qquad C_2 = \frac{\zeta \cdot \omega_n \cdot \Delta x_{p+\dot{x}_0}}{\omega_d}$$

$$x(t) = x_p + e^{-\zeta \cdot \omega_n \cdot t} \cdot \left(-\Delta x_p \cdot \cos \omega_d \cdot t + \frac{\zeta \cdot \omega_n \cdot \Delta x_p + \dot{x}_0}{\omega_d} \cdot \sin \omega_d \cdot t \right)$$

The initial conditions of the solution to the equation of motion are the deflection Δx_p and the initial velocity $\dot{x}_0$. As was done in Section 1.7, Formula 1.43 is converted into a compacted form, derived from the vector interpretation of oscillatory motions of Figure 1.9. Formula 1.29 is applied here to calculate x_a and φ as a function of C_1 and C_2. Therefore, the transient regime formula is:

Formula 1.44 Transient response to a step force input

$$x(t) = x_p + x_n(t) = x_p - x_a \cdot e^{-\zeta \cdot \omega_n \cdot t} \cdot \cos(\omega_d \cdot t + \varphi)$$

The family curves of a second-order system transient due to a unitary step input is shown in Figure 1.22.

The concept and formulas of the time constant T_z and settling time $t_{s2\%}$ apply to this transient regime because the step input does not modify the natural response.

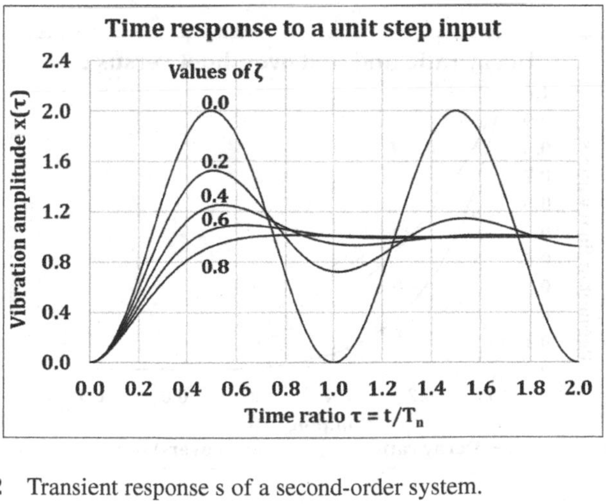

FIGURE 1.22 Transient response s of a second-order system.

It only inverts the natural oscillation curve and moves it vertically upwards (or downwards), a fixed displacement equal to the step deflection Δx_p, given by Formula 1.39.

The standard form of a system's transient response, with no initial velocity, to a unit step input, is below:

Formula 1.45 Standard form of the time second-order transient response

$$x(t) = 1 - \frac{e^{-\zeta \cdot \omega_n \cdot t}}{\sqrt[2]{1-\zeta^2}} \cdot \left(\sin \omega_n \cdot \sqrt[2]{1-\zeta^2} \cdot t + \tan^{-1} \frac{\zeta}{\sqrt[2]{1-\zeta^2}} \right)$$

This formula is graphically represented in chart 1.22 using the dimensionless time t/T_n.

1.8.3 Transient Response Overshoots to a Unit Step Input

A system with no initial velocity has the first peak at $t_p = T_d/2$. Therefore, replacing this time in Formula 1.43, the first overshoot in per unit of Δx_p is:

Formula 1.46 First overshoot in per unit of the permanent deflection Δx_p

$$s_1 = \frac{x_{peak} - x_p}{\Delta x_p} = e^{-\pi \cdot \frac{\zeta}{\sqrt[2]{1-\zeta^2}}}$$

Where Δx_p is the permanent deflection produced by the step force (calculated as x_p-x_0). The peak x_{peak} is the first maximum amplitude of x(t), which includes initial position x_0. This formula is plotted in Figure 1.23 along with the decay ratio curve (added only for comparison purposes). This plot demonstrates that the damping ratio ζ controls the natural response's decay ratio and the first overshoot to a step input.

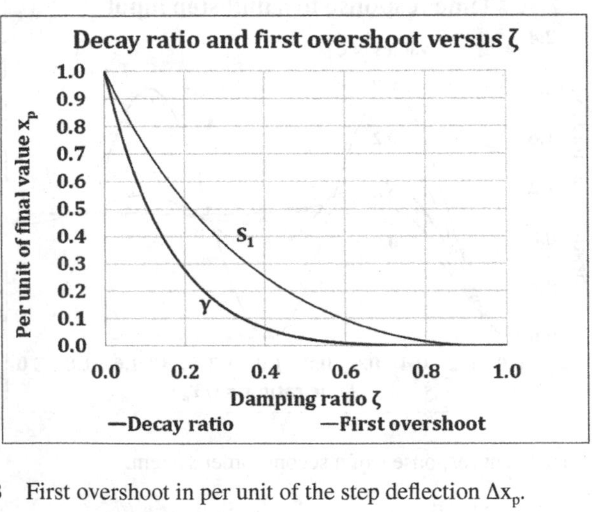

FIGURE 1.23 First overshoot in per unit of the step deflection Δx_p.

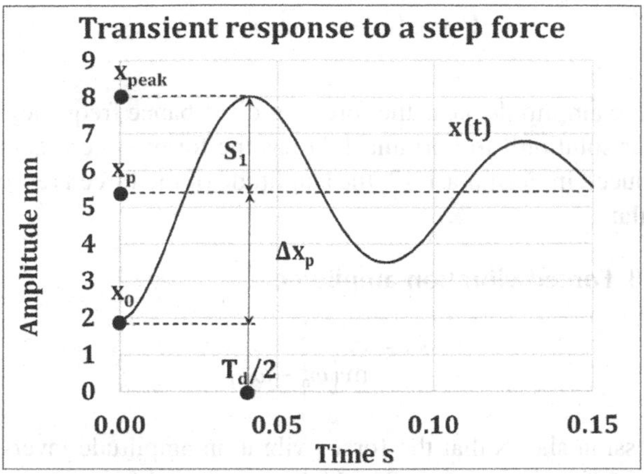

FIGURE 1.24 Graphical identification of transient response parameters.

Figure 1.24 shows the relations between initial deflection x_0, step deflection Δx_p, overshoot S_1, permanent deflection x_p and first peak (or maximum peak) x_{peak}.

As turbomachines' damping ratio is in the 0.015 to 0.125 range[10], the overshoot to step input must be expected between 67% and 91% of the permanent step Δx_p. These are significant figures that must be considered in the shaft design. Therefore, in second-order systems, the transient response to a step input system exhibits significant overshoots in the usual range of machinery damping ratio. Therefore, turbomachines need to be calculated to assess the material's stress during the overshoots. The first overshoot is the most important because it represents the maximum deflection. As the spring stress is proportional to its deflection, the maximum stress happens at the first peak.

1.9 TRANSIENT RESPONSE TO A HARMONIC FORCE INPUT

Several disturbance sources contribute to machine vibration. Most of them are harmonic. For example, if the shaft is unbalanced, the rotor vibrates with the same frequency as its angular velocity. In this case, the rotor unbalancing is a vibration source; consequentially, an accident is likely to occur due to resonance. Therefore, the system's reaction to harmonic perturbation is a valuable vibration engineering tool to identify resonance risks.

1.9.1 Conservative Vibrating System

The following ODE is the equation of motion of a frictionless machine excited by a harmonic force that may be produced by unbalancing (self-excited) or by an external source:

Formula 1.47 Equation of motion of a system with harmonic input

Applied step force = inertial + friction + rigidity reactions

$$F\cdot\cos(\omega_f\cdot t)=m\cdot\ddot{x}+k\cdot x$$

F is the force amplitude. ω_f is the forced or disturbance frequency of force F.

A particular solution[11] to Formula 1.47 has the form: $x_f \cdot \cos\omega \cdot t$. If this formula is introduced in Formula 1.47, the amplitude of the forced response is given by this formula:

Formula 1.48 Forced vibration amplitude

$$x_f=\frac{F}{m\cdot\left(\omega_n^2-\omega_f^2\right)}$$

This expression shows that the forced vibration amplitude inversely depends on the excitation force frequency. The higher the forced frequency, the lower the forced vibration amplitude.

The solution to the differential Formula 1.47 is the sum of the natural and the particular solution.

Formula 1.49 Natural and forced response to a harmonic input

$$x(t)=x_n(t)+x_f(t)=C_1\cdot\cos\omega_n\cdot t+j\cdot C_2\cdot\sin\omega_n\cdot t+x_f\cdot\cos\omega_f\cdot t$$

The initial conditions of this equation are the same as previously used, that is: $x(0)\neq 0$ and $\dot{x}(0)\neq 0$. Therefore, from the equation $x(0) = x_0 - x_f$, it arises that constant C_1 is equal to x_0, and from the equation $\dot{x}(0)=\dot{x}_0$. The result is $C_2=\omega_n/\dot{x}_0$. Replacing these constants in Formula 1.49, the general solution to the equation of motion 1.47, has the following form:

Formula 1.50 General solution to the equation of motion 1.47

$$x(t)=x_0\cdot\cos\omega_n\cdot t+\frac{\dot{x}_0}{\omega_n}\cdot\sin\omega_n\cdot t-x_f\cdot(\cos\omega_f\cdot t-\cos\omega_n\cdot t)$$

The term between parenthesis may be replaced by the product of two trigonometric expressions to demonstrate that the forced solution is the product of two harmonic signals of different frequencies.

Formula 1.51 The total response of a conservative system to a harmonic force

$$x(t)=x_a\cdot\cos(\omega_d\cdot t+\varphi)+x_f\cdot\sin\left(\frac{\omega_n-\omega_f}{2}\cdot t\right)\cdot\sin\left(\frac{\omega_n+\omega_f}{2}\cdot t\right)$$

The first component of Formula 1.51 is the system's natural response. The second term is the forced response, independent of the initial conditions[12]. The forced

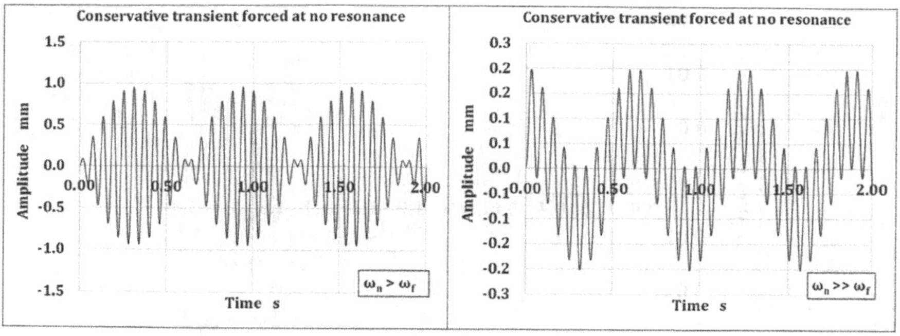

FIGURE 1.25 Forced vibration response to a harmonic force.

frequencies are the frequency difference ($\omega_n - \omega_f$) and the frequency sum ($\omega_n+\omega_f$). Therefore, the curve comprises one sinusoid of high-frequency, whose amplitudes follow a low-frequency sinusoid. See Figure 1.25.

Figure 1.25 shows the forced response for a low (left) and high (right) frequencies difference ($\omega_n - \omega_f$). The vibration curve pattern depends on the natural and forced frequency difference. The left figure pattern is known as beat frequency portrayed by an evolving sinusoid of frequency ($\omega_n - \omega_f$)/2. The beat frequency is usually observed in audio and sound vibrations. This phenomenon is also seen in electrical generators when an incoming generator is connected in parallel with other generators, and the incoming has a frequency different from the net[13].

The right side of Figure 1.25 illustrates the forced vibration for a significant frequency difference ($\omega_n-\omega_f$). The low-frequency wave carries a high-frequency wave without beating. In other words, the system acts as the amplitude modulation used in telecommunications, where one audio wave (ω_f) is carried by a high-frequency electromagnetic wave (ω_n).

1.9.1.1 Resonance of the Forced Response

As $x_f(t)$ has the difference ($\omega_n^2-\omega_f^2$) in the denominator, it is apparent that the resonance cannot be represented by said formula because there is an indetermination of the forced term of Formula 1.48 for $\omega_n = \omega_f$. In this case, the formula returns 0/0. This indetermination is prevented by applying L'Hopitale's rule.

Formula 1.52 L'Hopitale's rule applied to Formula 1.48

$$\lim_{\omega_n \to \omega_f} \frac{\left(\cos\omega_f \cdot t - \cos\omega_n \cdot t\right)}{\left(\omega_n^2 - \omega_f^2\right)} = \frac{\dfrac{d}{d\omega_f}\left(\cos\omega_f \cdot t - \cos\omega_n \cdot t\right)}{\dfrac{d}{d\omega_f}\left(\omega_n^2 - \omega_f^2\right)} = \frac{t \cdot \sin\omega_f \cdot t}{2 \cdot \omega_f}$$

This rule's application returns the following expression only valid for the resonance case of the forced vibration.

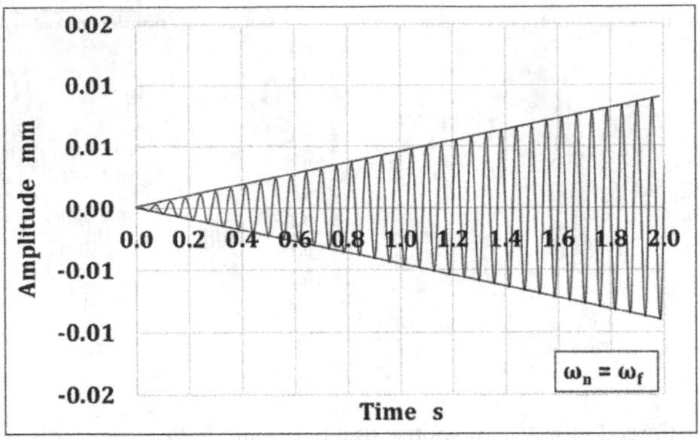

FIGURE 1.26 Forced response at resonance. Conservative system excited by a harmonic force.

Formula 1.53 Forced response at resonance

$$x_f(t) = \frac{F}{m} \cdot \frac{t \cdot \sin \omega_f \cdot t}{2 \cdot \omega_f}$$

The above formula does not include the indetermination of the third summand of Formula 1.50 for $\omega_n = \omega_f$. Therefore, the result is a growing function, which theoretically would reach an infinite value for an infinite time, which, of course, is impossible because frictionless systems do not exist in real life. Figure 1.26 shows the forced response at resonance. The wave peaks are limited by the straight lines of slope equal to F/m.

This plot concludes that resonance is manifested as a linearly growing amplitude of frequency equal to the input force. There is no doubt that this condition is undesirable and must be stopped immediately.

1.9.2 Non-Conservative Vibrating System

The equation of motion of a second-order system excited by an external harmonic is:

Formula 1.54 Differential form of the equation of motion for a system with harmonic input

$$F \cdot \cos(\omega_f \cdot t) = m \cdot \ddot{x}(t) + f \cdot \dot{x}(t) + k \cdot x(t)$$

Three components form the response of this case to a harmonic input: a) natural, b) transient forced, and c) permanent forced. The two forced responses are generated by the harmonic force applied to the system. Therefore, the system's responses can be summarized by the following formula:

$$x(t) = x_n(t) + x_f(t) \text{ where: } x_f(t) = x_t(t) + x_p(t)$$

The transient forced response $x_t(t)$ has the same damping constant T_z as the natural response, which means that both have the same settling time ($t_{s2\%}$ or $t_{s5\%}$). The permanent forced response has a constant amplitude and only disappears if the external force is removed. Formulas to calculate the transient and the permanent components are explained below in this section. The natural response is the same as the conservative case shown previously.

The general solution to the equation of motion 1.47 is:

Formula 1.55 General solution to the equation of motion 1.47

$$x(t)=e^{-\zeta\cdot\omega_n\cdot t}\cdot\left(C_1\cdot\cos\omega_d\cdot t+j\cdot C_2\cdot\sin\omega_d\cdot t\right)+x_t(t)+x_p(t)$$

The first summand is the damped natural response discussed in Section 1.7. The transient forced term $x_t(t)$ has the following form:

Formula 1.56 Transient forced response

$$x_t(t)=x_t\cdot\cos\left(\omega_f\cdot t+\varphi\right)$$

The phase angle φ appears because the friction reaction is a vector at 90° from the inertial and rigidity force vectors. Refer to Section 1.5.1.

If $x_t(t)$ is introduced in the equation of motion 1.47, the transient forced response's amplitude and phase formulas are found.

$$x_t=\frac{F}{m\cdot\sqrt[2]{\left(\omega_n^2-\omega_f^2\right)^2+4\cdot\zeta^2\cdot\omega_f^2}}\quad\varphi=\tan^{-1}\frac{2\cdot\zeta\cdot\omega_f}{\omega_n^2-\omega_f^2}$$

Applying the initial conditions used previously: $x(0) = 0$ and $\ddot{x}(0)=0$, the constants C_1 and C_2 are obtained; therefore the transient forced solution is:

Formula 1.57 General formula of the forced damped response

$$x_t(t)=-\frac{\frac{F}{m}\cdot e^{-\zeta\cdot\omega_n\cdot t}}{\left(\omega_n^2-\omega_f^2\right)^2+4\cdot\zeta^2\cdot\omega_n^2\cdot\omega_f^2}\cdot\left(C_a\cdot\cos\omega_d\cdot t+C_b\cdot\sin\omega_d\cdot t\right)$$

Where constants C_a and C_b are calculated with the following formulas:

Formula 1.58 Constants of the transient forced response

$$C_a=\omega_n^2-\omega_f^2\quad C_b=\frac{\zeta\cdot\omega_n}{\omega_d}\cdot\left(\omega_n^2+\omega_f^2\right)$$

Therefore, the amplitude and phase of the transient forced response in a compacted form are the following:

Formula 1.59 Amplitude and phase of the transient forced component

$$x_t = -\frac{\frac{F}{m} \cdot e^{-\zeta \cdot \omega_n \cdot t}}{\left(\omega_n^2 - \omega_f^2\right)^2 + 4 \cdot \zeta^2 \cdot \omega_n^2 \cdot \omega_f^2} \quad \varphi_t = \tan^{-1}\left(\frac{C_b}{C_a}\right)$$

Hence, the compacted form of the transient forced response formula is expressed:

Formula 1.60 Transient forced response

$$x_t(t) = x_f \cdot e^{-\zeta \cdot \omega_n \cdot t} \cdot \cos\left(\omega_f \cdot t + \varphi_t\right).$$

1.9.2.1 Permanent Forced Response

After the natural and transient forced components are damped, the applied harmonic force remains. This force excites the system and makes it vibrate until the input force is removed.

Formula 1.61 Permanent forced response

$$x_p(t) = x_p \cdot \cos\left(\omega_f \cdot t + \varphi_p\right)$$

However, frictional forces do not allow the system to react instantaneously and replicate the harmonic input excitation without delay. Therefore, there exists a phase delay φ included in the above formula, where the amplitude x_p and phase φ_p are given by:

Formula 1.62 Amplitude and phase of the non-conservative permanent response

$$x_p = \frac{F}{m \cdot \sqrt[2]{\left(\omega_n^2 - \omega_f^2\right)^2 + 4 \cdot \zeta^2 \cdot \omega_n^2 \cdot \omega_f^2}} \quad \varphi_p = \tan^{-1}\left(\frac{2 \cdot \zeta \cdot \omega_n \cdot \omega_f}{\omega_n^2 - \omega_f^2}\right)$$

See an example of a permanent forced response in the thin continuous curve of Figure 1.27. The frequency of the harmonic force frequency of this example is 60 rad/s.

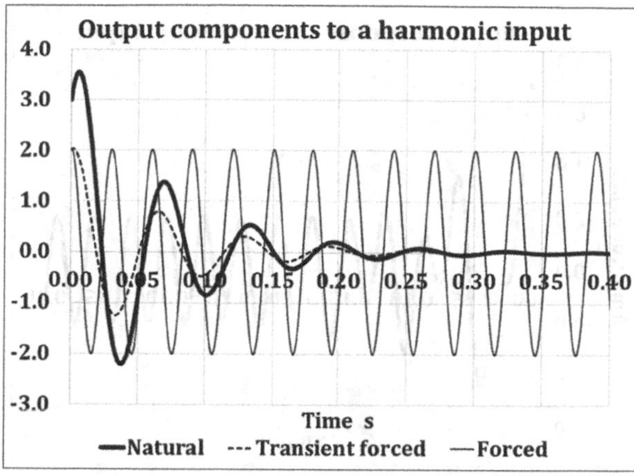

FIGURE 1.27 Vibration components.

1.9.2.2 Total Vibration

The total vibration is given by the sum of the three components discussed above. The mathematical representation of the three sums is the following formula:

Formula 1.63 Total vibration due to a harmonic force input

$$x(t) = x_a \cdot e^{-\zeta \cdot \omega_n \cdot t} \cdot \cos(\omega_n \cdot t + \varphi) + x_t \cdot e^{-\zeta \cdot \omega_n \cdot t} \cdot \cos(\omega_f \cdot t + \varphi_t) + x_p \cdot \cos(\omega_f \cdot t + \varphi_p)$$

Figure 1.28 represents the curve of the above formula. As there is a mix of three waves with different phases and frequencies (ω_n and ω_f), the first part of the response is inharmonic. This inharmonic regime disappears after the 2% settling time of the natural and damped forced components, which in this example is equal to 0.27 seconds for each of them. After the settling time has ended, only the constant forced vibration remains, with the same frequency as the exciting external source. It is recommended in practice to remove this permanent vibration due to fatigue risks. Natural and transient responses usually have a noticeably short duration, but in any case, they should be monitored and mitigated, for they may create risks for any machine.

1.9.3 Practical Assessment of a Transient Response

For example, consider the request to produce charts of components and total vibration against time for a 10 MW turbogenerator of 60 Hz (3,600 RPM). A technical report is requested with the most important conclusions. The rotor-shaft is a non-conservative second-order system, excited by the rotor-shaft due

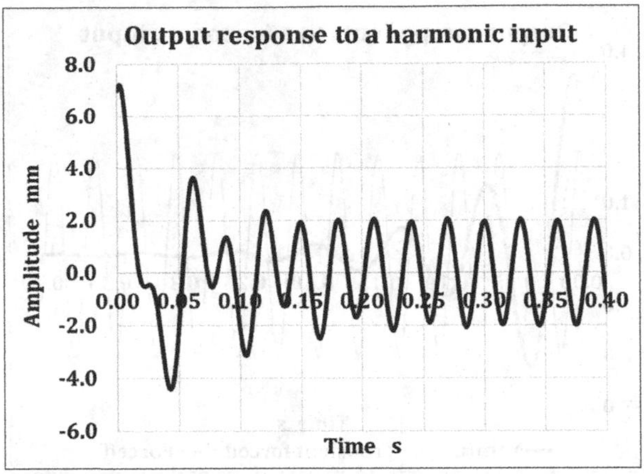

FIGURE 1.28 Total vibration.

to its imbalance. Adopt a peak-to-peak amplitude tolerance of 0.05 mm. See Figure 10.13.

Table 1.1 contains the technical specifications, formulas, and calculations of the transient response.

TABLE 1.1
Technical Specifications

Technical Specifications			
System Mass Weight	**W**	**2,000**	**kg**
The amplitude of the applied harmonic force	F	4,000	kg
Frequency of applied force	ω_f	377	rad/s
Natural frequency	ω_n	120	rad/s
Damping ratio	ζ	0.10	
Initial position x_i	x_0	0.50	mm
Initial velocity	v_0	10	mm/s
Response Parameters			
Mass	$m = W/g$	204	kg mass
Decay ratio	$\gamma(\zeta) = e^{-\frac{2\cdot\pi\cdot\zeta}{\sqrt[2]{1-\zeta^2}}}$	53.2%	per cycle
Damped frequency	$\omega_d = (1-\zeta^2)^{1/2}$	119	rad/s
Settling time	$t_{s2\%} = 1/(\zeta \cdot \omega_n)$	0.33	s
Natural Component			
Constant C_1	$C_1 = x_0$	0.50	mm
Constant C_2	$C_2 = \frac{\zeta\cdot\omega_n\cdot x_0 + \dot{x}_0}{j\cdot\omega_d}$	0.13	mm
Amplitude	$x_a = \sqrt[2]{C_1^2 + C_2^2} = \sqrt[2]{x_0^2 + \left(\frac{\zeta\cdot\omega_n\cdot x_0 + \dot{x}_0}{\omega_d}\right)^2}$	0.52	mm

(Continued)

TABLE 1.1 (Continued)

	Technical Specifications		
System Mass Weight	**W**	**2,000**	**kg**
Phase	$\varphi = \arctan\left(\frac{C_2}{C_1}\right)$	0.26	rad
Transient Forced Component			
Frequency ratio	$u_f = \omega_f / \omega_n$	3.14	
Constant C_a	$C_a = \omega_n^2 - \omega_f^2$	-0.0000078	$(s/rad)^2$
Constant C_b	$C_b = \frac{\zeta \cdot \omega_n}{\omega_d} \cdot \left(\omega_n^2 + \omega_f^2\right)$	0.0000010	$(s/rad)^2$
Amplitude x_t	$x_t = -\frac{\frac{F}{m} \cdot e^{-\zeta \cdot \omega_n \cdot t}}{\left(\omega_n^2 - \omega_f^2\right)^2 + 4 \cdot \zeta^2 \omega_n^2 \cdot \omega_f^2}$	0.1540	mm
Phase	$\varphi_t = \tan^{-1}\left(\frac{C_b}{C_a}\right)$	-0.123	rad
Permanent Forced Component			
Amplitude	$x_p = \frac{\frac{F}{m}}{\sqrt[2]{\left(\omega_n^2 - \omega_f^2\right)^2 + 4 \cdot \zeta^2 \cdot \omega_n^2 \cdot \omega_f^2}}$	0.1532	mm
Phase	$\varphi_p = \tan^{-1}\left(\frac{2\zeta \cdot \omega_n \cdot \omega_f}{\omega_n^2 - \omega_f^2}\right)$	-0.063	rad

1.9.3.1 Technical Assessment Summary

The amplitude tolerance is 0.05 mm peak to peak, which at 3,600 RPM is between fair and good. See the Vibration Severity Table in Chapter 10. However, as the amplitudes are $x_a = 0.520$ mm, $x_t = 0.154$ mm, and $x_p = 0.153$ mm, the permanent response peak is three times the allowable tolerance.

The natural component exceeds the peak tolerance in the first 0.01 seconds, and then, the damped forced component has peaks higher than tolerance during the first 0.007 seconds. If the rotating machine is subject to frequent startups during service, these transient peaks may affect the shaft's mechanical integrity. Therefore, some of the techniques described in Part III Chapters are recommended to reduce vibration amplitudes.

Figure 1.29 shows the oscillograms of the output harmonic components and the total vibration, the same as would appear in a vibrometer's display.

Figure 1.29 demonstrates that resonance is a dangerous condition that requires an urgent machine stoppage. In this case, it is recommended to implement a

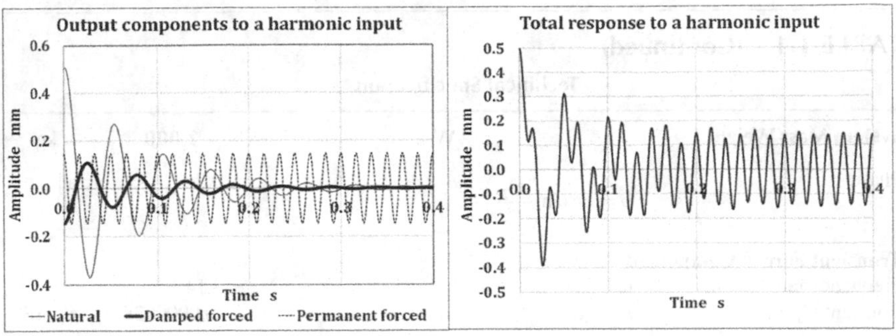

FIGURE 1.29 Vibration amplitude. Components and total.

solution before a new startup. Modern machines are provided with a control system that allows setting the alarm stop levels and prevent dangerous conditions.

1.10 FREQUENCY RESPONSE

The frequency response studies the system's amplitude and phase as a function of the frequency. This study, based on a harmonic excitation input, returns essential information about system performance. In Vibration Engineering, the frequency response enables finding resonance risks and provides valuable information for the design of vibration absorbers, isolators, and active feedback systems.

The frequency response may be obtained by a test, where the system is subject to a harmonic input force, and the output is recorded or theoretically derived based on the equation of motion. In this section, a second-order system's frequency response is derived and discussed to reach conclusions about its resonance and critical parameters.

Control systems engineers use the frequency response to determine whether a control system is stable and design compensators to obtain the desired performance indexes of stability and accuracy. One of the most common techniques to portray the frequency response is using the Bode plot. This plot represents the amplitude in decibels, and phase versus frequency, on a logarithmic scale. See in Chapter 11 the use of this plot to assess the performance and design of feedback systems for vibration control.

1.10.1 Frequency Response of Second-Order Systems

In linear systems, the frequency response is the ratio of vibration amplitude x_a over the shaft static deflection x_0 as a function of frequency ω. The static deflection is the shaft deflection that would produce the amplitude of the harmonic force excitation under static conditions, that is, with no vibration. The formula to calculate the static deflection is $x_0 = F/k$, where x_0 is the static deflection, F is the force amplitude of the harmonic excitation, and k is the shaft stiffness coefficient for symmetrical concentrated load F on the shaft. The static deflection is a constant value; therefore, it does not depend on the excitation frequency.

The x_a/x_0 ratio is designated Q(ω), usually named the magnification factor. In Vibration Engineering. Control, engineers prefer the term frequency response modulus.

Formula 1.64 Magnification factor or frequency response modulus

$$Q(\omega)=\frac{x_a(\omega)}{x_0}$$

x_a (ω) is the vibration amplitude against frequency ω.

The magnification factor Q(ω) formula is derived based on the equation of motion 1.2. The input force is harmonic of frequency ω, represented by the exponential formula $F \cdot e^{j\omega t}$, where F is the input force's amplitude. According to Euler's formulas, the input force may be expressed as F · (cos ω · t + j · sin ω · t). Nonetheless, the derivation that follows uses complex exponential expressions because they have some advantages. The differential equation of motion is:

Formula 1.65 Equation of motion with harmonic input

$$F\cdot e^{j\cdot\omega\cdot t}=m\cdot\ddot{x}(t)+f\cdot\dot{x}(t)+k_w\cdot x(t)$$

The integral of the above equation and its derivatives is:

Formula 1.66 Particular integral and its derivatives

$$x(t)=x_a\cdot e^{j\cdot\omega\cdot t}\ \dot{x}(t)=x_a\cdot j\cdot\omega\cdot e^{j\cdot\omega\cdot t}\ \ddot{x}(t)=-x_a\cdot\omega^2\cdot e^{j\cdot\omega\cdot t}$$

The imaginary unit j means that the velocity vector is 90° ahead of the vibration amplitude vector. See Section 1.5.1. Formulas 1.66 are replaced in the equation of motion 1.65 to see that the vibration amplitude x_a is a function of the frequency, the natural frequency and the damping ratio. After these replacements are done, the following formula of forces equilibrium is obtained:

Formula 1.67 Applied force and reactions in complex notation

$$F=\left(k-m\cdot\omega^2+j\cdot f\cdot\omega\right)\cdot x_a(\omega)$$

Formula 1.67 is represented by the triangle forces of Figure 1.30.

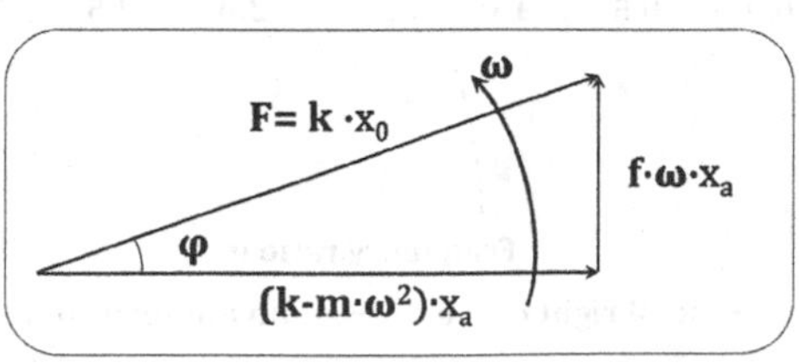

FIGURE 1.30 Vector interpretation of applied and reaction forces.

The vibration's frequency response $x_a(\omega)$ is cleared from Formula 1.67, where the $k \cdot x_0$ product replaces the force amplitude F.

Formula 1.68 Frequency response of the vibration amplitude

$$x_a(\omega) = \frac{k \cdot x_0}{k - m \cdot \omega^2 + j \cdot f \cdot \omega}$$

The real part of this function is the inertial and rigidity reactions, and the imaginary part is the friction reaction force. The above is a complex expression that can be split into one real part and one imaginary part by multiplying and dividing it by the denominator conjugate. After this operation, the magnification factor and phase's frequency response is given by the following formulas:

Formula 1.69 Frequency response of a mechanical system

$$|Q(\omega)| = \left|\frac{x_a(\omega)}{x_0}\right| = \frac{k}{\sqrt[2]{\left(k - m \cdot \omega^2\right)^2 + (f \cdot \omega)^2}} \quad \varphi(\omega) = -\tan^{-1}\left(\frac{f \cdot \omega}{k - m \cdot \omega^2}\right)$$

In the remainder of this book, the vertical bars notation of absolute values is dropped off. Therefore, $Q(\omega)$ stands for the frequency response modulus.

Formulas 1.69 are converted to dimensionless expressions introducing the frequency ratio $u = \omega/\omega_n$ and the damping ratio ζ. With the adoption of these dimensionless parameters, it is possible to plot the universal curves of Figures 1.32 and 1.33. These curves are also applicable in rotating systems except for self-excited

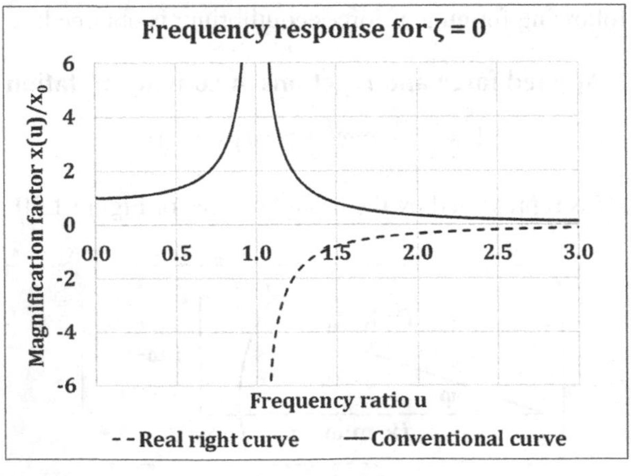

FIGURE 1.31 Frequency response of a frictionless second-order system.

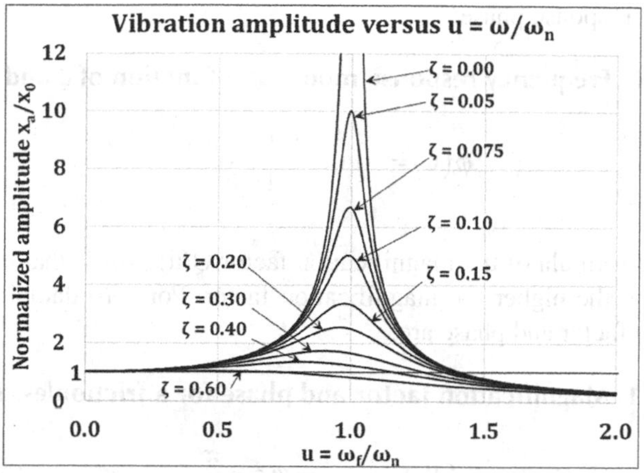

FIGURE 1.32 Magnification factor versus frequency.

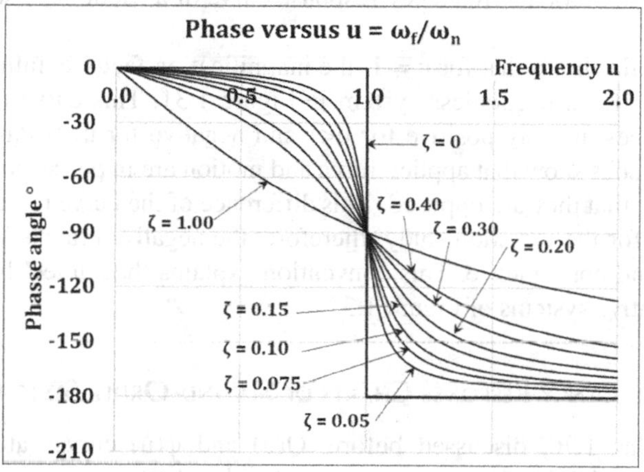

FIGURE 1.33 Curves of phase versus frequency.

and base-excited vibration. See these cases in Section 1.11. Formulas used to create these figures are the following:

Frequency ratio:

$$u = \frac{\omega}{\omega_n}$$

Magnification factor:

$$Q(u) = \frac{x_a(u)}{x_0} = \frac{1}{\sqrt[2]{\left(1-u^2\right)^2 + \left(2\cdot\zeta\cdot u\right)^2}}$$

Frequency response phase:

Formula 1.70 Frequency response model as a function of ζ and u

$$\varphi(u) = -\tan^{-1}\left(\frac{2\cdot\zeta\cdot u}{1-u^2}\right)$$

The above formula of the magnification factor Q(u) proves that the lower the damping ratio, the higher the magnification factor. For zero damping ratio, the magnification factor and phase are:

Formula 1.71 Magnification factor and phase for a frictionless system

$$Q(u) = \frac{1}{1-u^2} \quad \varphi = -\frac{\pi}{2}$$

Figure 1.31 shows the frequency response curve of a frictionless second-order system.

This formula shows that for u = 1, the magnification factor is infinite. See the curve of Q(u) for a frictionless system in Figure 1.31. This curve is composed of two branches, namely positive for u<1 and negative for u>1 (dashed curve). Positive ordinates show that applied force and motion are in phase. Negative ordinates indicate that they are opposed. This difference of the curve representation is not essential for the vibration study; therefore, the negative branch is positive by convention and convenience. This convention explains the curves' bell shape of non-conservative systems of Figure 1.32.

1.10.2 Frequency Response Charts of Second-Order Systems

With Formulas 1.70, discussed before, Q(u) and φ(u) curves are plotted in Figures 1.32 and 1.33. These curves are considered the most important vibration engineering plots due to their universal validity and intense application in practice. They allow visualizing resonance problems in many types of machines, structures, and devices.

The damping ratio is the parameter of each curve. This set of curves helps to quickly determine the resonance peak at any frequency and damping ratio as a frequency ratio function. At resonance, the phase is -90° for any $\zeta \neq 0$.

For the typical damping ratio range of rotating machines (0.015 to 0.125), the magnification factor at the resonance is in the 33 to 4 range. It means that 1 mm of the shaft static deflection x_0 is converted into 33 to 4 mm due to centrifugal forces generated during rotation. In most cases, this deflection increment is not acceptable.

Machines and constructions are not precisely second-order systems. For example, when machines are implemented with vibration absorbers, they form a 4th order system. Nonetheless, their behavior is assessed using the concepts, figures and formulas of second-order systems as the first approach to vibration problems.

1.10.3 Resonance Parameters

Resonance is a harmful condition to the machine's mechanical integrity and usually requires an in-depth understanding of the root causes to adopt preventative measures. There is an elementary and typical example of resonance prevention in a small bridge walked by a military regiment. If the soldiers' steps coincide with the bridge's natural frequency, resonance may happen, and the bridge may collapse. Thus, in some cases, military parades on bridges are replaced by a free walk. This affirmation is not an overstated anecdote because the Tacoma Narrows bridge in Washington state was inaugurated in 1940 and fell four months later. The fatal accident was due to von Karman vortices and other reasons that excited the bridge at its natural frequency. On that opportunity, Von Karman knew that the governor publicly promised: "We are going to build the exact same bridge, exactly as before." Von Karman, famous for his ironic sayings, sent the following telegram to the governor: "If you build the exact same bridge as before, it will fall in the exact same river exactly as before."

The resonant frequency formula ω_r is cleared from the derivative of Formula 1.70 of Q(u) with respect to u and equating this expression to zero. The result is:

Formula 1.72 Derivation of the resonant frequency ratio

$$\frac{dQ(u)}{du} = 0 \rightarrow u_r = \sqrt[2]{1 - 2\cdot\zeta^2}$$

There are three characteristic frequencies in second-order systems: the natural frequency ω_n, the damped frequency ω_d, and the resonant frequency ω_r. The natural frequency ω_n is the highest, followed by the damped frequency ω_d. The resonant frequency ω_r is the lowest. Nonetheless, they are of the same order within the machines' typical ζ range. In this range, the maximum differences between the natural frequency and the resonant and damping frequencies are -2.3% and -8.1%, respectively.

The following set of formulas summarize the relations between the three characteristic frequencies.

Formula 1.73 Damping and resonance frequencies as a function of the natural frequency

$$u_d = \sqrt[2]{1 - \zeta^2} \quad u_r = \sqrt[2]{1 - 2\cdot\zeta^2}$$

$$\omega_d = \omega_n \cdot u_d \quad \omega_r = \omega_n \cdot u_r$$

u_d is the damped frequency ratio
u_n is the natural frequency ratio

Replacing these frequencies in Formula 1.70 of Q(u), the magnification factors at those three frequencies are returned.

Formula 1.74 Magnification factors at their characteristic frequencies

$$Q_n\left(u_n\right)=\frac{1}{2\cdot\zeta}\;Q_d\left(u_d\right)=\frac{1}{\zeta\cdot\sqrt{4-3\cdot\zeta^2}}\;Q_r\left(u_r\right)=\frac{1}{2\cdot\zeta\cdot\sqrt{1-\zeta^2}}$$

$Q_n(u_n)$ is the magnification factor at the natural frequency u_n
$Q_d(u_d)$ is the magnification factor at the damped frequency u_r
$Q_r(u_r)$ is the magnification factor at the resonant frequency u_r

Figure 1.34 shows that the magnification factors' curves are almost superimposed, especially for damping ratios lower than 0.2, where the rotating machine's damping ratio range is placed. Therefore, in rotating machines, it is accepted, with little error, that $Q_n \simeq Q_d \simeq Q_r$.

Table 1.2 shows that the resonance magnification factor is close to unity for damping ratios higher than 0.6. In a nutshell: the values of damping ratio higher

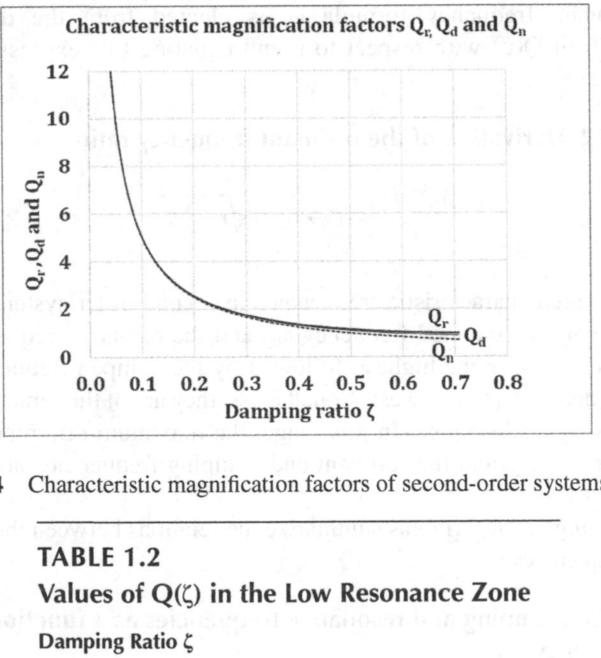

FIGURE 1.34 Characteristic magnification factors of second-order systems.

TABLE 1.2
Values of Q(ζ) in the Low Resonance Zone

Damping Ratio ζ					
0.600	**0.620**	**0.640**	**0.660**	**0.680**	**0.707**
Resonance magnification factor Q_r					
1.042	1.028	1.017	1.008	1.003	1.000

than 0.6 poorly amplify the vibration amplitude. In this zone, the magnification factor Q is considered equal to 1.

Therefore, the zone of negligible resonance is at values of ζ higher than 0.60, and there is no resonance at values of ζ higher than 0.707.

1.11 FUNDAMENTAL VIBRATION FORMS

Mechanical vibration is a dynamic oscillation of small amplitude occurring in simple daily life devices as well as complex and sophisticated machinery such as laboratory instrumentation, industrial equipment, aircraft, ships, cars, structures, and many other construction and manufactured products. Vibration is a phenomenon produced for many reasons. Based on vibration causes, four fundamental vibration forms are identified, namely:

1. Form 1. External excitation
2. Form 2. Self-excited
3. Form 3. Base-excited
4. Form 4. Transmitted force

The equations of motion return formulas for these vibration modes describing the system performance, which are useful for the first approach to a vibration problem, determining potential causes, assessing risks, and how to mitigate them. The frequency response of these vibration modes reveals significant differences between them, discussed in the following sections.

Any of these four fundamental vibration forms can guide the design of workable solutions. The vibration modes' models of this section assume that the system is SDOF; thus, it is straightforward to produce a consistent model using standard spreadsheets based on second-order system formulas. After that, if the problem requires a more in-depth analysis, it is recommended to use formulas of multi-degree of freedom (MDOF) or continuous systems or finite element models. In general, the use of the last method is done by specialists. John Vance[14] recommends engineers start solving a vibration problem with the simplest model. After that, if this simple model is inaccurate or does not correctly describe the observed vibration and does not provide directions to design a solution, the model must be upgraded to a more sophisticated mathematical structure.

1.11.1 Externally Excited Mode

The physical configuration of this fundamental vibration form is depicted in Figure 1.35. It vibrates because it is excited by a vertical harmonic force applied to the mass.

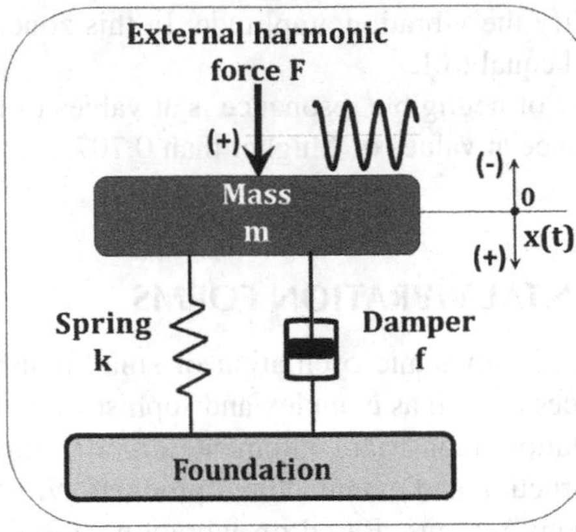

FIGURE 1.35 Physical configuration of the externally excited mode.

The frequency response expression is Formula 1.70, reproduced below for the reader's convenience.

Formula 1.75 Frequency response of the externally excited mode

$$Q(u) = \frac{x_a(u)}{x_0} = \frac{1}{\sqrt[2]{\left(1-u^2\right)^2 + \left(2\cdot\zeta\cdot u\right)^2}}$$

At frequencies higher than resonant frequencies, the magnification factor tends to zero. Figure 1.36 shows the Q(u) curves for several values of the damping ratio. On each curve, there is a crossover frequency u_x at which it is Q = 1. Thus, beyond this crossover frequency, the static deflection x_0 is attenuated instead of amplified.

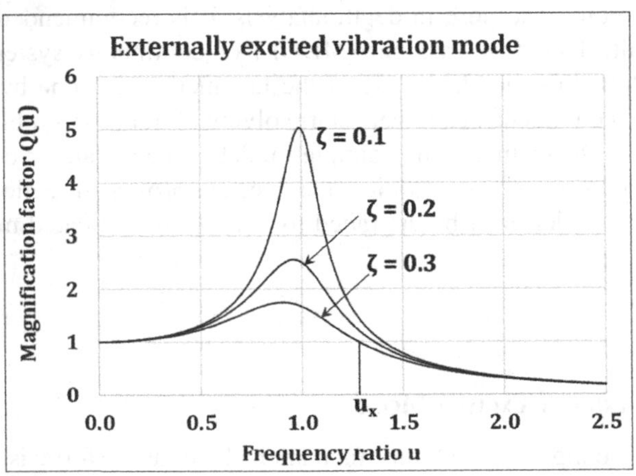

FIGURE 1.36 Frequency response of the externally excited mode.

The formula of u_x is cleared from Formula 1.75 for Q(u) = 1 resulting in a function of the damping ratio.

Formula 1.76 Crossover frequency at Q = 1 of externally excited mode

$$u_x(\zeta) = \sqrt[2]{2\cdot(1-2\cdot\zeta^2)}$$

At frequencies higher than u = 0.707, the crossover frequency turns imaginary; therefore, the magnification factor has no physical meaning at frequencies higher than that frequency. The modulus of Q(u) descends after the resonance peak. At an infinite frequency, Q(u) tends to zero.

The most important conclusions about the externally excited form are:

- At zero frequency, the amplitude x_a and the static deflection x_0 coincide. Both are equal to 1. It means that there is no vibration.
- For frequencies between zero and the resonant peak, Q(u) is higher than 1. Therefore, the static deflection x_0 is amplified in that frequency range. That is why Q(u) is known as the magnification factor.
- The frequency response curve presents a peak at $u_r = \sqrt[2]{1-2\cdot\zeta^2}$. For $\zeta \geq 0.707$, there is no resonance peak.
- At frequencies higher than the resonant frequency, the magnification descends, and after the crossover point, it is lower than 1. It means that the system does not magnify but attenuates the input disturbance.

It is unusual in practice to deal with a machine repeatedly knocked on its mass, as suggested by this vibration mode. However, there are industrial applications where the machine is externally forced to vibrate. That is the case of vibrating troughs and screens used in many industries where bulk material must be conveyed or classified according to size. This equipment is often used in mining, food, paper, plastic, metalworking, recycling, pharmacy, sawmills, cement, etc.

1.11.2 Self-Excited Mode

Turbomachines and reciprocating motors are often affected by rotor imbalance. Therefore, the proper form to investigate their behavior is the self-excited mechanical configuration of Figure 1.37. The system is a rotor-shaft mounted on bushings, pedestals, and foundations. A foundation supports the assembly through dampers and springs.

An unbalanced rotor-shaft means that its gravity center is not exactly on its gyration axis. The distance from this axis to the gravity center is the gyration radius of a centrifugal force produced by the unbalanced mass m. As the rotor gyrates, this force has no constant modulus for reasons discussed in Chapter 5 (refer to lateral vibration). Therefore, the centrifugal force produces a self-excited

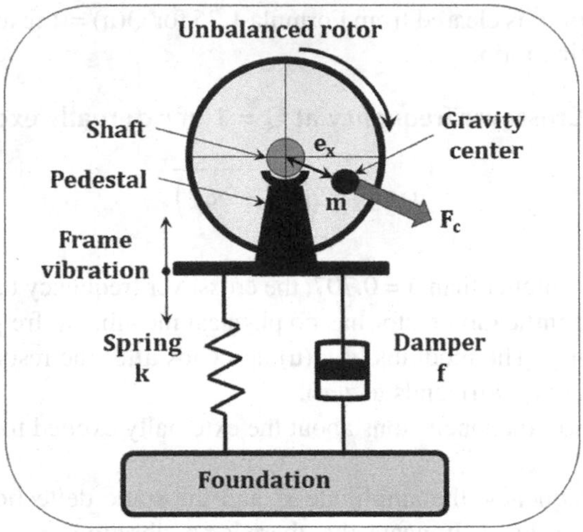

FIGURE 1.37 Physical configuration of self-excited mode.

vibration that affects the whole machine. In Figure 1.37, e_x is the gravity center offset at standstill condition, named eccentricity.

The centrifugal magnification factor $Q_c(u)$ is obtained from the centrifugal force equilibrium with the rotor-shaft rigidity and friction reactions. The following formula expresses this equilibrium. It is a complex expression based on the vector interpretation of forces of Section 1.5.1.

From the forces equilibrium formula, the deflection produced by the centrifugal force is cleared, and the result is:

Formula 1.77 Force's equilibrium and centrifugal deflection

$$m\cdot(e_x + \delta_0 + \delta_c)\cdot\omega^2 = (k + j\cdot f\cdot\omega)\cdot\delta_c$$

$$\delta_c = \frac{m\cdot(e_x + \delta_0)\cdot\omega^2}{k - m\cdot\omega^2 + j\cdot f\cdot\omega}$$

The shaft bending produces the deflection δ_0 due to the shaft and rotor weight at static conditions. As shown in Chapter 5, the above formula considers that the centrifugal force is pointing downward, where the maximum centrifugal force is produced. At any other angular position of the centrifugal force, deflection δ_c is lower. Therefore, the above formula returns the maximum deflection produced during one shaft turn.

It is reminded that the imaginary term of Formula 1.77 means that the friction vector is at 90° of the rigidity and inertia vectors. Taking the modulus of $Q_c(u)$ and clearing the $\delta_c/(e_x + \delta_0)$ ratio and replacing ω by $\omega_n \cdot u$, the magnification factor of the centrifugal deflection versus the frequency ratio u is returned.

Formula 1.78 Magnification factor of the centrifugal deflection

$$Q_c(u) = \frac{\delta_c(u)}{e_x + \delta_0} = \frac{u^2}{\sqrt[2]{\left[1-u^2\right]^2 + 4\cdot\zeta^2\cdot u^2}}$$

Figure 1.38 is the graphic representation of this magnification factor against u for a self-excited system

The magnification factor is lower than 1 between u = 0, and the crossover frequencies u_s indicated in Figure 1.38, where the subscript s stands for the self-excited mode. At high velocities, the frequency response modulus tends to 1 because the rotor-shaft inertia motion cannot replicate high frequencies and then, it tends to return to its standstill deflection.

The following formula gives the crossover frequency at $Q_c = 1$.

Formula 1.79 Crossover frequency at $Q_c = 1$

$$u_s = \frac{1}{\sqrt[2]{2\cdot\left(1-2\cdot\zeta^2\right)}}$$

It can be demonstrated that the crossover frequency u_s is the inverse of u_x ($u_s = 1/u_x$). However, as both crossover frequencies belong to two different vibration modes (externally excited and self-excited), it is impossible to infer an easy physical consequence from this mathematical property. The crossover frequencies u_s and u_x are plotted on the left of Figure 1.39 as a function of the damping ratio. The right figure compares the externally excited and self-excited modes and indicates u_s, u_x and the attenuation zone. The base excited form is described below.

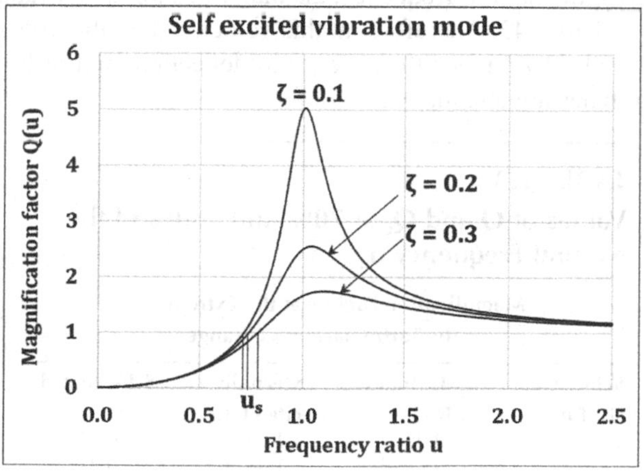

FIGURE 1.38 Magnification factor of the self-excited mode.

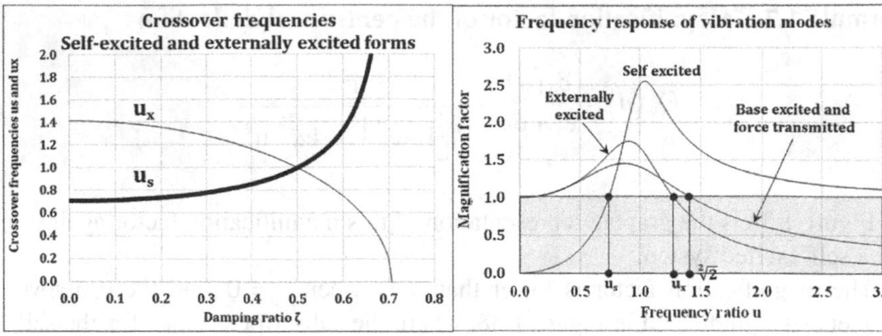

FIGURE 1.39 Crossover frequencies at Q = 1.

For $u<u_s$, in this case, the offset formed by the eccentricity plus the static deflection sum is attenuated. For $u > u_s$, the magnification factor $Q_c(u)$ is higher than 1; therefore, the shaft offset is magnified. In this range, the resonant frequency occurs. Therefore, it is recommended to refrain from operating at frequencies higher than 70% of the resonant frequency, which is not always possible. Were it is not feasible to operate at velocities lower than 70% of the resonance, the best solution is balancing the shaft rotor in a specialized workshop.

1.11.2.1 Note About the Recommended Velocities Range

The usual recommended velocities range for rotating machines is less than 70% or higher than 130% of the natural frequency. However, the magnification factor has a significant value at these velocities, as shown in Table 1.3.

This table shows that in the machines' damping ratio range, the magnification factor for an externally-excited mode, at a velocity of 70% of the natural frequency, is in the 1.86 to 1.94 range. Therefore, the vibration amplitude is almost double the static deflection x_0 at velocities lower than the natural frequency. At a speed of 130% of the natural frequency, the magnification factor range has lower values, from 1.31 to 1.42. Therefore, at this velocity, the vibration amplitude is between 31% and 42% higher than the static deflection, and at higher speeds, vibration amplitudes are attenuated.

TABLE 1.3
Values of Q and Q_c at 70% and 130% of the Natural Frequency u_n

	Magnification Factors at the Extremes of the Turbomachines ζ Range		
Velocity in % of u_n	ζ Machines Range	Externally Excited Q(u)	Self-Excited $Q_c(u)$
70%	0.050	1.94	0.95
	0.125	1.86	0.91
130%	0.050	1.42	2.40
	0.125	1.31	2.21

The self-excited form attenuates the amplitudes at 70% of the natural frequency but significantly amplifies vibration amplitude at 130% of the natural frequency. In the last case, amplitudes are more than double the $e_x+\delta_b$ offset. Therefore, an unbalanced rotor gyrating at velocities higher than the natural frequency produces a more significant vibration amplitude than an externally excited mode. This significant difference between the externally excited vibration amplitudes and the self-excited modes is clearly shown in the right side of Figure 1.39. Note that while the externally excited curve is in the attenuation zone, the self-excited is always in the amplification zone.

Therefore, due to the rotor-shaft imbalance, a machine can vibrate at velocities beyond the natural or resonance frequencies. The importance of the periodic rotor inspection and the tracking of its amplitude vibration is derived from this property. The solution for this case is always to order the rotor balanced in a specialized workshop.

In summary: high amplitude vibration at low velocities is mainly produced by external excitation and at high speeds due to rotor imbalance.

1.11.3 Base-Excited Form

This fundamental vibration form is based on the configuration illustrated in Figure 1.40. The base vibration is produced by other structures or machines that propagate their alternating forces through the ground. There are two different displacements: the base displacement $x_b(t)$ is the input, and the mass displacement $x_m(t)$ is the output. Therefore, the magnification factor is the ratio x_m/x_b as a function of the exciting frequency. Therefore, this case is not an SDOF system. Vibration is transmitted from the base to the mass through a spring and damper set.

The frequency response curve of the base-excited system is in Figure 1.41. It must be noted the attenuation zone for frequencies higher than $\sqrt[2]{2}$.

In this case, the base applies a force transmitted to the mass through the spring-damper set. This force is equal to the spring coefficient times the base

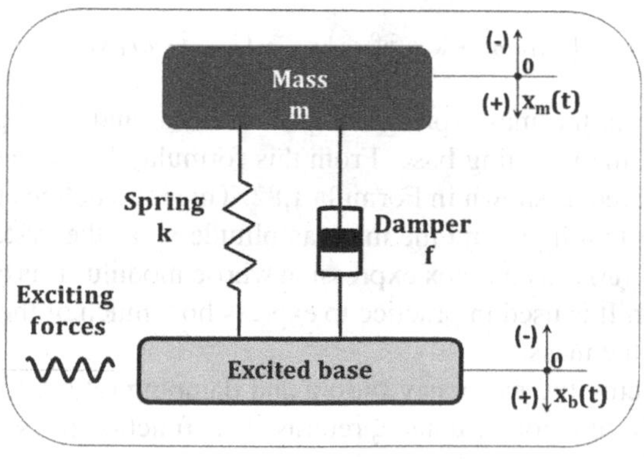

FIGURE 1.40 Physical configuration of base-excited mode.

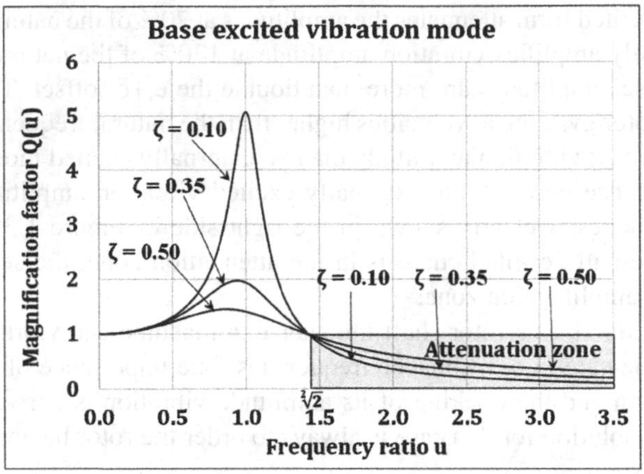

FIGURE 1.41 Frequency response of base-excited mode.

displacement x_b plus the damper coefficient times the base displacement velocity v_b. The reaction force has three components: the mass times its acceleration a_m, the damper friction coefficient times the mass velocity $v_{m,}$ and the spring rigidity coefficient times the mass displacement x_m.

The following is the expression of the equation of motion of the base-excited mode:

Formula 1.80 Equation of motion of the base-excited mode

$$m \cdot \ddot{x}_m + f \cdot \dot{x}_m + k \cdot x_m = f \cdot \dot{x}_b + k \cdot x_b$$

Therefore, the equation of motion is explained in Section 1.5.1; this equation is converted into the following complex expression:

Formula 1.81 Vector expression of the equation of motion

$$\left(-m \cdot \omega^2 + k + j \cdot f \cdot \omega\right) \cdot x_m = \left(k + j \cdot f \cdot \omega\right) \cdot x_b$$

The left term is the mass-spring-damper reaction, and the right term is the applied force by the vibrating base. From this formula, the vibration amplitudes ratio x_m/x_b is cleared as shown in Formula 1.82. This ratio defines the base amplitude x_b as the system input and the mass amplitude x_m as the system output. The result of $x_m(\omega)/x_b(\omega)$ is a complex expression whose modulus r_x is called the transmissibility factor. It is used in practice to express how much of the base vibration is transferred to the mass.

After introducing the frequency ratio u and damping ratio ζ formulas[15] in the above equation, a function of u and ζ returns. This function physically represents the base-excited mode's frequency response.

Formula 1.82 Amplitude ratio or frequency response of the base-excited mode

$$r_x(u) = \frac{x_m(u)}{x_b(u)} = \sqrt[2]{\frac{1+4\cdot\zeta^2\cdot u^2}{\left(1-u^2\right)^2+4\cdot\zeta^2\cdot u^2}}$$

The chart in Figure 1.41 shows that both displacements, mass and foundation are equals at zero frequency. Like in the externally forced mode, the amplitude grows until it reaches the maximum value at the resonant frequency. In the descending branches, all curves cross at the point of Q = 1 and $u = \sqrt[2]{2}$. It means that at that frequency, for any damping ratio value, the base and the mass displacements are the same. For values of frequency higher than $\sqrt[2]{2}$ all curves are below 1; therefore, vibration amplitudes are attenuated in that zone. See the attenuation zone in Figure 1.41. In this zone, the lower the damping ratio, the lower the amplitude. It means that adding friction to mitigate vibration is useless because the system behaves in the opposite direction.

1.11.4 Transmitted Force Mode

A typical engineering subject is the transmitted force calculation to a rigid foundation for designing or evaluating the base breakage risks. See Figure 1.42. The foundation is supposed perfectly rigid; therefore, it is motionless and cannot vibrate. Due to the foundation's perfect rigidity, forces received by the foundation are directly transmitted to the ground, also assumed to be perfectly rigid. Therefore, in Figure 1.42, the mass system has only one vertical vibration displacement.

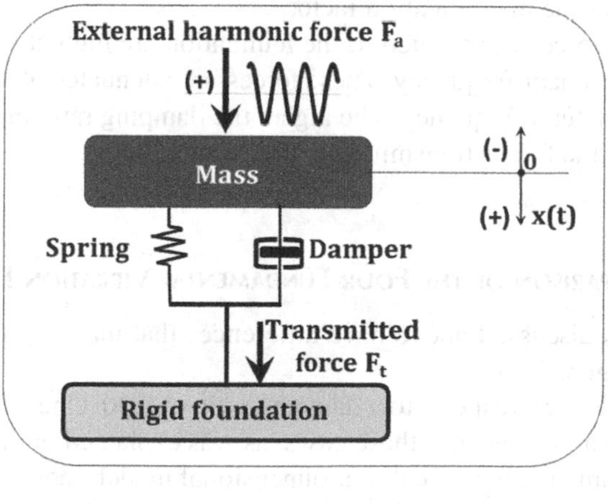

FIGURE 1.42 Physical configuration of the force transmitted system.

The ratio between the transmitted and the applied force is derived in Chapter 8. The result is Formula 8.8, reproduced below.

Formula 1.83 Transmissibility ratio or frequency response of the force transmitted mode

$$r_t(u) = \frac{F_t(u)}{F_a(u)} = \sqrt[2]{\frac{1+4\cdot\zeta^2\cdot u^2}{\left(1-u^2\right)^2+4\cdot\zeta^2\cdot u^2}}$$

The denominator under the square root is the expression of the mass-damper-spring reaction to the applied force $F_a \cdot \sin(\omega \cdot t)$. The numerator represents the force transmitted to the foundation. Physically, this ratio is the portion of the applied force transmitted to the foundation and the ground. In practice, the maximum allowable ratio is usually specified for machines that vibrate or may vibrate during their operation.

Formula 1.83 is identical to Formula 1.82; however, the physical systems that they represent are different. In the base-excited mode, the system input and output are the amplitudes vibration. In the force-transmitted mode, the system input and output are the applied and transmitted forces, respectively. As the magnification factor formulas of both modes are identical, the force-transmitting frequency response curve is also represented by the chart in Figure 1.41.

All considerations made for the base-excited form with respect to the attenuation zone hold valid for the transmitted force mode. Critical frequency, $u = \sqrt[2]{2}$ at which all curves cross in Q = 1 is the frequency of equal forces. At this point, the transmitted force is equal to the applied force. For values of frequency higher than $u = \sqrt[2]{2}$, the transmitted force is lower than the applied force. In this attenuation zone, same as in the base-excited mode, the higher the damping ratio, the higher the magnification factor. Therefore, it does not make sense to increase the attenuation zone's damping ratio to reduce the transmitted force because the system reaction increases the magnification factor.

In summary: forces transmitted to the foundation are high at velocities below the machine's resonant frequency. These forces are attenuated at velocities much higher than the natural frequency. The higher the damping ratio in the attenuation zone, the higher the forces transmitted to the foundation.

1.11.5 Comparison of the Four Fundamental Vibration Forms

The four modes discussed above have differences that may be visualized in the right chart in Figure 1.39.

The three curves correspond to a damping ratio of 0.30. One of the curves represents modes 3 and 4 because these cases, as was explained in the previous section, have the same mathematical non-dimensional model, though they represent different physical systems.

The most important conclusion of this figure is:

- The four modes have different resonant frequencies.
- The base-excited and force transmitted curves have a lesser peak amplitude and tend to zero at a slower rate than the externally excited mode.
- At high velocities, only the self-excited curve is not in the attenuation zone. After the resonance peak, it has the highest amplitude, followed by the base-excited and force transmitted curves, which are attenuated.
- At low velocities, the self-excited form is the only curve in the attenuation zone.
- An unbalanced rotor is unnoticeable at velocities lower than the frequency u_s, but it is significant at velocities equal or higher than the resonant frequency.

NOTES

1 Recommended books for Chapter 1: 1. Mechanical Vibrations by S. Graham Kelly. Schaum's Outline Series. McGraw Hill, 1996 – 2. Advanced Vibration Analysis. By S. Graham Kelly. CRC Press. Taylor and Francis Group, 2007 – 3. Fundamentals of Vibration Engineering by Isidor Bykhovsky. MIR Publishers, Moscow. 1st published 1972 – 4. Mechanical Vibrations by J.P. Den Hartog. Dover Publications, Inc. New York, 1985, 4th edition – 5. Mechanical Vibration by Haym Benaroya, Mark Nagurka and Seon Han. CRC Press. Taylor and Francis Group, fourth edition 2018.

2 See https://www.researchgate.net/publication/310627137_Bladed_wheels_damage_detection_through_Non-Harmonic_Fourier_Analysis_improved_algorithm. May 2017. Mechanical Systems and Signal Processing 88:1-8. Paolo Neri. Universitá di Pisa

3 In the rest of the text the word set is elided. Therefore, only the term rotor-shaft is used to design this mechanical set.

4 See Mechanical Vibration by Haym Benaroya, Mark Nagurka and Seon Han. CRC Press. Taylor and Francis Group, fourth edition 2018., Section 6.2.

5 ODE stands for Ordinary Differential Equation

6 For a detailed discussion of first and second-order systems formulas see on internet MIT site, 2.151 Advanced System Dynamics and Control, by Derek Rowell, 10-Oct-2004, derivation of formula (65) in page 24. https://stuff.mit.edu/afs/athena/course/2/2.151/www/Handouts.html, Review: First and Second-order Systems Response

7 Consult the book Théorie et Calcul des Asservissements Linéaires, by Jean Charles. Gille, Paul Decaulne, and Marc Pelegrin. Dunod, 1967 See figure 6.19.

8 The Grainger College of Engineering https://courses.engr.illinois.edu/ece486/fa2020/laboratory/docs/lab2/estimates.html. This paper is based on the following references: Gene F. Franklin, J. David Powell, and Abbas Emami-Naeini, Feedback Control of Dynamic Systems, 4th ed., Prentice Hall, 2002. Benjamin C. Kuo, Automatic Control Systems, 7th ed., Prentice Hall, 1995.

9 Note: as $\gamma(\zeta)$ is lower than 1, the above formulas always return a positive value of settling time.

10 See Torsional Analysis of Variable Frequency Drives by Fred R. Szenasi, PE, Senior Project Engineer, Engineering Dynamics Inc. San Antonio, Texas. There exist other technical publications that mention other ζ values for rotating machines. As most of them have similar figures, in this text it is adopted the 0.015 to 0.125 range. There exist other technical publications that mention other ζ values for rotating machines. As most of them have similar figures, in this text it is adopted the 0.015 to 0.125 range.

11 See Fundamentals of Vibration Engineering, by I Isidor Bykhovsky, MIR Publishers, Moscow. 1st published 197, Chapter 2, Section 7.

12 sidor Bykhovsky in his book Fundamentals of Vibration Engineering, calls the forced terms as excited natural vibration.

13 Consult Mechanical Vibrations by JP. Den Hartog, Dover Publications, Inc. New York, 1985, 4th edition. Section 1.6.

14 See Machinery vibration and rotordynamics by John Vance, Fouad Zeidan and Brian, Murphy. John Wiley and Sons, 2010. Chapter 1, Section: Using Simple Models for Analysis and Diagnostics.

15 To do this, the ratio x_m/x_b is cleared from Formula 1.81, and after divide the result by k and introduce the following formulas: $\left[f = f_c \cdot \zeta = 2 \cdot \sqrt[2]{k \cdot m} \cdot \zeta\right]$, $\left[k = m \cdot \omega_n^2\right]$ and $[u = \omega/\omega_n]$, to obtain: $\frac{x_m}{x_b} = \frac{1 + j \cdot 2 \cdot \zeta \cdot u}{1 - u^2 + j \cdot 2 \cdot \zeta \cdot u}$. Formula 1.83 is the modulus of this complex expression.

2 Dynamics of Rotating SDOF Systems

2.1 INTRODUCTION TO TORSIONAL VIBRATION

This chapter discusses the simple SDOF method of lumped systems. It is an extension of force and deflection formulas of linear systems applied to rotating systems. Most linear systems concepts and formulas hold for rotating systems by substituting linear nomenclature with angular nomenclature.

Shaft torsional vibration is a reciprocating angular motion around the gyration axis, characterized by small amplitude and high frequency. Initial conditions different to zero (position or velocity or both) and abrupt load torque changes produce torsion angle oscillations. During this vibration, there is an energy exchange between the rotating mass and the shaft rigidity. Figure 2.1 is a geometric description of turbine shaft torsion, where the torque is applied in the rotor blades. This figure is also useful in understanding torsional vibration. As in linear systems, the applied torque is equilibrated by the rotor-shaft assembly's inertia and rigidity reactions.

An abrupt load change produces torsional vibrations. For example, power plants are affected by abrupt load changes generated by the network demand. These abrupt changes are caused by a) short circuit currents, b) circulation of harmonics due to variable velocity drives, c) long transport lines with capacitors forming resonant RLC circuits, d) rectifier stations, and e) phase imbalance. These causes and several others produce unbalanced phase torques, which generate torsional vibration. These electrical phenomena produce torsional interactions between the grid and the turbogenerator set. In naval propulsion plants, the propeller's imbalance, or non-uniform water flow, may also produce abrupt load changes and generate torsional vibration. In drive-shafts, worn couplings and misaligned joints and other unbalanced parts are also a vibration source. Torsional vibrations are significantly harmful, but they can be predicted in the design stage and, consequentially, mitigating measurements adopted.

How extensive is the torsional vibration harm? It has a significant extension because it encompasses the shaft and wheels and other parts, such as the flywheel, gearbox, couplings, clutches, propeller, ship hulls, basements, and pedestals. This vibration is an unavoidable propagation of failures in the machine.

The applied torque equals the sum of the rated torque and transient resisting torques produced by the load changes. These transient torques generate stress added to the stress created by the rated turbomachine torque; then, it is important to predict the oscillating torques under transient conditions. In the peaks, they may exceed the allowable stress and fatigue limit of the rotor-shaft steel.

DOI: 10.1201/9781003175230-3

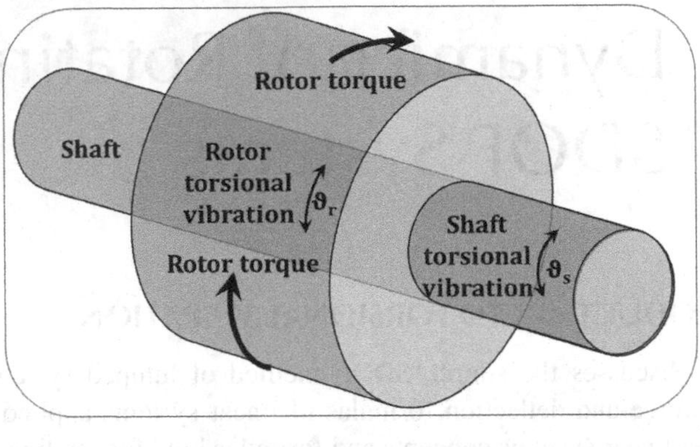

FIGURE 2.1 Sketch of a turbine rotor-shaft under torsional vibration.

The first step to assess torsional vibration risk is to know the rotor-shaft's natural frequencies. If the shaft speed coincides with some component's natural frequency (shaft, wheels, bearings, blades, and other parts), it is said the turbomachine operates at a critical speed. There is more than one critical speed because not all turbomachine parts have the same natural frequency.

The aftermath of torsional vibration is the shaft fatigue and eventual breakage and wear of the gear and other parts, which may generate their failure. This deterioration is severe because, for example, teeth breakage precludes any machine operation. Experience indicates that severe torsional vibration is detected by noise and excessive coupling wear; however, the best practice is to anticipate problems with a sound monitoring system and specialized instrumentation.

Once the natural frequencies are determined, the next step is to calculate the rotor-shaft's forced vibration response and torsional stress efforts. This study should also encompass the identification of disturbing elements. For example, malfunction of couplings is a source of torsional vibration; therefore, it is good to check if coupling specifications match the turbomachine's manufacturer specifications.

2.2 TORSIONAL VIBRATION OF SDOF SYSTEMS

As commented in Chapter 1, two physical properties define a second-order system performance: the natural frequency ω_n and the damping ratio ζ. These two properties are defined in this section for rotating systems.

The SDOF model is a simplified vision of reality. However, its simplicity provides a fast understanding of the problem and guides to plan vibration mitigation. A more sophisticated method, like continuous system formulas or finite element models, can be used if the lumped model does not provide accurate results or guidance to solve the problem. As in linear systems, the applied and reaction torques equilibrium is represented by the differential equation of motion. As this equation is analogous to the linear equation of motion, rotating systems' performance is explained based on analogous concepts and formulas used in Chapter 1.

2.2.1 Torsional System Response

The technical definition of transient and frequency response parameters for linear systems has already been discussed in Chapter 1. This discussion holds for rotating systems. In the torsional vibration of SDOF systems, three properties are analogous to linear systems: the stiffness coefficient, the mass moment of inertia, and the damping factor. The angle ϑ is analogous to the displacement x.

2.2.1.1 Natural Frequency of Rotating Systems

The damped frequency, resonant frequency, and magnification factors have the same form as linear systems. See Section 1.6 for the natural frequency formula of rotating systems and its physical parameters: k_t, I_p, and J_p. Applying Formulas 1.12 of k_t and 1.13 of J_p in Formula 1.11, the natural frequency formula of rotating systems is a function of the stress propagation velocity.

Formula 2.1 Natural frequency of a circular shaft

$$\omega_n = \frac{1}{L}\sqrt[2]{\frac{G}{\rho}} = \frac{c}{L}$$

Where c is the propagation speed of torsional stress in m/s. In steel, the typical propagation velocity of torsional waves (shear waves) is 3,100 m/s; therefore, if this figure replaces the speed c in Formula 2.1, it returns a good rule of thumb to calculate the natural frequency in shafts.

Formula 2.2 Approximate formula for shafts natural frequency

$$\omega_n \cong \frac{3{,}100}{L}$$

If discrete systems formulas cannot be used, then the system's natural frequencies are calculated with formulas of continuous systems discussed in Chapter 3.

The above formulas prove that shafts' natural frequency is inversely proportional to the shaft length only. Therefore, it does not depend on the shaft diameter! Conversely, linear systems' natural frequency does depend on the diameter because it is proportional to the D/L^2 ratio.

The above formula suggests the following acceptable rule of thumb to calculate RPM's natural frequency: $N_n \cong 30{,}000/L$. Therefore, shafts of 8.2 m length in machines of 3,600 RPM are prone to resonate in an auto-excited mode. In 3,000 RPM turbomachines, this length is 9.9 m.

2.2.1.1.1 *Natural Frequency of a Rotor-Shaft Assembly*

The torsional natural frequency of the rotor-shaft assembly is similar to Formula 4.13 for lateral vibration. Therefore, the polar mass moment of inertia of the rotor is introduced, and the result is the following formula:

Formula 2.3 Torsional natural frequency of the rotor-shaft

$$\omega_{n\,\text{shaft-rotor}} = \sqrt[2]{\frac{k_t}{J_{\text{shaft}} + J_{\text{rotor}}}}$$

2.2.1.2 Damping Ratio ζ

The following formula returns the critical friction coefficient of torsional motion.

Formula 2.4 Critical friction coefficient

$$f_c = 2 \cdot \sqrt[2]{k_t \cdot J_p}$$

From this formula arises the damping ratio ζ, which identifies whether the shaft reaction to torque changes is an oscillatory or a non-oscillatory response.

Formula 2.5 Damping ratio or critical damping ratio

$$\zeta = \frac{f}{f_c} = \frac{f}{2 \cdot \sqrt[2]{k_t \cdot J_p}}$$

As all turbomachines are designed with friction forces as low as possible to prevent energy losses and mechanical damage, the damping ratio value is much lower than 1, predicting an oscillating time response. It was said before that the damping ratio range of turbomachines is 0.050 to 0.125.

2.2.2 Transient Response With a Step Torque Input

The equation of motion of the rotating system is given by the equilibrium between the applied torque and the resisting torques of inertia, friction, and torsional rigidity. The solution to this equation is the torsional angle ϑ as a function of time.

Formula 2.6 Equation of motion of torsional vibration

$$T_q(t) = J_p \cdot \ddot{\vartheta}(t) + f_t \cdot \dot{\vartheta}(t) + k_t \cdot \vartheta(t)$$

This equation is solved with the same procedure explained in Section 1.5 for linear systems. Therefore, the derivation procedure of the equation of motion of linear systems holds for rotating systems.

Figure 2.2 shows a shaft driven by a motor or turbine on the right extreme, where the torque is applied, and a load placed on the left extreme. This case could be a turbogenerator shaft. The turbine is on the right extreme, and the generator on the left extreme. Both turbine and load are not shown in the figure. The system's initial condition is the angle ϑ_0, produced by a torque of constant amplitude at the normal machine operation, with no vibration. If T_{q0} is the torque demanded by the load at the initial condition, then the static angular deflection is given by formula $\vartheta_0 = T_{q0}/k_t$.

Then, the rotating angle ϑ_0 is equivalent to the deflection x_0 of linear systems. Figure 2.2 graphically displays the natural torsional vibration produced by an initial condition of $\vartheta_0 = -0.24°$. In a cartesian plot, this vibration is plotted in the chart in Figure 2.3, where the 2% settling is 0.11 s.

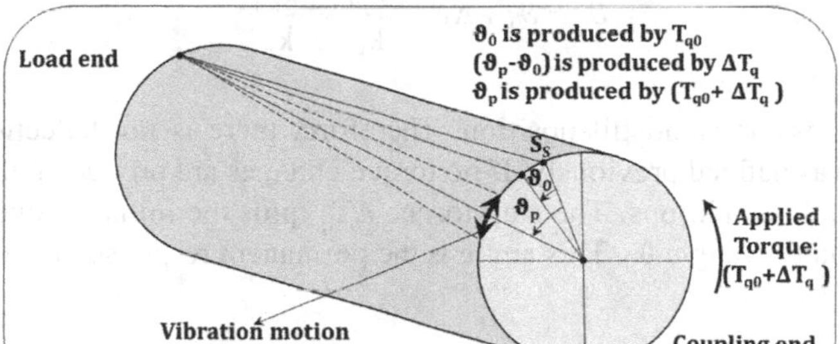

FIGURE 2.2 Torsional vibration produced by a torque change.

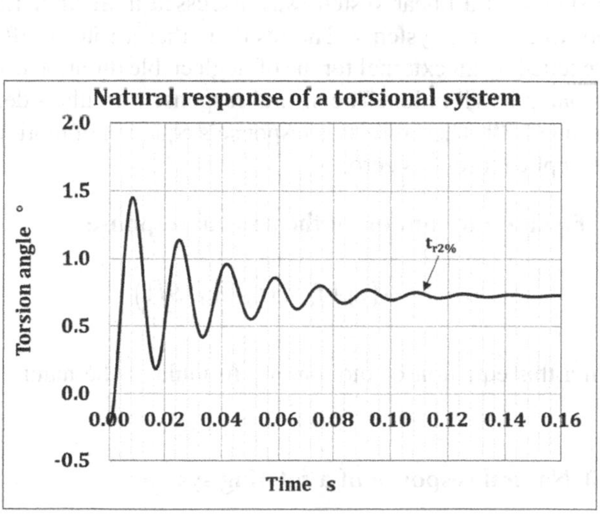

FIGURE 2.3 Natural response of a torsional system.

At static conditions, a step torque is applied by a load change. This torque produces a permanent deflection ϑ_p. However, the shaft reaches this permanent deflection after the step torque's natural frequency response is damped.

Formula 2.7 Total torsional angle

$$\vartheta(t) = \vartheta_p + \vartheta_n(t)$$

Where ϑ_p is the permanent angular deflection and $\vartheta_n(t)$ is the natural time response, which is considered zero for any $t > t_{s2\%}$. The permanent or final deflection is formed as indicated by Formula 2.8.

Formula 2.8 Permanent torsional angle after a step torque.

$$\vartheta_p = \vartheta_0 + \Delta\vartheta = \frac{T_{q0}}{k_t} + \frac{\Delta T_q}{k_t}$$

Point S_s is the standstill position. Therefore, there is no deflection angle. Angle ϑ_0 was defined previously. If no torque changes are produced, the system is under static conditions. The step torque ΔT_{q0} plus the initial torque T_{q0} produce the rotating angle ϑ_p. This angle is the permanent response due to the total torque $T_{q0} + \Delta T_{q0}$.

2.2.2.1 Torsional Natural Response

The natural response of a linear system was discussed in Section 1.5. The same concepts apply to rotating systems. The result is that a rotor-shaft with initial conditions or excited by an external torque of neglectable duration oscillates until friction forces damp its vibration. The natural response formula is derived as was done with Formula 1.30. Therefore, this response's equation of motion is Formula 2.9, where the applied torque is zero.

Formula 2.9 Equation of motion of the natural response

$$0 = J_p \cdot \ddot{\vartheta}(t) + f_t \cdot \dot{\vartheta}(t) + k_t \cdot \vartheta(t)$$

After solving this equation of motion, the formula of the natural response is obtained:

Formula 2.10 Natural response of a rotating system

$$\vartheta_n(t) = \vartheta_a \cdot e^{-\zeta \cdot \omega_n \cdot t} \cdot \cos(\omega_d \cdot t - \varphi)$$

The amplitude ϑ_a and phase φ depend on the initial conditions, as seen in Section 1.5:

Formula 2.11 Amplitude and phase of the natural response

$$\vartheta_a = \sqrt[2]{\vartheta_0^2 + \left(\frac{\zeta \cdot \omega_n \cdot \vartheta_0 + \dot{\vartheta}_0}{\omega_d}\right)^2} \quad \varphi = \arctan\left(\frac{\zeta \cdot \omega_n \cdot \vartheta_0 + \dot{\vartheta}_0}{\vartheta_0 \cdot \omega_d}\right)$$

See in Figure 2.2 that the natural response vibration is around the angle ϑ_p. The thickest arrows of Figure 2.2 represent this vibration. The only difference between this chart and the same chart in linear systems is the ordinate scale. In rotating systems, this scale represents torsion angles and not linear deflections. The decay curve formula is like the formula of linear systems.

Formula 2.12 Torsional decay curve of the natural response

$$d(t) = \vartheta_a \cdot e^{-\zeta \cdot \omega_n \cdot t}$$

2.2.2.2 Transient Response to a Step Torque

If the load experiences a sudden change as a step torque, the rotating system reaction is given by the solution to the equation of motion 2.6. The result is derived with the same procedure used in Section 1.7. The formula that describes the transient response is:

Formula 2.13 Transient damped response

$$\vartheta(t) = \vartheta_0 + \Delta\vartheta_p - \vartheta_a \cdot e^{-\zeta \cdot \omega_n \cdot t} \cdot \cos(\omega_d \cdot t + \varphi)$$

The two following expressions give the amplitude ϑ_a and phase φ.

Formula 2.14 Amplitude and phase of the torsional vibration angle

$$\vartheta_a = \sqrt[2]{\Delta\vartheta_p^2 + \left(\frac{\zeta \cdot \omega_n \cdot \vartheta_0 + \omega_0}{\omega_d}\right)^2} \quad \varphi = \tan^{-1}\left(\frac{\zeta \cdot \omega_n \cdot \vartheta_0 + \dot{\vartheta}_0}{\vartheta_0 \cdot \omega_d}\right)$$

The decay curve formula is:

Formula 2.15 Decay curve of the step response

$$d(t) = \vartheta_0 + \Delta\vartheta_p + \vartheta_a \cdot e^{-\zeta \cdot \omega_n \cdot t}$$

The application of these formulas returns Figure 2.4. It shows the case of a 400 MW, 3,600 RPM (377 rad/s) turbine in a power plant experiencing a sudden

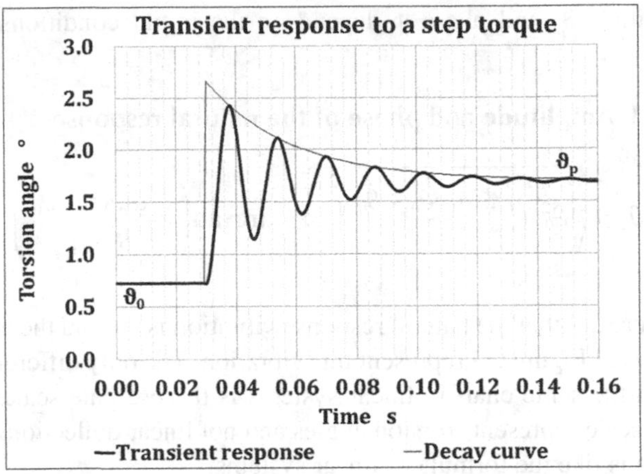

FIGURE 2.4 Transient regime of a rotating system to a step input.

load change from 150 to 350 MW. The shaft length and diameter are 4 m and 345 mm, respectively. The damping ratio used in this simulation is 0.09. The natural frequency is $\omega_n \cong 3{,}100/L = 775$ rad/s. There is no self-excitation risk because of the significant difference between the natural frequency and the rated velocity.

Many types of torque disturbances and many workable solutions to the equation of motion 2.6 describe the transient response. These solutions are valid only if the shaft yield strength is higher than the stress produced by the total torque; therefore, the equation of motion assumes that no permanent (or plastic) deformation happens during the natural response and the torsional vibration.

2.2.3 Velocity Transient of a Turbine-Generator Set

If the fuel (or steam) valve opening of a steam turbine-generator set is suddenly increased (step input), the turbine will apply a step torque ΔT_q to the generator. Due to the rotor-shaft inertia of the assembly, the shaft velocity cannot reproduce the step torque. It means that there is a velocity transient governed by the turbine's speed control system, which modulates the steam or the steam neck valve. This control system neither prevents nor provides any mitigation to the torsional vibration because its purpose is to re-establish the turbine operating velocity after any torque change.

Therefore, after the step torque is applied, two superimposed oscillations arise. They are the rotor-shaft velocity transient, governed through the steam neck valve and the uncontrollable shaft torsional vibration. The frequency tolerance in an electrical net is ±0.5%. As the frequency is proportional to the generator speed, this tolerance is the same for the velocity.

Refer to Figure 2.5; note that the time axis is out of scale. The thicker curve is the rotor-shaft velocity transient produced by a sudden torque increase. As the load torque boosts, the turbine velocity comes down from 3,600 RPM to 3,585

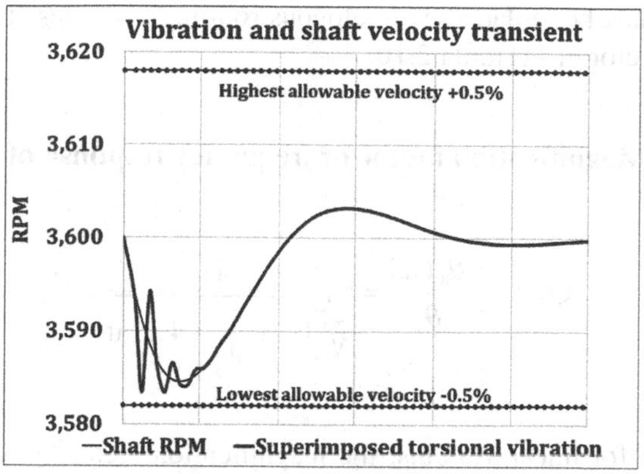

FIGURE 2.5 Turbogenerator's vibration and velocity transient.

RPM. In the meantime, the speed controller opens the steam neck valve, and the velocity tends to recover its earlier speed of 3,600 RPM.

There is a slight overshoot before reaching the rated velocity because the velocity control is a high-order closed-loop system tuned with no little resonance. In a turbine-generator set, the energy storing elements are the turbine rotor-shaft inertia, the shaft rigidity, the mechanical generator inertia, the generator winding inductance and the steam valve inertia. It means that the turbogenerator set is a 5th order system. However, components with a short time constant, like the fuel valve, may be neglected in the system's study.

The velocity transient is damped by the speed controller adjustment, as shown by the fine curve of Figure 2.5. The torsional vibration has a shorter settling time than the velocity system; therefore, it is damped before the rated velocity is achieved again. The thick line shows the torsional vibration superimposed on the shaft transient. Figure 2.5 gives an idea of how fast the shaft torsional vibration is compared to the rotor's velocity oscillations controlled by an automatic governor.

2.2.4 Frequency Response

The reader is referred to Section 1.10, where the frequency response technique is described. Nonetheless, a short recapitulation is reproduced below.

In a turbomachine, resonance is produced when its angular velocity or an external disturbance frequency coincides with the machine's natural frequency. The consequence of this coincidence is a resonant vibration, which puts the equipment in danger of permanent damages if not mitigated on time.

The frequency response formula is obtained from the solution to the equation of motion 2.6 using the frequency ratio u instead of the frequency ω. The modulus of the solution to this equation divided by ϑ_0 is the frequency response or magnification factor. The procedure of this formula's derivation is the same sequence followed by Formulas 1.64 to 1.70. Of course, in this case, the linear variables of said formulas must be replaced by analogous rotating variables. The result is the magnification factor of Formula 2.16.

Formula 2.16 Magnification factor or frequency response of a rotating system

$$Q(u) = \frac{\vartheta_a(u)}{\vartheta_0} = \frac{1}{\sqrt[2]{\left[1-u^2\right]^2 + 4\cdot\zeta^2\cdot u^2}}$$

As was done for linear systems, the magnification factor Q_r is approximated to 1/(2 · ζ) at values of ζ under 0.15. Therefore, for the ζ turbomachines range (0.015 to 0.125), the formula of Q_r returns values in the 4 to 33 range. This range indicates that the resonant regime may be violent and increase torsional stress to values beyond the steel yield strength during rotation.

Mitigation of torsional resonance is done with couplings implemented with springs with Q values as low as 4. However, gears are in the high range of Q because of their rigidity. Therefore, mitigation measurements are not that effective in gears, and consequently, teeth are at risk of wear and breakage in extreme vibration conditions.

The magnification factor of Formula 2.16 has already been discussed. Revisit Section 1.10 and especially Figures 1.32 and 1.33. At the resonant frequency u_r, the above formula gives the maximum magnification factor Q_r. The formulas to calculate this resonance peak are:

Formula 2.17 Resonant frequency ratio and magnification factor

$$u_r = \sqrt[2]{1-2\cdot\zeta^2} \quad Q_r(u_r) = \frac{1}{2\cdot\zeta\cdot\sqrt[2]{1-\zeta^2}}$$

A rotating system's frequency response has a vector interpretation that helps visualize the applied torque and the resisting torques as in linear systems. See Figure 2.6. As was discussed in Section 1.10, the oscillation output is not in phase

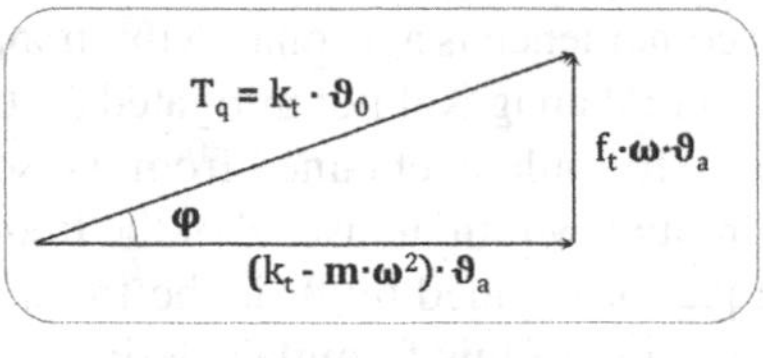

FIGURE 2.6 Vector interpretation of the frequency response of a rotating system.

with the input signal. Therefore, the phase angle between input and output harmonic signals is calculated with the following formula:

Formula 2.18 Phase angle of shaft torsional oscillations

$$\varphi(\omega) = \arctan\left[\frac{2\cdot\zeta\cdot u}{1-u^2}\right]$$

At resonance peak, it is $\omega \simeq \omega_n$; then, the phase angle is −90°. The maximum phase angle is −180°, which happens at an infinite frequency, which, of course, is an unrealistic case. In higher-order systems (more than two energy storage elements), the phase angle is higher than the previous values. At infinite frequency, every added order contributes with a maximum phase of -90°. For example, if one mass and one spring are added to the original second-order system, a 4th order system is formed, whose maximum theoretical phase angle is minus 360°.

2.2.5 Torsional Stress Under Vibration

The shaft torsional stress is proportional to the torsional deflection ϑ, measured at the torque application section. The torque is equal to the rigidity coefficient k_t times the shaft deflection angle. If the resisting torque abruptly changes, torsional vibration is induced, as shown in Figure 2.5; therefore, the stress is amplified by the magnification factor. It means that the rotating angle is no longer equal to the static angle ϑ_0 at a standstill but equal to $Q(u) \cdot \vartheta_0$.

Formula 2.19 Torsional shear stress versus frequency

$$\tau(u) = \frac{k_t \cdot Q(u) \cdot \vartheta_0}{Z_t}$$

This formula indicates that the shear stress changes with frequency according to one of the curves of Figure 1.32, whose parameters are the damping ratios. Therefore, there is a shear stress peak at a resonant frequency whose magnitude depends on the damping ratio calculated with the following formula:

Formula 2.20 Torsional shear stress under resonance conditions

$$\tau_r = \frac{k_t \cdot Q_r \cdot \vartheta_0}{Z_t}$$

This formula's result should be compared with the allowable shear stress to determine the shaft safety factor under resonance.

2.2.6 Cumulative Fatigue Generated by Turbomachines Startup

Though this section is not exactly a vibration topic, the steel fatigue imposed by vibration is included here as a warning to engineers about the vibration aftermath on the mechanical integrity of machines subject to vibration.

Fatigue stress is the stress at which steel collapses if it is subject to a harmonic force. The collapse stress value depends on the applied force's stress and the accumulated number of cycles. The higher the number of cycles endured by the piece, the lower the stress of collapse. This phenomenon is represented by the Wohler curve that is specific for each steel type. See an example in the sketch of Figure 2.7. Under the fatigue limit S_f, there is no collapse regardless of the accumulated number of cycles. Above this stress, the metal piece reaches the breakage zone, depending on the cycles' accumulated number. This condition is an impending rupture that occurs at any stress above the fatigue limit stress S_f. Figure 2.7 shows this phenomenon for the stress S_x. The piece collapses when the number of accumulated cycles is N_x. If the applied force is not harmonic, it collapses at its ultimate tensile strength (UTS). In this case, the breakage is only produced by static stress. Therefore, the fatigue breakage is strongly dependent on the cycle accumulation.

During startup, significant stress is endured by the turbomachines shaft. There is a combination of bending and torsional stress, the last being the most important because the startup process imposes sudden and significant torque changes on the shaft. Every startup produces torsional vibration and reduces the future allowable cycles of startups; therefore, it is important to estimate the accumulated cycles and stress produced by successive startups and record their progress in time. If the startup is not moderated by some technical means, transient torsional vibrations may produce unacceptable transitory stress. However, as the startup is a transient process, overpassing once the fatigue limit does not forcefully result in undesired aftermath because the fatigue damage depends on the accumulated cycle quantity. Typical alloy steels' fatigue limit is equal to 50% of the UTS (Ultimate Tensile Strength), and the flat part of the Wohler curve starts between 0.1 and 0.5 million of accumulated cycles.

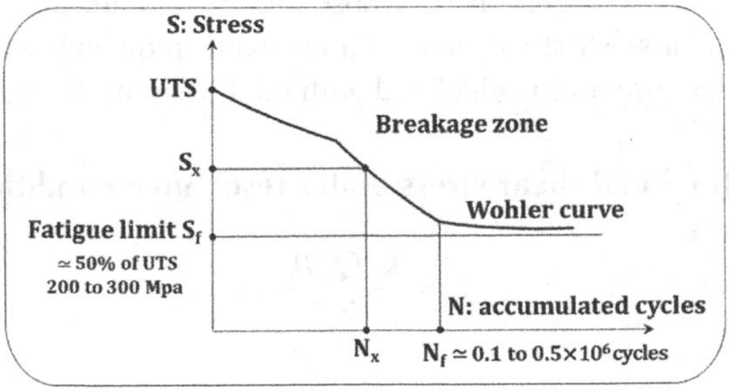

FIGURE 2.7 S-N curve.

The accumulative stress theory is applied to estimate how many cycles are still left before reaching the S-N (Stress versus Number of Accumulated Cycles) curve. This calculation is done by performing a transient regime analysis to estimate the fraction of the total available cycles left before reaching the allowable stress in the S-N curve. S-N curves of most shafting steels are published and are used to calculate how many cycles can be endured at certain stress before failure[2].

Military standard MIL-STD 167 recommends that the torsional allowable fatigue stress is calculated by dividing the steel UTS over 25. This recommendation purports to subject the piece to stress lower than the fatigue limit S_f. For example, the chromium-nickel steel A-422 UTS is 483 MPa. Therefore, the allowable fatigue stress is 483/25 = 19 MPa. This value barely stands for 9% of the steel yield strength. Nonetheless, it is recommended to de-rate this allowable stress with a safety factor and a stress concentration factor (SCF). Therefore, the allowable fatigue stress is given by the following formula:

Formula 2.21 Allowable shear stress as per MIL-STD 167

$$S_{\text{allowable fatigue stress}} = \frac{S_{UTS}}{25 \cdot SF \cdot SCF}$$

SCF is the Stress Concentrated Factor, which de-rates the allowable fatigue stress due to mechanical discontinuities like a keyway in a shaft. A typical safety factor (SF) for the shaft diameter calculation is 5. However, in the allowable stress calculation, it is sometimes common to use a safety factor of 2. Under typical torsional conditions, the shear yield strength is 57.7% ($SYS = YS / \sqrt[2]{3}$) of the tensile yield strength. This limit is expected to be lower in keyways or any other shape discontinuities where the shear stress is concentrated.

2.2.7 Multidisciplinary Assessment of Torsional Vibration

This example is a multidisciplinary report because vibration and its impact on the shaft stress are analyzed. It is requested to calculate the stress and safety factor produced by the natural response's peak and disturbances.

2.2.7.1 Technical Scenario

This case refers to a 500 MW turbine-generator set with a 5 m shaft length between the turbine and the generator. Figure 2.8 shows that the initial torsional deflection is close to 0.4° with a delivered power of 200 MW. At the instant t = 0.1 s, a 250 MW step demand happens, which increases the delivered power to 450 MW and the resisting torque to 121.7 t.m. The step torque and the consequent torsional vibration, measured in the field, are shown in Figure 2.8. From a practical point of view, this sudden and significant load increment is not a frequent event, but in any case, it has been adopted in this model to show the potential impact of step demands.

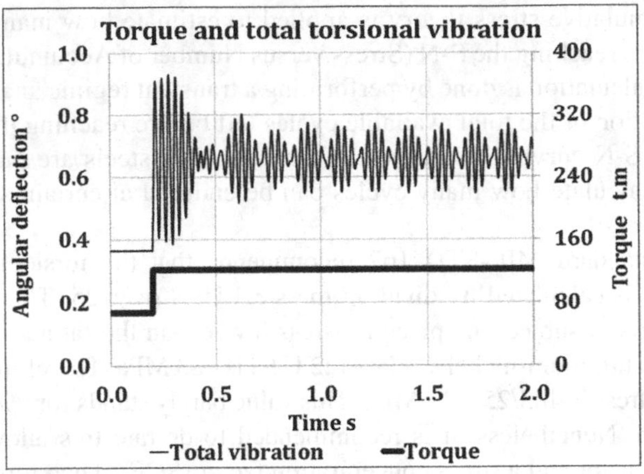

FIGURE 2.8 Total vibration as seen by an instrument in the field and step torque estimation.

According to the Fourier theory, a step torque is formed by infinite harmonic components. Some component's frequency may be equal to any of the turbomachine's resonance frequencies. Therefore, it is important to measure the component's frequency to assure that no resonance risks happen in a turbomachine.

The case under study assumes that the turbomachine endures a step torque due to a sudden load change, and at the same time, two external sources of vibration excite the turbomachine. The chart in Figure 2.9 shows one low-frequency disturbance of 180 rad/s (LF disturbance) and another high-frequency disturbance of 205 rad/s (HF disturbance) that initiate exciting the turbomachine at t = 0.2 s. See the oscillogram of these two vibrations in Figure 2.9.

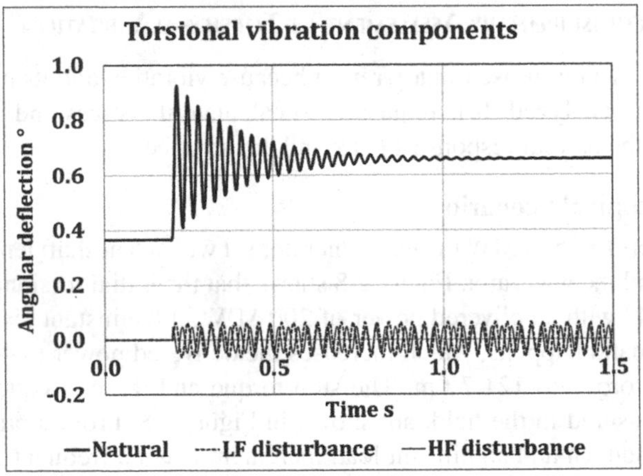

FIGURE 2.9 Startup and disturbances vibration.

Due to the two disturbances, a static, permanent regime does not happen. Instead, the mix of two harmonic components generates a permanent response of torsional oscillations. As was commented before, this oscillation produces fatigue and peaks of torsional stress. As the disturbances have a significant frequency difference, they form a beat pattern described in Section 1.9 and depicted on the right side of Figure 1.25. See Figure 2.8.

2.2.7.2 Calculation Model

The calculation model to assess the stress of a 500 MW turbine's shaft during the transient response and under a hypothetical resonance condition is programmed in the following spreadsheet. The formulas used in calculating each of the steps are in the left column, and the results are in the second column. Some results are represented with two different physical units for convenience reasons. A technical report describing the most important conclusions inferred from the model output is included after the spreadsheet.

Block 1. Technical specifications				
Turbine rated characteristics				
Rated power P	500	MW		
Rotational velocity RPM	3,600	RPM	377	rad/s
Shaft steel properties. AISI 302. 18Cr 9Ni 70Fe				
Elasticity modulus E (Young)	193	GPa	1.97E+10	kg/m²
Shear modulus of rigidity G	77.2	GPa	7.87E+09	kg/m²
Yield strength YS	330	MPa	33,650,635	kg/m²
Shear yield strength SYS	191	MPa	19,428,203	kg/m²
Steel specific weight γ kg/m³	7,860	Steel density ρ	801	kg mass/m³
Rotor-shaft properties				
Length L	5	m		
Diameter	500	mm	0.500	m
r_J ratio = J_{total}/J_{shaft}	10	Times J_{shaft}		
Damping coefficient	0.026	Typical range: ζ = 0.015 to 0.125		
Torsional safety factor SF	5	SCF is not included		
Stress Concentration Factor SCF	1.3			
Power demand and disturbances				
Initial power P_0 at t = 0	250	MW		
Step power demand ΔP at t = t_s	200	MW		
Step power starts at instant t_s	0.2	s		
The mechanic equivalent of heat J	102	kg.m/kJ		

Low disturbance frequency ω_L	30	rad/s		
High disturbance frequency ω_H	350	rad/s		
Initial vibration velocity ω_0	0	rad/s	There is no initial vibration	
Calculation procedure and results				
Block 2. Shaft strength properties				
Shaft-wheels strength properties				
Shaft weight $W = (\pi \cdot D^2/4) \cdot L \cdot \gamma$	7,717	kg		
Shaft mass $m = W/g$	787	kg mass		
Area polar moment of inertia $I_p = \pi \cdot D^4/32$	0.00614	m^4		
Torsional stiffness coefficient $k_t = G \cdot I_p/L$	9,660,654	kg.m/rad		
Torsional section modulus $Z_t = I_p/(D/2)$	0.02454	m^3		
Shaft polar mass moment of inertia $J_{shaft} = m_s \cdot D_{shaft}{}^2/8$	24.6	kg.m.s²/rad or kg mass.m²		
Total polar mass moment of inertia $J_{total} = k_J \times J_{shaft}$	246	kg.m.s²/rad or kg mass.m²		
Block 3. Deflections and stresses under stable conditions				
Deflection and stress under stable conditions (Initial means before the step torque)				
Initial torque $T_{q0} = P_0 \cdot J/\omega_{shaft} \cdot 1000$	67,641	kg·m		
Initial rotating angle $\vartheta_0 = T_{q0}/k_t$	0.0070	rad	0.40	°
Static deflection produced by the step torque				
Step torque $\Delta T_q = \Delta P \cdot J/\omega_{shaft} \cdot 1000$	54,113	kg.m		
Step angle $\Delta\vartheta_{Tq} = \Delta T_q/k_t$	0.0056	rad	0.32	°
Final or permanent torque, deflection and stress				
Permanent torque $T_{qp} = T_{q0} + \Delta T_q$	121,754	kg.m		
Permanent rotating angle $\vartheta_p = T_{qp}/k_t = \vartheta_0 + \Delta\vartheta_{Tq}$	0.0126	rad	0.72	°
Stress under the permanent regime $\tau_p = SCF \cdot T_{qf}/Z_t$	6,448,891	kg/m²	63.2	MPa
Safety factor = SYS/τ_{final}	3.0	Includes SCF		
Block 4. Transient and frequency response parameters				
Transient due to step torque				
$C_1 = -\Delta\vartheta_{Tq}$	−0.00560	rad	−0.32	°
$C_2 = (-\zeta \cdot \omega_n \cdot \vartheta_0 + \omega_0)/\omega_d$	−0.00011	rad	−0.01	°
Amplitude of natural response $\vartheta_a = (C_1{}^2 + C_2{}^2)^{1/2}$	0.00560	rad	0.32	°

Phase angle in natural response $\varphi = \arcsin [C_2/(C_1^2 + C_2^2)^{1/2}]$	−0.020	rad	−8.15	°
Characteristic frequencies				
Natural frequency $\omega_n = (k_t/J_{total})^{1/2}$	198.24	rad/s	1,893.1	RPM
Damped frequency $\omega_d = \omega_n \cdot (1 - \zeta^2)^{1/2}$	198.21	rad/s	1,892.7	RPM
Resonant frequency $\omega_r = \omega_n \cdot (1 - 2 \cdot \zeta^2)^{1/2}$	198.17	rad/s	1,892.3	RPM
The decay constant and settling time				
Decay time constant $T_z = 1/(\zeta \cdot \omega_n)$	0.25	s		
Settling time 2% $t_{s2\%} = t_{Tq+}4 \cdot T_z$	8.11	s		
Block 5. Deflection, stress and safety factor at the first peak of the natural regime				
Damping period $T_d = 2 \cdot \pi/\omega_d$	0.0317	s		
First peak time in the natural response $\omega_0 = 0$; then, $t_{peak} = t_s + T_d / 2$ $\omega_0 > 0$; then, $t_{peak} = t_s + \left(f + \operatorname{atan}\left(\zeta / \left(1-\zeta^2\right)^{1/2}\right)\right) / \omega_d$	0.116	s		
First peak deflection $\vartheta(t_{peak})$ in the natural response $\vartheta_{peak} = \vartheta_0 + \Delta\vartheta_p - \vartheta_a \cdot e^{(-\zeta \cdot \omega_n \cdot t_{peak})} \cdot \cos(\omega_d \cdot t_{peak} + \varphi)$	0.0179	rad	1.02	°
Shear stress at peak deflection of the natural response $\tau_{peak} = SCF \cdot k_t \cdot \varphi_{peak}/Z_t$	9,140,484	kg/m2	89.6	MPa
Safety factor at peak deflection $SF_n = SYS/\tau_n$	2.1	Includes SCF		
Block 6. Torsional stresses and safety factors due to disturbances				
Stress due to low-frequency disturbance after the natural regime				
Frequency ratio $u_L = \omega_L/\omega_n$	0.15			
Magnification factor $Q_L = 1/\left[\left(1-u_L^2\right)^2 + \left(2\zeta \cdot u_L\right)^2\right]^{1/2}$	1.02			
Torsional angle $\vartheta_L = Q_L \cdot \Delta\vartheta_{Tq}$	0.006	rad	0.33	°
Additional shear stress $\tau_L = SCF \cdot k_t \cdot \vartheta_L/Z_t$	2,933,292	kg/m²	28.8	MPa
Stress due to high-frequency disturbance after the natural regime				
Frequency ratio $u_H = \omega_H/\omega_n$	1.77			
Magnification factor $Q_H = 1/[(1 - u_H^2)^2 + (2 \cdot \zeta \cdot u_H)^2]^{1/2}$	0.47			
Torsional angle $\vartheta_H = Q_H \cdot \Delta\vartheta_{Tq}$	0.003	rad	0.15	°
Additional shear stress $\tau_H = SCF \cdot k_t \cdot \vartheta_H/Z_t$	1,353,149	kg/m²	13.3	MPa

Combined stress and safety factor after the natural response is damped				
Torsional combined stress $\tau_c = \tau_{final} + \tau_L + \tau_H$	10,735,332	kg/m²	105.3	MPa
Combined safety factor SF = SYS/τ_c	1.81	Includes SCF		
Peak produced by the natural response and the disturbances at t = t_{peak}				
Partial and total peaks at t_{peak} $\vartheta_{max} = \vartheta_{peak} + \vartheta_L \cdot \cos[\omega_L \cdot (t_{peak} - t_{Tq})] + \vartheta_H \cdot \cos[\omega_H \cdot (t_{peak} - t_{Tq})]$	1.02 Natural	0.29 Low frequency	0.11 High frequency	1.43 Total at t_{peak}
Stress and SF at t_{peak}				
Combined maximum stress $\tau_{max} = SCF \cdot k_t \cdot \vartheta_{max}/Z_t$	125	MPa		
$SF_{max} = SYS/\tau_{max}$	1.52	Includes SCF		
Stress and safety factor under resonance during the natural regime				
Magnification factor under resonance $Q_r = 1/[2 \cdot \zeta \cdot (1 - \zeta^2)^{1/2}]$	25.0			
Resonance angle $\vartheta_r = Q_r \cdot \Delta\vartheta_{Tq}$	0.14	rad	8.02	°
Shear stress $\tau_r = SCF \cdot k_t \cdot \vartheta_r/Z_t$	71,668,676	kg/m²	702.8	MPa
Safety factor SF = YS/τ_r	0.47	Includes SCF		

2.2.7.3 Technical Summary

Block 3. Deflection and stress under stable conditions

Before the step power is applied, the generator delivers 200 MW, and the torque is 67.6 t.m. Due to this torque, the torsional deflection is 0.4°.

The step torque is 54.1 t.m, which produces an additional torsional deflection of 0.32°.

The total torque has escalated to 121.7 t.m, and the total final deflection is 0.72°. The stress after the natural response is damped equals 63.2 MPa, and the safety factor is 3.0. However, the step torque produces dynamic efforts in the natural response, whose peaks are discussed below.

Block 4. Transient and frequency response parameters

The damping ratio is 0.020 that is within the usual range of turbomachines. However, this value anticipates a significant overshoot discussed below. This damping ratio also says that each torque change produced by the electrical demand on a 3,600 RPM turbine accumulates 25 cycles, which increases the steel fatigue. It is recommended to do a detailed study of total accumulated cycles per year due to transients, permanent vibration or other causes and compare with the AISI 302 steel Wohler curve specified for the shaft. The total duration of the natural response at $t_{s2\%}$ is 8.11 s.

The natural frequency is 198.2 rad/s. Damped ω_d and resonance ω_r frequencies are equivalent to this natural frequency due to the damping coefficient's low value.

Block 5. Stress and safety factor at maximum overshoot of the natural response

The first peak of the natural response is at t = 0.116 seconds. The first peak at that instant is 1.02°. The peak stress produced by this deflection is 89.6 MPa. As the shear yield strength is 191 MPa, the peak deflection's safety factor is 2.9, which is considered low. However, as the step torque assumed in this example is too high, this SF could eventually be accepted. The plant technical team should assess the severity of fatigue if this overshoot is frequent.

Block 6. Torsional stresses and safety factors due to disturbances

The low-frequency disturbance has a magnification factor of 1.02. Therefore, the deflection produced by this disturbance is 0.33°, and the additional stress is 28.8 MPa. This stress, plus the stress produced by the step torque, is equal to 92 MPa, which corresponds to a safety factor of 191/92 = 2.08.

The high-frequency disturbance has a magnification factor of 0.47; therefore, this vibration is attenuated and not magnified. The additional torsion angle is 0.15°, and the additional stress is 13.3 MPa.

The combined stress equals 105.3 MPa. The safety factor is 191.0/105.3 = 1.81, which may be considered acceptable; however, if additional dynamic efforts arise, this safety factor could be insufficient because the shaft stress may reach the plastic deformation zone.

Warning. Stress and safety factor under resonance

At resonance, the magnification factor is 25.0. It means that the final angular deflection would increase by an unacceptable angle of 8.02°. The additional stress is 702.8 MPa higher than the steel UTS. Under this condition, the shaft collapse is inevitable. In conclusion, this turbomachine cannot operate at resonance (as all the other existing turbomachines!). A partial solution would be to install a flexible coupling, which can be tuned to a frequency other than resonance or use control techniques described in part III of this book.

NOTES

1 Recommended books for Chapter 2: 1. Mechanical Vibrations by S. Graham Kelly. Schaum's Outline Series. McGraw Hill, 1996 – 2. Advanced Vibration Analysis. S. Graham Kelly. CRC Press. Taylor and Francis Group, 2007 – 3. Fundamentals of Vibration Engineering by Isidor Bykhovsky. MIR Publishers, Moscow. 1st published 1972 – 4. Mechanical Vibrations by J. P. Den Hartog. Dover Publications, Inc. New York, 1985, 4th edition – 5. Mechanical Vibration by Haym Benaroya, Mark Nagurka and Seon Han. CRC Press. Taylor and Francis Group, fourth edition 2018.

2 Consult Atlas of Fatigue Curves of the American Society of Metals. Editor: H.E. Boyer. ASM, 1986

3 Dynamics of Linear and Rotating MDOF and Continuous Systems

3.1 INTRODUCTION TO MDOF AND CONTINUOUS SYSTEMS

Inertia, rigidity, and damping properties of any solid body type are not lumped properties as assumed in SDOF systems' models; instead, they are distributed on every differential part of the body. Nonetheless, the theoretical basis discussed in Chapters 1 and 2 holds for modelling discrete multi-degree of freedom and continuous systems described in this section.

3.1.1 Discrete Multi-Degree of Freedom Systems

Discrete systems are divided into several elemental parts for modelling purposes, each having one of the three physical properties used in SDOF models: mass, rigidity, and damping. This assumption suggests that free bodies' mechanically interconnected form a unique body composed of many SDOF systems. After each model part is created, a unique interconnected model is built using a matrix method. Each part is modelled based on D'Alembert's principle or Newton's second law. This matrix method demonstrates that discrete multi-degree of freedom systems have as many natural frequencies as the masses' quantity integrating the system.

Although there are infinite natural frequency values, only frequencies close to the machine speed or close to any of the system's resonance frequencies are of interest. These risks are sometimes surprising because the resonance phenomenon is unpredictable; it occurs at more than one vibration mode[2] caused by external random excitations or initial position and velocity conditions.

In multi-degree and continuous systems, a vibration mode is a harmonic component of vibration. According to the Fourier series, systems vibrate theoretically up to infinite components of different frequencies. The sum of all vibration modes is not a harmonic signal. Every vibration mode corresponds to a term of the Fourier series. The first term of the series has the highest amplitude, and the lowest frequency, called the fundamental frequency. Frequencies of all other components are multiple of the fundamental. The vibration mode number identifies the harmonic component of the Fourier series. The term with the fundamental frequency is mode 1. For example, vibration mode 3 is the third harmonic of vibration and has three times the fundamental frequency.

DOI: 10.1201/9781003175230-4

If a system is disturbed by harmonic inputs, the closest disturbance frequency to any of the resonance frequencies predominates on the others. All harmonic components contribute with their vibration modes to the total vibration. However, high-order harmonics have lower amplitudes than the fundamental (mode 1); therefore, most high-order-harmonics are usually neglected in the first approach to a vibration problem.

3.1.2 Continuous Systems

Machines' shafts and rotors have more than one natural frequency due to infinite elemental energy storages distributed in the machine, like rotating masses and rigid shafts. These machines can be studied as a multi-degree of freedom (MDOF) system; however, better accuracy is obtained by adopting continuous system models. As machines' physical properties are continuously distributed, they form infinite elemental vibrating systems. Calculus mathematical tools allow integrating these elemental systems and derive formulas that calculate the most significant natural frequencies in practice.

Under usual operating conditions, like a machine startup, resonance at different speeds may happen. For the sake of a machine or structure's technical integrity, resonance causes are identified in the design stage or at the field during commissioning tests. An important aspect of the design stage is calculating the stress during the transient regime at different frequencies and evaluating fatigue risks on the machine lifespan.

3.1.2.1 Stress Waves and Propagation Velocity

In continuous systems, a sudden effort, like a strong shock or explosion applied to any part of their structure, produces maximum stress at the application point. The applied effort generates an elemental displacement of molecules due to their elasticity, which pushes the next particles, and these particles push other particles and so on. The result is that the stress propagates to the rest of the structure as a high-speed wave.

There are two types of travelling waves in a continuous vibrating system: longitudinal (L-wave) and transverse (T-wave). In L-waves, the displacement of the particle is parallel to the wave's displacement. L-waves are produced by initial conditions or by flexural efforts. In stainless steel, the typical propagation velocity of L-waves is close to 5,000 m/s. Instead, in T-waves, the molecules and wave displacements are perpendicular to the propagation velocity. These waves are produced by initial condition causes or by torsional efforts. The typical propagation velocity of T-waves in stainless steel is 3,100 m/s. The propagation velocity of L-waves depends on Young's modulus and density. Instead, T-waves velocity depends on the shear modulus and density.

This chapter describes how to calculate vibration wave propagation in a beam or a shaft. The natural frequencies' calculation of beams subject to flexural vibration (or lateral vibration) and shafts under torsional vibration use the Euler-Bernoulli equation and the wave equation, respectively.[3]

The propagation velocity of elastic stress waves depends on the structure's two physical properties: Young's modulus E (for flexural stress) or the shear modulus G (for torsional stress) and the material density. Wave velocities must not be confused with the pushed molecules velocities, which only move at a short distance range. The velocity of these molecules is lower than the travelling waves. The theory of elastic waves inside solids demonstrates that the following formulas return their propagation velocity.

Formula 3.1 Stress propagation velocity of flexural and torsional waves

$$c = \sqrt[2]{\frac{E}{\rho}} \quad c = \sqrt[2]{\frac{G}{\rho}}$$

L-waves T-waves

This formula stands for the stress propagation velocity. Young's modulus E and shear modulus G are expressed in kg/m^2 and ρ in kg mass; then velocity c is returned in m/s. Typical velocity mentioned before of 5,000 m/s for L-waves in stainless steel assumes E = 197 GPa and ρ = 802 kg mass/m^3. Instead, the 3,100 velocity of T-waves in stainless steel assumes G = 75.6 GPa and ρ = 802 kg mass/m^3. These figures are often used in practice for machine natural frequency estimates, as explained in this chapter.

The stress wave propagation must not be confused with vibration waves, which propagate at a velocity called phase velocity, which depends not only on E or G and ρ but also on the beam's geometry. Instead, the stress propagation time between the beam end where the effort is applied, and the other extreme is:

Formula 3.2 Propagation time of elastic stress waves at a distance L

$$t_p = \frac{L}{c}$$

For example, because of the different propagation velocities that L-waves and T-waves have, in a 4 m length machine shaft, the torsional stress propagation time is 1.3 ms. In the same length shaft, the flexural stress propagation time is 0.6 ms.

3.2 LINEAR MULTI-DEGREE OF FREEDOM SYSTEMS

The continuous system models have a high degree of accuracy to predict the system's natural frequencies. However, they may have a mathematical complexity because their fundamental equations need partial derivatives that are not always readily produced. This disadvantage is overcome using a multi-degree system formed by elemental second-order systems to calculate the most important natural frequencies that may affect the system's behavior. This method returns acceptable values of the system's natural frequencies, though it does not use continuous physical properties as real bodies have.

A mass not linked to any other part of the system can be represented as a free body with several forces acting on it. D'Alambert's principle states that the inertial reaction to an external force may be replaced by a group of passive forces that do not depend on mass acceleration. These forces are due to the system's rigidity and friction; therefore, D'Alembert's expression for a linear system is:

Formula 3.3 Principle of D'Alembert

$$m \cdot \ddot{x}_i = \sum_{i=1}^{D} f_i \cdot \dot{x}_i + \sum_{i=1}^{S} k_i \cdot x_i$$

D stands for dampers quantity and S for springs quantity.

Newton's second law states that the inertial forces are equal to the body mass times its acceleration. And D'Alembert's principle completes this law stating that inertial forces are in equilibrium with all other passive forces acting upon the body. Therefore, the mass is represented with a free body diagram, as shown in Figure 3.1. The inertial force F_i is $(m_i \cdot \ddot{x}_i)$. The other forces are passive due to the rigidity and damping system's reactions.

The applied force is F_i. The generalized coordinate x_i determines the mass position, velocity, and acceleration. The mass is mechanically connected, on both sides, to springs and dampers. The terms $(x_i - x_{i-1})$ and $(x_i - x_{i+1})$ are the springs net stretching or compression, and the terms $(\dot{x}_i - \dot{x}_{i-1})$ and $(\dot{x}_i - \dot{x}_{i+1})$ represent the velocity difference between mass i and the contiguous masses m_{i-1} and m_{i+1}. Applying the second Newton's law and D'Alembert's principle, the following expression of forces equilibrium is returned.

Formula 3.4 Forces equilibrium in a free-body mass

$$m_i \cdot \ddot{x}_i = -k_i \cdot (x_i - x_{i-1}) + k_{i+1} \cdot (x_i - x_{i+1}) - f_i \cdot (\dot{x}_i - \dot{x}_{i-1}) + f_{i+1} \cdot (\dot{x}_i - \dot{x}_{i+1}) + F_i$$

The application of this formula requires careful use of subscripts. Note that, for a mass m_i, generalized coordinates x_i are always in the first term of expressions

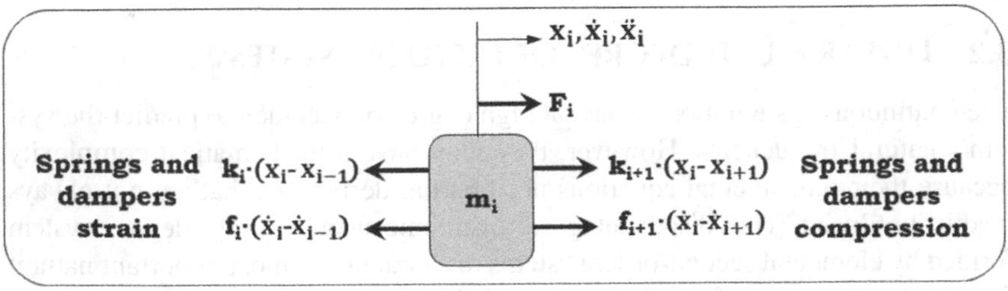

FIGURE 3.1 Applied and reaction forces on a free-body mass of a multi-degree of freedom system.

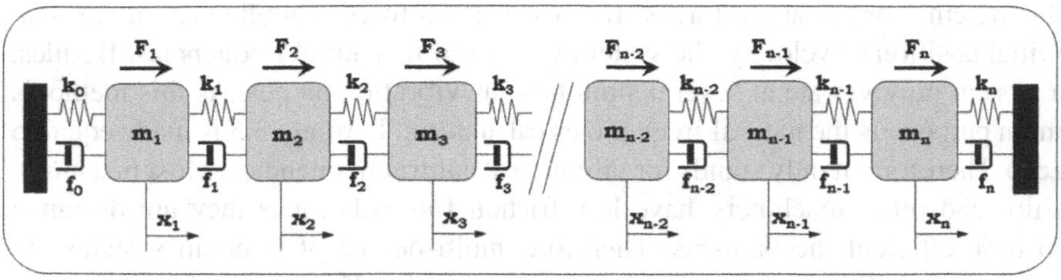

FIGURE 3.2 System's boxes diagram with one restriction on each side.

between parentheses. Figure 3.2 shows the complete boxes diagram of a multi-degree of freedom system with two restrictions on both extremes. Restrictions are the elements linked to the ground where the extreme spring-damper sets are embedded. Notwithstanding these restrictions, the system can vibrate if there are proper initial conditions of position and velocity or an external excitation.

Formula 3.4 demonstrates that rigidity and damping reaction forces depend on the mass coordinate and other masses' coordinates; thus, the mass m_i is subject to reaction forces that depend on the coordinates of masses mi-$_1$ and m_{i+1}. When this happens, the system is considered coupled because one mass's motion depends on the other masses' motion.

What is the main purpose of modelling a multi-degree of motion system? For vibration engineering, it is mainly the natural frequencies determination. The designer uses these frequencies to prevent vibration mode frequencies close to any of the natural frequencies. The designer must adopt techniques to have a safety margin between the disturbance frequency and the closest natural frequency. Technical publications mention safety margins in the 5% to 20% range. However, the decision depends on the expected frequency variations. If these variations are small, then a lower safety margin may be adopted. Random frequency disturbance variations are usually a source of surprising resonance situations that must be prevented by adopting higher safety margins. In general, the first natural frequencies are the most likely to produce resonance accidents; therefore, the designer or the field engineer must be aware of the system's natural frequencies and potential disturbance frequencies' proximity.

3.2.1 Matrix Model of Multi-Degree Systems

A system formed by n masses has n differential equations. The display and manipulation of n differential equations may be cumbersome; therefore, they are represented in matrix form.

Formula 3.5 Matrix formulation of the equation of motion

$$\mathbf{M}\cdot\ddot{\mathbf{x}}+\mathbf{F}\cdot\dot{\mathbf{x}}+\mathbf{K}\cdot\mathbf{x} = \mathbf{F(t)}$$

Bold and capital letters imply that these symbols are a matrix or a vector. If the applied forces are harmonic, the solution to this set of equations shows that the system vibrates at the forces' frequency. If no force is applied and there is an initial position or velocity, the system vibrates at its natural frequencies. Besides, a system may vibrate at several simultaneous vibration modes. As this method's main purpose is the natural frequencies calculation, Formula 3.5 is made equal to zero. Therefore, it only stands for vibration at natural frequencies. Most machines, ships and other machinery have low friction forces because they are designed to have efficient mechanisms. Therefore, multi-degree of freedom systems are mostly studied with a zero-friction matrix (**f** = 0). The matrixial equations of motion of the undamped natural response:

Formula 3.6 Matrixial equations of motion of undamped natural response

$$\mathbf{M}\cdot\ddot{\mathbf{x}}+\mathbf{K}\cdot\mathbf{x} = 0$$

M and **K** are square diagonal matrices of order n (nxn). Acceleration $\ddot{\mathbf{x}}$ and generalized coordinates **x** are vector matrices of one column (n×1). **M** is the inertial or masses matrix, and **K** is the rigidity or springs matrix. It must be understood that the term spring includes the bodies' rigidity like shafts or beams. Variable n is the system's quantity of inertial elements (linear masses and rotating wheels). The expanded matrix of the equations of motion 3.6 is displayed below.

Inertial Matrix M						Vector $\ddot{x}_i$		Rigidity Matrix K						Vector x_i		
m_1	0	0	…	0		$\ddot{x}_1$		k_1+k_2	$-k_2$	0	…	0		x_1		0
0	m_2	0	…	0		$\ddot{x}_2$		$-k_2$	k_2+k_3	$-k_3$	…	0		x_2		0
0	0	m_3	…	0	×	$\ddot{x}_3$	+	0	$-k_3$	k_3+k_4	…	0	×	x_3	=	0
…	…	…	…	…		…		…	…	…	…	…		…		…
…	…	…	…	…		…		…	…	…	…	…		…		…
0	0	0	…	m_n		$\ddot{x}_n$		0	0	0	…	$k_{n-1}+k_n$		x_n		0

This representation is seldom used in the study of multi-degree of freedom systems because of the matrices' nomenclature convenience of Formula 3.6. Expanded matrices are used in systems with no more than two or three degrees of freedom.

By analogy with friction-less SDOF systems, it is supposed that the free-vibration motion is simple harmonic. This assumption returns the following matrix formulas as solutions to the differential Formulas 3.4 and 3.6.

Formula 3.7 Solution to the matrix differential Formula 3.6

$$\mathbf{x}(\mathbf{t}) = \mathbf{x_a}\cdot \sin\omega_n\cdot \mathbf{t}$$

$$\ddot{\mathbf{x}}(t) = -\mathbf{x}_a \cdot \omega_n^2 \cdot \sin\omega_n \cdot t = -\mathbf{x}(t) \cdot \omega_n^2$$

Vibration amplitudes $\mathbf{x}_a$ represents the system's time-invariant shape.

Matrix Formulas 3.7 of **x**(t) and $\ddot{\mathbf{x}}(t)$ are replaced in the matrix equation of motion 3.6 to derive the natural frequencies formula of multi-degree of freedom systems. This replacement returns the two matrices' product: the amplitudes vector $\mathbf{x}_a$ and the matrix frequency equation, shown between brackets in Formula 3.8.

Formula 3.8 Matrix solution to the equation of motion 3.6

$$\left[-\mathbf{M} \cdot \omega_n^2 + \mathbf{K}\right] \cdot \mathbf{x}_a = \mathbf{0}$$

As solution 3.8 states that the system is vibrating, the amplitudes vector $\mathbf{x}_a$ cannot be zero. Therefore, to satisfy this solution, the determinant of the matrix term between brackets must be zero. This equality to zero is the condition of no trivial solution ($\mathbf{x}_a = 0$) to the matrix differential equation of motion 3.6. This determinant is called the characteristic determinant.

Formula 3.9 Condition of no trivial solution to Formula 3.6

$$\mathbf{det}\left[-\mathbf{M} \cdot \omega_n^2 + \mathbf{K}\right] = \mathbf{0}$$

$$\mathbf{M} \cdot \omega_n^2 = \mathbf{K}$$

This equation's unknowns are the square natural frequencies ω_n^2. The inertial and rigidity matrices are usually known. The characteristic determinant's expansion produces an equation in ω_n^2 named frequency equation or characteristic equation, or eigenequation. ω_n^2 is cleared by a matrix method, as shown in Formula 3.10. This formula is derived by multiplying both sides of $\mathbf{M} \cdot \omega_n^2 = \mathbf{K}$ by the inverse rigidity matrix $\mathbf{M}^{-1}$.

Formula 3.10 Solution to the matrix frequency equation

$$\mathbf{I} \cdot \omega_n^2 = \mathbf{K} \cdot \mathbf{M}^{-1}$$

The matrix ratio Adj M/det M gives the inverse matrix M^{-1}. In Formula 3.10, matrix **I** is the identity diagonal square matrix. The diagonal components of **I** are equal to 1, and all the others are zero. The $\mathbf{I} \cdot \omega_n^2$ product returns another n × n diagonal square matrix, whose diagonal elements are the square natural frequencies, and the rest are zero. The right term of Formula 3.10 is the rigidity matrix times the inverse of the inertial matrix. The result is the dynamic matrix $\mathbf{A} = \mathbf{K} \cdot \mathbf{M}^{-1}$, an n × n matrix. The square frequencies ω_n^2 are called the eigenvalues, and the amplitudes of vector $\mathbf{x}_a$ are the eigenvectors.

The above procedures produce the following general frequency equation, whose solution returns the system's natural frequencies.

Formula 3.11 Frequency equation of an n/2-degrees of freedom system

$$a_n\cdot\omega_{nn}^{n} + a_{n-1}\cdot\omega_{nn-2}^{n-2} + \cdots + a_4\cdot\omega_{n4}^{4} + a_2\cdot\omega_{n2}^{2} + a_0 = 0$$

Coefficients a_i are functions of the rigidity coefficients k_i and masses m_i. The exponents of ω_n are only even numbers. This equation has n roots, out of which n/2 are real and positive numbers. These positive numbers are the natural frequencies' values.

3.2.2 Natural Frequencies of a System with Three Degrees of Freedom

This section discusses matrices' application to the natural frequencies calculation of a system with three degrees of freedom. Figures 3.3, 3.4, and 3.5 displays the system's schemes for three different cases: a) the system is restricted on both extremes, b) the system is restricted in only one extreme, and c) the system has no restrictions.

These restrictions influence the rigidity matrix configuration shown in each figure. Rigidity matrices are also applicable to systems with one or two degrees of freedom. For example, if m_3 and k_4 are set equal to zero in the first case, two degrees of freedom system with two restrictions are returned. The two restrictions case is illustrated here. Three equations can be written based on the matrix equation of motion.

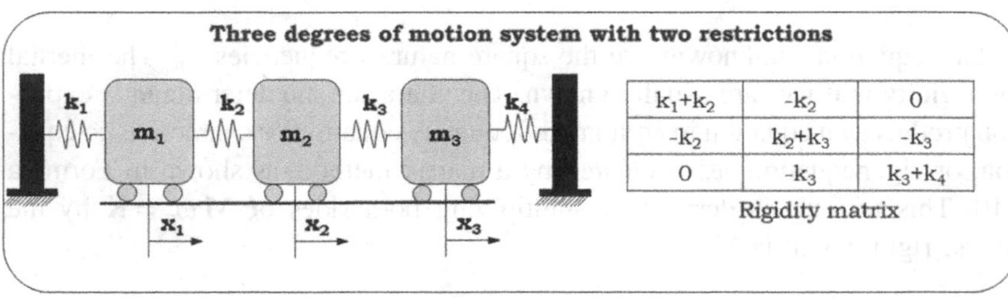

FIGURE 3.3 Rigidity matrixes of three degrees of freedom system with two restrictions.

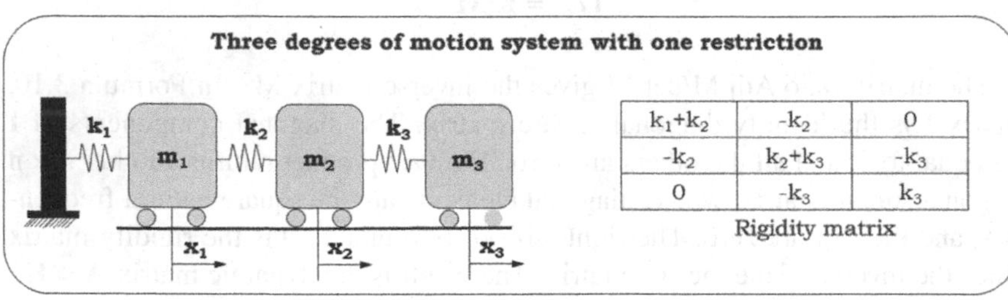

FIGURE 3.4 Rigidity matrixes of three degrees of freedom system with one restriction.

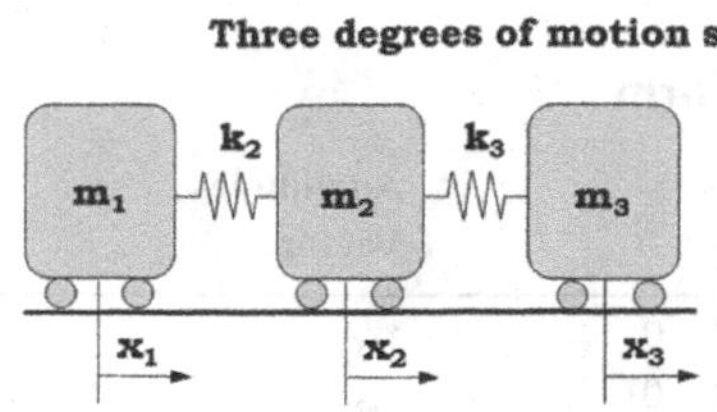

k_2	$-k_2$	0
$-k_2$	k_2+k_3	0
0	$-k_3$	k_3

Rigidity matrix

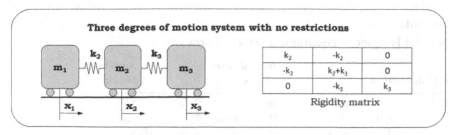

FIGURE 3.5 Rigidity matrixes of three degrees of freedom system with no restrictions.

Formula 3.12 Equations of motion for three degrees of freedom system with two restrictions

$$m_1 \cdot \ddot{x}_1 - (k_1 + k_2) \cdot x_1 + k_2 \cdot x_2 = 0$$

$$m_2 \cdot \ddot{x}_2 + k_2 \cdot x_1 - (k_2 + k_3) \cdot x_2 + k_3 \cdot x_3 = 0$$

$$m_3 \cdot \ddot{x}_1 + k_3 \cdot x_2 - (k_3 + k_4) \cdot x_3 = 0$$

These expressions show that each inertial force depends on its coordinate and also on contiguous coordinates. For example, the rigidity reaction of m_1 depends on two coordinates: x_1 and x_2. Therefore, as commented before, there is a coupling between mass m_1 and mass m_2.

The solution to these differential equations is the vector **x**(t) of Formula 3.7, which may take two forms: trigonometric or exponential with complex exponent according to Euler's formulas of complex numbers.

Formula 3.13 Solutions to equations of motion 3.12

$$\mathbf{x}(t) = \mathbf{x_a} \cdot \sin \omega_n \cdot t$$

$$\mathbf{x}(t) = \mathbf{x_a} \cdot \frac{e^{j \cdot \omega \cdot t} - e^{-j \cdot \omega \cdot t}}{2 \cdot j}$$

Both forms are harmonic functions, predicting that the motion is vibratory, and the frequency must be the natural frequency because there are no external forces applied to the system. As was said before, $\mathbf{x_a}$ is the wave amplitude of vibration and stands for the vibrating system's shape. For beams, this shape is shown in Figure 3.11 for vibration modes 1, 2 and 3.

After replacing the trigonometric solution 3.13 in the differential equations of motion 3.12, natural frequency equations 3.14 are obtained.

Formula 3.14 Natural frequency equations

$$(k_1 + k_2) \cdot x_{a1} - k_2 \cdot x_{a2} = m_1 \cdot \omega_n^2$$

Formula 3.15

Natural Frequency Equations in Matrix Form

Rigidity Matrix K				Inertial forces Matrix $M\cdot\omega_n^2$				Amplitudes Vector		
k_1+k_2	$-k_2$	0		$m_1\cdot\omega_n^2$	0	0		x_{a1}		0
$-k_2$	k_2+k_3	$-k_3$	−	0	$m_2\cdot\omega_n^2$	0	x	x_{a2}	=	0
0	$-k_3$	k_3+k_4		0	0	$m_3\cdot\omega_n^2$		x_{a3}		0

Formula 3.16

Characteristic Determinant of Frequency Equations

$k_1+k_2-m_1\cdot\omega_n^2$	$-k_2$	0		0
$-k_2$	$k_2+k_3-m_2\cdot\omega_n^2$	$-k_3$	=	0
0	$-k_3$	$k_3+k_4-m_3\cdot\omega_n^2$		0

$$-k_2\cdot x_{a1}+\left(k_2+k_3\right)\cdot x_{a2}-k_3\cdot x_{a3}=m_2\cdot\omega_n^2$$

$$-k_3\cdot x_{a2}+\left(k_3+k_4\right)\cdot x_{a3}=m_3\cdot\omega_n^2$$

The matrix form of these equations is:

Summing the rigidity and inertial forces matrices, the matrix frequency equation is returned[4]. As it was seen before, the characteristic determinant of this equation must be zero.

Expanding this determinant, the frequency equation, a sixth-degree algebraic equation for three degrees of freedom systems, is returned.

Formula 3.17 Frequency equation of three degrees of freedom system

$$a_6\cdot\omega_n^6+a_4\cdot\omega_n^4+a_2\cdot\omega_n^2+a_0=0$$

This equation has six roots; three of them are positive and real. These roots are the system's natural frequencies.

3.3 ROTATING MULTI-DEGREE OF FREEDOM SYSTEMS

High-order systems have more than one degree of freedom. For example, in rotating machines, a fifth-order system is formed by five wheels, each having one degree of freedom. These high-order systems have a significant quantity of physical properties and the state variables of position, velocity, and acceleration.[5] As noted in Section 3.2, algebraic expressions used to model high-order systems are extensive and may be tedious when applied to high-order systems modelling. Instead, matrices notation is a compact way to write the physical properties and manipulate them according to pre-established physical laws. Therefore, the calculation of properties and state variables of high-order systems are easier to carry out with the Matrix Algebra rules.

The most common matrices of physical properties in Vibration Engineering are inertia, friction, and rigidity. These three matrices are square; they have the same number of rows and columns (n×n). As they convert forces or torques into deflections, they are also called transfer matrices. Deflections (linear or angular), velocity and acceleration, are state variables represented by (n×1) vectors. The modelling procedure of a linear system with matrices was discussed in Section 3.2. All concepts and formulas of that section hold for rotating systems. However, some changes must be introduced. Instead of mass, the polar mass moment of inertia must be used, designated J_p, and the linear rigidity coefficient is changed by k_t, the torsional coefficient of rigidity.

The equation of motion of a frictionless high-order system is represented with the following matrix equation (Revisit Formula 3.6). The solution to this equation is the natural system response

Formula 3.18 Matrix equation of motion of an undamped rotating system

$$\mathbf{J_p} \cdot \ddot{\vartheta} + \mathbf{K_t} \cdot \vartheta = \mathbf{0}$$

The application of this equation to circular uniform shafts has the following definitions.

i identifies a system element
$\mathbf{J_p}$ is the wheels' polar mass moment of inertia matrix (n×n). $J_{pi} = m_i \cdot D_i^2 / 8$
ϑ is the angular acceleration vector (n×1).
$\mathbf{K_t}$ is the angular rigidity coefficient matrix (n×n). $k_{ti} = T_{qi}/\vartheta_i$
ϑ is the angular deflection vector (n×1)

By analogy with linear systems, the solution to the equation of motion is:

Formula 3.19 Solution to the equation of motion of a rotating system

$$\vartheta(t) = \vartheta_\mathbf{a} \cdot \sin(\omega_n \cdot t + \varphi_i)$$

$$\ddot{\vartheta}(t) = -\vartheta_\mathbf{a} \cdot \omega_n^2 \cdot \sin(\omega_n \cdot t + \varphi_i) = -\vartheta(t) \cdot \omega_n^2$$

Replacing these formulas in the equation of motion 3.18, the matrix expression of this solution is:

Formula 3.20 Matrix solution to the equation of motion 3.18

$$\left[-\mathbf{J_p} \cdot \omega_n^2 + \mathbf{K_t}\right] \cdot \vartheta_\mathbf{a} = \mathbf{0}$$

The term between brackets is the matrixial expression of the characteristic determinant. Section 3.2 explained that this determinant must be equal to zero to avoid the trivial solution of $\boldsymbol{\vartheta}_a = 0$. The amplitudes vector $\boldsymbol{\vartheta}_a$ cannot be physically zero because formulas 6.23 anticipate that the system is vibrating. Therefore, the matrix frequency equation is:

Formula 3.21 Matrixial expression of the characteristic determinant

$$\mathbf{det}\left[-\mathbf{J}_p \cdot \omega_n^2 + \mathbf{K}_t\right] = \mathbf{0}$$

The natural frequencies quantity is equal to the system's degrees of freedom system; however, the frequency equation degree is double the multi-degrees of freedom. For example, the tenth-order system of Figure 3.6 has five natural frequencies, calculated by a tenth-degree equation derived from the characteristic determinant. Each set of this scheme is a mass-spring element. The spring represents the shaft rigidity.

This system could represent an air compressor used in gas turbines. The rigidity matrix must be derived based on a wheel-free body diagram, as in linear systems. If the configuration under study is different, for example, if friction in bearings is not neglectable; then, the friction term $\mathbf{F}.\omega_n$ should be included in the equation of motion 3.18.

Formula 3.22
Expansion of the Frequency Equation for a Five-Degrees of Freedom System

Rigidity Matrix K_t

k_{t1}	$-k_{t1}$	0	0	0
$-k_{t1}$	$k_{t1}+k_{t2}$	$-k_{t2}$	0	0
0	$-k_{t2}$	$k_{t2}+k_{t3}$	$-k_{t3}$	0
0	0	$-k_{t3}$	$k_{t3}+k_{t4}$	$-k_{t4}$
0	0	0	$-k_{t4}$	$k_{t4}+k_{t5}$

-

Inertial Forces Matrix $J \cdot \omega_n^2$

$J_{p1} \cdot \omega_n^2$	0	0	0	0
0	$J_{p2} \cdot \omega_n^2$	0	0	0
0	0	$J_{p3} \cdot \omega_n^2$	0	0
0	0	0	$J_{p4} \cdot \omega_n^2$	0
0	0	0	0	$J_{p5} \cdot \omega_n^2$

= 0

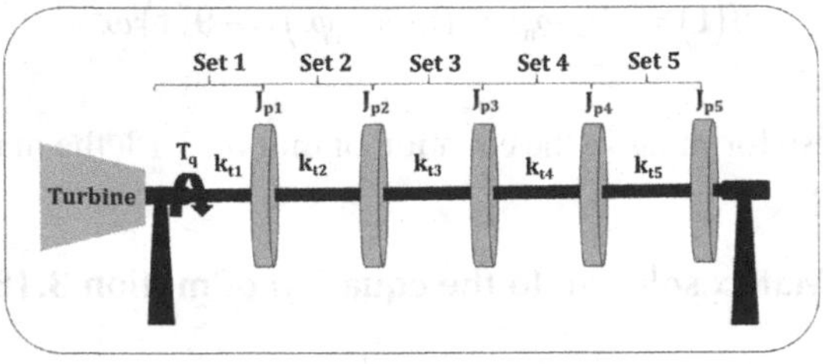

FIGURE 3.6 Tenth-order system driven by a turbine.

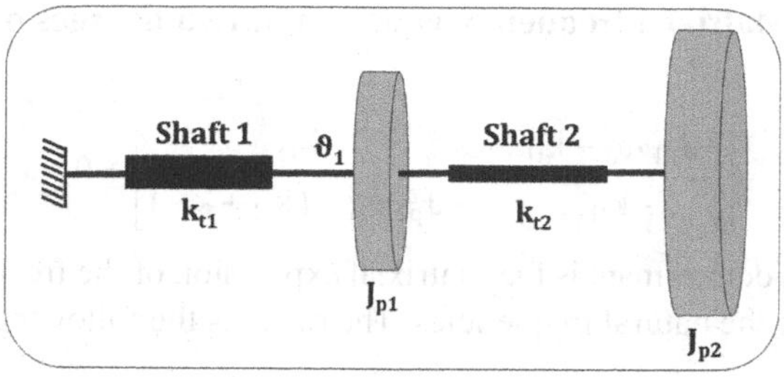

FIGURE 3.7 Schematics of two degrees of freedom system.

This approach to torsional oscillations[6] creates a mechanical model by virtually cutting the shaft into smaller shafts, each with a wheel of small width. Each has inertia and rigidity properties that may be or not equal to the others. Therefore, this method allows for modelling a stepped shaft supporting wheels of different diameters.

3.3.1 Natural Frequencies of Two Degrees of Freedom System

The application of the frequency equation to two degrees of freedom system is explained in this section.

It is requested to calculate the system's natural frequencies as a function of the shaft rigidity and the wheels' polar moments of inertia shown in Figure 3.7.

There does not exist friction in the system; therefore, its oscillations are undamped. As the natural frequency formula is required, no external torques must be included in the equation of motion. As the system has no external excitation, it vibrates at its natural frequency. Therefore, the equations of motion are:

$$J_{p1}\cdot\ddot{\vartheta}_1 + k_1\cdot\vartheta_1 + k_2\cdot(\vartheta_1 - \vartheta_2) = 0$$

$$J_{p2}\cdot\ddot{\vartheta}_2 + k_2\cdot\vartheta_2 + k_2\cdot(\vartheta_2 - \vartheta_1) = 0$$

These equations are statically coupled because each one has two independent variables ϑ_1 and ϑ_2 because the spring reaction is proportional to the net stretching $(\vartheta_1 - \vartheta_2)$ or $(\vartheta_2 - \vartheta_1)$.

The matrixial frequency equation of this case is the following expression.

Formula 3.23 Matrixial frequency equation of two degrees of freedom system

$$-\begin{bmatrix} J_{p1}\cdot\omega_n^2 & 0 \\ 0 & J_{p2}\omega_n^2 \end{bmatrix} + \begin{bmatrix} k_{t1} & -k_{t1} \\ -k_{t1} & k_{t1} + k_{t2} \end{bmatrix} = 0$$

Therefore, the frequency equation is the sum of two matrices: the inertial $\mathbf{J_p}$ and the rigidity $\mathbf{K_t}$. This sum returns the following matrix[7]:

Formula 3.24 Matrixial frequency equation of two degrees of freedom system

$$\begin{bmatrix} -J_{p1}\cdot\omega_n^2 + k_{t1} & -k_{t1} \\ -k_{t1} & -J_{p2}\cdot\omega_n^2 + (k_{t1}+k_{t2}) \end{bmatrix} = 0$$

This matrix's determinant is the matrixial expression of the frequency equation used to calculate the natural frequencies. The result is the following fourth-degree equation.

Formula 3.25 Frequency equation of two degrees of freedom system

$$\omega_n^4 - \left(\frac{k_{t1}}{J_{p1}} + \frac{k_{t2}}{J_{p2}} + \frac{k_{t1}}{J_{p2}}\right)\cdot\omega_n^2 + \frac{k_{t2}\cdot k_{t2}}{J_{p1}\cdot J_{p2}} = 0$$

Therefore, the natural frequencies formula is:

Formula 3.26 Natural frequencies formula

$$\omega_{n1,2} = \sqrt[2]{\frac{1}{2}\cdot\left(B \pm \sqrt[2]{B^2 - 4\cdot\frac{k_{t1}\cdot k_{t2}}{J_{p1}\cdot J_{p2}}}\right)}$$

Where the term B is calculated with the following formula.

Formula 3.27 Natural frequencies of a fourth-order system

$$B = \frac{k_{t1}}{J_{p1}} + \frac{k_{t2}}{J_{p2}} + \frac{k_{t1}}{J_{p2}}$$

Only real and positive results are valid. The application of this formula is easy if a spreadsheet is used. This expression is more complicated for order systems higher than two; therefore, in those cases, to avoid cumbersome algebraic manipulations, it is recommended to use mathematical software to manipulate matrices and calculate determinants.

3.3.2 Practical Assessment of Natural Frequencies

Let two degrees of freedom system be formed by a stepped shaft and two different-sized wheels gyrating at 3,600 RPM. The system's natural frequencies need to be calculated by implementing the model in the spreadsheet reproduced below. All used formulas are on the left column in the order of calculation. This spreadsheet shows the formulas to be used and the calculation sequence.

Torsional Natural Frequencies		
Two Degrees of Freedom		
Input data		
Steel density ρ	802	kg mass/m³
Shear stress modulus G	77.20 7.9E+09	GPa kg/m²

(Continued)

(Continued)

Torsional Natural Frequencies Two Degrees of Freedom		
Property	**Wheel-shaft set 1**	**Wheel-shaft set 2**
Shaft partial lengths L_i m	1.00	2.00
Shaft diameter D_s m	0.30	0.35
Wheels average thickness of t_w m	0.025	0.025
Wheels diameter D_w m	1.50	2.00
Calculations		
Shafts rigidity		
Mass $m_{si} = \pi \cdot D_s^2/4 \cdot L \cdot \rho$ kg mass	56.69	154.32
Polar mass moment of inertia $J_p = m_s/8 \cdot D^2$ kg.m.s²/rad	0.64	2.36
Area moment of inertia $I_p = \pi \cdot D_s^4/32$ m⁴	0.00080	0.00147
Rigidity coefficient $k_t = G \cdot I_p/L_i$ kg.m/rad	6,260,127	5,798,829
Shaft natural frequency $\omega_{nshaft} = (k_t/J_p)^{1/2}$ rad/s	3,133	1,567
Wheels polar mass moment of inertia		
Mass $m_{wi} = \pi \cdot D_w^2/4 \cdot t_w \cdot \rho$ kg mass	36.00	64.00
Polar mass moment of inertia $J_p = m_{wi}/8 \cdot D_w^2$ kg.m.s²/rad	10.1	32.0
Natural frequencies		
Constant $B = k_{t1}/J_{p1} + k_{t2}/J_{p2} + k_{t2}/J_{p1}$ [1/s²]	1.3E+06	
System's natural frequency $(\omega_n)^2$ [rad/s]²	1.2E+06	8.1E+04
System natural frequencies ω_n [rad/s]	**1,099**	**285**
System natural frequencies ω_n [RPM]	**10,495**	**2,722**

As the rated velocity is 3,600 RPM and natural frequencies are 2,722 and 10,495 RPM, the turbomachine is subject to resonance only during the startup process when passing through the 2,722 RPM. As the rated velocity is higher than the lower resonant frequency, it does not experience resonance problems. There are no risks with the second resonance because it is much higher than the rated velocity.

3.4 THE EULER-BERNOULLI EQUATION

The Euler-Bernoulli equation is the cornerstone of continuous beam vibration. It is derived based on the beam's geometric properties related to the efforts supported by a beam: flexural moment and shear. The Euler-Bernoulli beam is an ideal model that describes how beams with small curvature behave under bending and axial forces[8]. This model assumes that beams have continuous and homogeneous physical properties, meaning that rigidity and inertia are evenly distributed along the beam length. The Euler-Bernoulli equation is based on two assumptions, a) after bending, the beam's cross-sections are plane and perpendicular to the beam's geometric axis; and b) the beam's slopes are small. These conditions hold after the beam is bent[9]. The Euler-Bernoulli equation proves that the beam axis under load adopts a curved shape, depending on the support types and vibration mode. See beam shapes for diverse types of support and vibration mode in Figure 3.11.

Shear and torsional deflections are neglected for the beam shape determination because beams have a right cross-section much smaller than their length. It is accepted that a beam has a D/L ratio lower than 0.15. The study of these shapes dated from the 18th century when Daniel Bernoulli was living in Russia. There, he formulated the differential equation of a vibrating beam's motion. Later, Leonhard Euler joined him in Russia and adopted this equation in his elastic beams' shape theory under several load types. Since then, the Euler-Bernoulli equation is the mathematical basis of continuous beam models. It is a simple and exact approach to many engineering problems that involve beams' vibration.

3.4.1 Deflections and Efforts at Beam's Supports

This section explains some concepts and formulas of Mechanics of Materials (or Strength of Materials) regarding the beam's technical conditions at its extreme supports, essential for the solution of the Euler-Bernoulli equation developed in Section 3.4.3. These formulas refer to the relationship between the deflection's spatial derivatives and the bending moments and shear forces at the supports.

A distributed load along a beam is designated q and measured in units of force per length unit. This load produces a deflection δ(x), which generates a beam's curvature dependent on the support type at the beam's extremes and the vibration mode. Figure 3.8 shows a beam bent by a uniform load, where x is the space coordinate along the beam axis from any of the beam ends. In this illustration, the left end is used as the coordinate's origin, therefore at the left end, x = 0, and at the right end, x = L, being L the beam's length.

1. The beam's curve slope at any point is: $S(x) = \frac{d\delta(x)}{dx}$. This derivative is an important formula used to determine the beam's bending moment and shear force at each support in the points that follow x.
2. At any point of the beam, the bending moment is proportional to the deflection's second derivative with respect to x. The product $E \cdot I_z$ is the proportionality coefficient, representing the bending stiffness. I_z is the cross-section beam's moment of inertia. For uniform circular beam or shaft, its formula is: $I_z = I_y = \frac{\pi \cdot D^4}{64}$. z and y are perpendicular axes on the cross-section's plane. Then: $M(x) = E \cdot I_z \cdot \frac{d^2\delta}{dx^2}$.
3. Torsional or vertical efforts produce the shear force. Torsional efforts are more important in a machine's shafts. This force is proportional to the derivative of M(x) with respect to x or the third derivative of the deflection with respect to x. Then: $V(x) = \frac{dM(x)}{dx} = E \cdot I_z \cdot \frac{d^3\delta}{dx^3}$

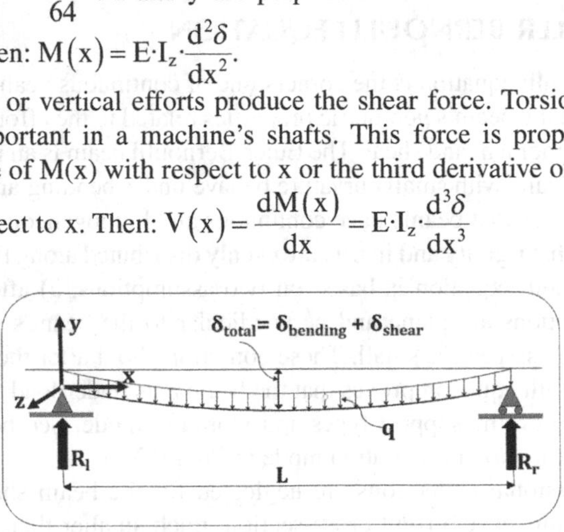

FIGURE 3.8 Beam load and deflection.

4. Load q is proportional to the fourth derivative of the beam's deflection with respect to x.$q = E \cdot I_z \cdot \frac{d^4\delta}{dx^4} = \frac{d^2M(x)}{dx^2} = \frac{dV(x)}{dx}$. This formula is pivotal in the derivation of the Euler-Bernoulli equation, used to calculate the beam's natural frequencies.

In summary: the first derivative of the beam's deflection with respect to x is the beam's curve slope. The second derivative is proportional to flexural moments, the third derivative is proportional to shear forces, and the fourth derivative is proportional to the load per unit length.

In the derivation of the Euler-Bernoulli equation, it is assumed that the beam is vertically vibrating with no external excitation, then it vibrates at its natural frequencies.

3.4.1.1 Boundary Conditions at Beam Supports

Support type determines if the deflection δ(x) or its derivatives are zero at the beam ends. The boundary conditions are derived from the support reaction to external efforts. These technical characteristics make each support type unique. The value of derivatives with respect to x mathematically identifies the support reaction with respect to x of the deflection δ(x) at the beam end. The following description of the boundary conditions refers to the right end at x = L. However, they are the same for the left support (x = 0) [10].

- Hinged support at x = L:
  The beam is only free to revolve around the support's joint point in a vertical plane. It means that no vertical movement and no reaction moment exists at the support, then δ(L) = 0 and M = 0. According to the formula of item 3, the second derivative with respect to x of the deflection is zero; then, the boundary conditions are:

$$\delta(L) = 0 \text{ and } \left.\frac{d^2\delta}{dx^2}\right|_{x=L} = 0$$

- Free support at x = L:
  This condition is at the right end of a cantilever beam with the left end clamped. The right end is free to revolve and move vertically, which means that at this end, a reaction moment and a vertical shear force do not exist. Therefore, the second and third derivative of δ(L) with respect to x is zero; hence the boundary conditions are:

$$\left.\frac{d^2\delta}{dx^2}\right|_{x=L} = 0 \text{ and } \frac{d^3\delta}{dx^3} = 0$$

- Clamped support at x = L:
  The clamped condition prevents the beam from gyrating and moving in a vertical plane at the right end. Therefore, the reaction moment and shear force are not zero. However, as the beam is forcefully horizontal because it is clamped inside the support, the deflection and the slope are zero. Therefore, the first derivative of δ(x) with respect to x is zero, and the boundary conditions are:

$$\delta(L) = 0 \text{ and } \left.\frac{d\delta(x)}{dx}\right|_{x=L} = 0$$

3.4.2 Derivation of the Euler-Bernoulli Equation

The beam vibration theory is based on the equilibrium between applied efforts and the beam inertia reaction. The Euler-Bernoulli equation represents this equilibrium. This section describes this equation's derivation for a frictionless beam or shaft with a uniform load q, freely vibrating· As the beam is frictionless, it vibrates with infinite components at their natural frequency each.

According to Newton's second law, if the beam is freely vibrating, its mass reacts to a uniform load q with a distributed force proportional to the vibration acceleration. Additionally, the Mechanics of Materials science demonstrates that the uniform load q is proportional to the fourth derivative of the deflection with respect to x. Therefore, these equalities are combined in the following formula.

Formula 3.28 Applied and reaction forces per unit length of a vibrating beam

$$q = -\mu \cdot \frac{d^2\delta}{dt^2} = E \cdot I_z \cdot \frac{d^4\delta}{dx^4}$$

The mass per unit length is $\mu = m/L$, E is the Young's module, and I_z is the beam's cross-section moment of inertia. From this equality, the Euler-Bernoulli equation is obtained.

$$\frac{\partial^4\delta(x,t)}{\partial x^4} + \frac{\mu}{E \cdot I_z} \cdot \frac{\partial^2\delta(x,t)}{\partial t^2} = 0$$

Formula 3.29 The Euler-Bernoulli equation for a beam lateral vibration

This equation units are $|1/m^3|$, and the term $\mu/E \cdot I_z$ is in $|s^2/m^4|$. The first term $(\partial^4\delta/\partial x^4)$ is the spatial distribution of rigidity forces along the beam due to loading. The second term is the maximum value of the inertial load.

3.4.3 Solution to the Euler-Bernoulli Equation

The general solution to the Euler-Bernoulli equation is the product of two different equations, one is a function of x, and the other is a function of time.

Formula 3.30 Compacted form of the solution to the Euler-Bernoulli equation

$$\delta(x,t) = \delta(x) \cdot \delta(t)$$

Functions $\delta(x)$ and $\delta(t)$ are the sums of harmonic waves each. The first is a spatial wave, and the second is the classical wave as a function of time.

By replacing this solution in the Euler-Bernoulli equation, the following expression is obtained.

Formula 3.31 Substitution of Formula 3.30 in the Euler-Bernoulli equation

$$\delta(t)\cdot\frac{d^4\delta(x)}{dx^4}+\frac{\mu}{E\cdot I_z}\cdot\delta(x)\cdot\frac{d^2\delta(t)}{dt^2}=0$$

Grouping temporal and space terms on both sides of the equals sign, the result is that these groups are equal to a constant, designated g (do not confuse this constant with the gravity acceleration). The physical interpretation of this constant is explained in the next section.

Formula 3.32 Constant terms of Formula 3.31

$$\frac{1}{\delta(x)}\cdot\frac{d^4\delta(x)}{dx^4}=-\frac{\mu}{E\cdot I_z}\cdot\frac{1}{\delta(t)}\cdot\frac{d^2\delta(t)}{dt^2}=g$$

From this expression, two differential equations are inferred: one is a function of time, and the second is a function of the spatial coordinate x.
Spatial differential equation Temporal differential equation

Formula 3.33 Differential equations whose solutions are δ(x) and δ(t)

$$\frac{d^4\delta(x)}{dx^4}-g\cdot\delta(x)=0\qquad\frac{d^2\delta(t)}{dt^2}+g\cdot\frac{E\cdot I_z}{\mu}\cdot\delta(t)=0$$

These equations are applicable for one vibration mode. Therefore, in successive formulas, the subscript j identifies the vibration mode. The structure of these two formulas predicts that their solutions are formed by harmonic series. One is a spatial series, and the other is a temporal series.

3.4.3.1 Solution to the Spatial Equation

The general solution to the spatial Formula 3.33 is:

Formula 3.34 General solution to the spatial differential Formula 3.33

$$\delta(x)=A_1\cdot e^{g\cdot x}+A_2\cdot e^{-g\cdot x}+A_3\cdot\cos\left(\sqrt[4]{g}\cdot x\right)+A_4\cdot\sin\left(\sqrt[4]{g}\cdot x\right)$$

δ(x) represents the beam shape when it vibrates at a specific natural frequency; therefore, deflections δ(x) are the beam vibration amplitudes at any section of the coordinate x for a specific vibration mode. Therefore, the condition of the solution δ(x) to the spatial differential equation is that it must be a function of x whose fourth derivative with respect to x is equal to the same function δ (x). A group of formulas that accomplish this condition is $e^{\lambda.x}$, $e^{-\lambda.x}$, sinh λ.x, cosh λ.x, sin λ.x, cos λ.x. However, only sin λ.x satisfies one of the boundary conditions discussed previously: the hinged support. The rest of the listed functions of x are not applicable because they don't verify any beam supports' boundary condition. Other methods have been carried out to determine δ(x) for supports other than hinged. However, these solutions require extensive derivations that are out of the scope of this book. As an alternative, some authors, like Den Hartog[11], prefer to apply the approximate Rayleigh method.

In summary: only the hinged-hinged beam has an exact solution. Therefore, the only valid component of Formula 3.34 is $A_4 \cdot \sin\left(\sqrt[4]{g} \cdot x\right)$, where $\sqrt[4]{g}$ is the inverse of a length (1/m). This concept indicates $\sqrt[4]{g}$ is the spatial frequency of δ(x) designated with λ. Hence the spatial solution to the Euler-Bernoulli for vibration mode j is:

Formula 3.35 Solution to the spatial Euler-Bernoulli

$$\delta_j(x) = A_j \cdot \sin(\lambda_j \cdot x)$$
$$\lambda_j = \sqrt[4]{g}$$

The boundary condition for this equation is that for x = 0 or x = L, it is δ(x) = 0 (see Section 3.4.1.1). Therefore, at x = L, the deflection is equal to $B_j \cdot \sin(\lambda_j \cdot L) = 0$. This expression is zero only if the angle $(\lambda_j \cdot L)$ is equal to zero or some multiple of the number π; therefore, $\lambda_j \cdot L = 0, \pi, 2 \cdot \pi, 3 \cdot \pi$ and so on up to $j \cdot \pi$. Hence, the formula of λ_j is:

Formula 3.36 Spatial frequency as a function of vibration mode j and beam length

$$\lambda_j = j \cdot \frac{\pi}{L}$$

As this is a hinged-hinged beam, at the support extremes x = 0 or x = L, the beam has a slope different to zero $(d\delta(x)/dx|_{x=0} \neq 0)$. This slope is an arbitrary boundary condition that prevents the existence of a trivial solution. The slope formula is: $d\delta j(x)/dx = A_j \cdot \lambda_j \cdot \cos(\lambda_j \cdot x)$. For x = 0 this derivative is $A_j \cdot \lambda_j$. Therefore, the spatial amplitude A_j is equal to the slope at the supports over the spatial frequency, and the integral of the space equation with initial conditions is:

Formula 3.37 Solution to the spatial Euler-Bernoulli equation with initial conditions

$$\delta_j(x) = \frac{\left|d\delta_j / dx\right|_{x=0}}{\lambda_j} \cdot \sin(\lambda_j \cdot x) = A_j \cdot \sin(\lambda_j \cdot x)$$

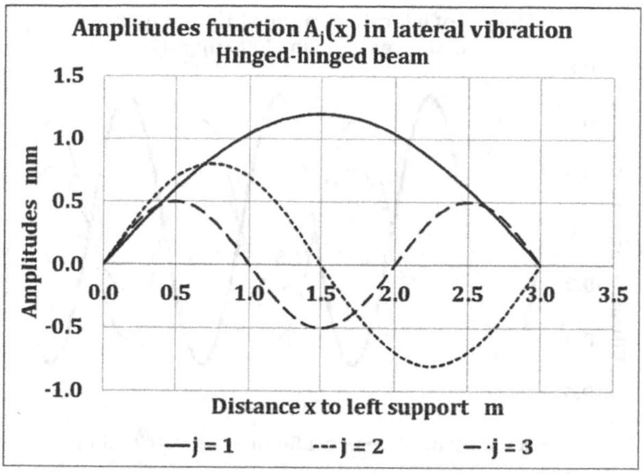

FIGURE 3.9 Amplitude's function $A_j(x)$ for the first three vibration modes.

Formula 3.37 is the solution to the spatial equation for a hinged-hinged beam. Physically, the product $A_j \cdot \sin(\lambda_j \cdot x)$ is the vibration amplitude function, whose values only depend on the beam's point where vibration amplitudes are studied. The abscissa x is the distance from that point to the left support. Therefore, Formula 3.37 represents the spatial standing sinusoidal waves that determine the vibration amplitude of $\sin(\omega_{nj} \cdot t)$ at any abscissa x. See Figure 3.9, representing the spatial amplitude of vibration for the first vibration modes.

There is an analogy between the spatial and temporal domains, summarized in the following table.

L_j is the spatial period of the harmonic j, and the number of cycles in the beam is L/L_j or j/2.

When the function $A_j(x)$ is zero, vibration mode j does not exist, meaning that the beam does not vibrate with the j mode at those points. However, it does not mean that the beam does not vibrate at that point because it docs at other vibration modes. Figure 3.10 shows vibration mode 3 at three different positions on the beam, at 0.3 m, 1.5 m and 2.85 m. This figure shows that sections near the supports have lesser amplitude than the vibration at the middle of the beam.

3.4.3.2 Beam's Vibration Shapes

Every vibration mode produces a specific beam shape. Therefore, the beam's shape is the combination of shapes generated by the most significant vibration mode. Theoretically, there are as many shapes as vibration modes. See in Figure 3.11 the shapes adopted by the beam at three distinct vibration modes.

These figures suggest that the support type influences the shape of each vibration mode. The most typical supports are hinged-hinged, free (cantilever), and clamped-clamped.

In summary: there are infinite natural vibration modes along a beam. All the beam points vibrate but not with the same amplitude and frequency. In the

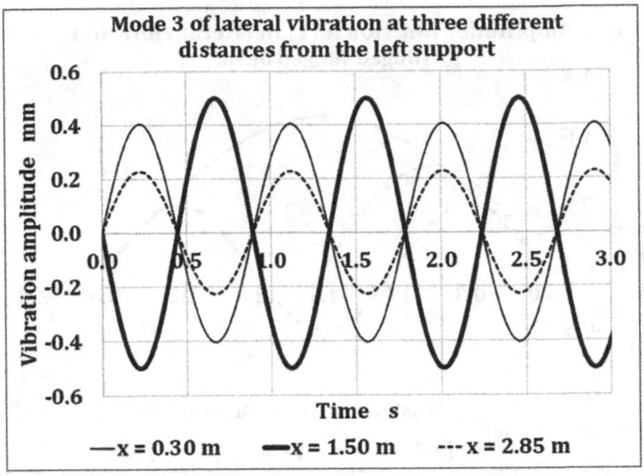

FIGURE 3.10 Vibration curves of mode 3 at different distances from the left support.

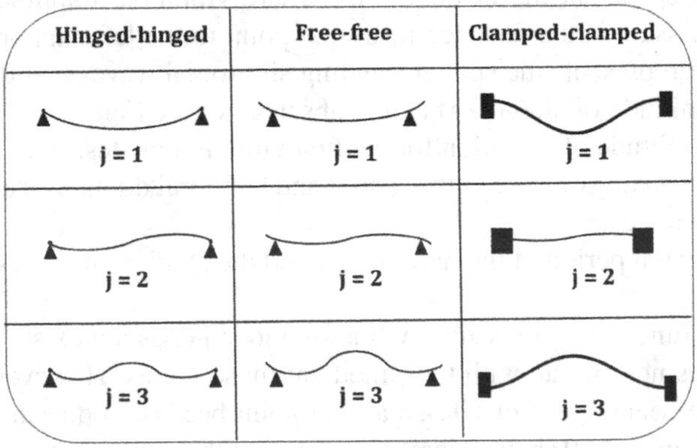

FIGURE 3.11 Beams' shapes for several types of support and vibration modes.

amplitude j curve's intersections with the time axis, the shaft does not vibrate at the mode j. As spatial frequencies are different, the zero points for one mode are not forcefully zero for the others.

3.4.3.3 Solution to the Temporal Equation

According to Calculus theory, differential equations of the form y'' (x) + a · y(x) = 0 have the following solution:

$$y(x) = C_1 \cdot \cos\left(\sqrt[2]{a}\, x\right) + C_2 \cdot \sin\left(\sqrt[2]{a} \cdot x\right)$$

Therefore, applying this solution to temporal Formula 3.33, the following temporal formula is obtained.

Formula 3.38 General solution to temporal Formula 3.33

$$\delta(t) = C_1 \cdot \cos\sqrt[2]{\Omega} \cdot t + C_2 \cdot \sin\sqrt[2]{\Omega} \cdot t$$

If this solution is compared with the temporal Formula 3.33, it is concluded that $\Omega = g \cdot \frac{E \cdot I_z}{\mu}$. Formula 3.38 suggests that Ω is the square of the vibratory frequency ω_n. As $g = \lambda_j^4$, this constant is replaced in the natural frequency formula as shown below.

Formula 3.39 Natural frequency of vibration mode j

$$\omega_{nj} = \sqrt[2]{\Omega} = \sqrt[2]{g \cdot \frac{E \cdot I_z}{\mu}} = \lambda_j^2 \cdot \sqrt[2]{\frac{E \cdot I_z}{\mu}}$$

Initial conditions of the temporal Formula 3.33 are: at t = 0, $\delta(t) = 0$ and $\dot{\delta}(0) = \dot{\delta}_0$. Therefore, the temporal solution to the frictionless beam vibration is:

Formula 3.40 Solution to the temporal Euler-Bernoulli equation

$$\delta_j(t) = C_j \cdot \sin(\omega_{nj} \cdot t)$$

This solution must not be trivial; then, constant C_j must be different from zero. This condition is accomplished if the initial velocity is different from zero. Therefore, as $\frac{d\delta(t)}{dt} = C_j \cdot \omega_{nj} \cdot \cos(\omega_{nj} \cdot t)$, at t = 0, the initial system velocity is $\frac{d\delta_j(t)}{dt} = \dot{\delta}_{j0}$. This velocity is an arbitrary initial condition. Replacing it in the velocity formula, the results are, $\dot{\delta}_{j0} = C_j \cdot \omega_{nj}$, because cos 0 = 1. Therefore, the integral of the differential solution is:

Formula 3.41 Solution to the temporal Euler-Bernoulli equation with initial conditions

$$\delta_j(t) = \frac{\dot{\delta}_{j0}}{\omega_{nj}} \cdot \sin(\omega_{nj} \cdot t)$$

This formula is always valid because the vibration is a motion with a velocity different from zero.

3.4.3.4 General Solution of the Euler-Bernoulli Equation

The general solution for vibration mode j is the product $\delta_j(x) \cdot \delta_j(t)$; therefore, the total solution is obtained by summing all these products from j = 1 to infinite, being j the infinite series of whole numbers. Hence, the formula of deflection waves is:

Formula 3.42 General solution to the Euler-Bernoulli equation for a hinged-hinged beam

$$\delta(x,t) = \sum_{j=1}^{\infty}\left(\frac{S_0}{\lambda_j}\cdot\frac{\dot{\delta}_{j0}}{\omega_{nj}}\right)\cdot\sin(\omega_{nj}\cdot t) = \sum_{j=1}^{\infty} A_j(x)\cdot\sin(\omega_{nj}\cdot t)$$

Formula 3.42 for one vibration mode can be compacted with this expression: $A_j(x) \cdot \sin(\omega_{nj} \cdot t)$. The amplitude A_j depends on the distance to the left support and the vibration mode under study.

In practice, it is not necessary to solve the Euler-Bernoulli equation to settle the beam's vibration issues because Formula 3.42 provides the information often needed in practice. This formula was derived for a conservative system; therefore, the system vibrates at its natural frequencies ω_{nj}, where j identifies the vibration mode. Of course, in practice, the above sum is never done for infinite vibration modes, but only for the first and most significant modes, for example, for j = 1 to 5.

3.4.4 Natural Frequencies with the Euler-Bernoulli Equation

This section derives the natural frequency formulas of a hinged-hinged beam on the base of the Euler-Bernoulli equation.

As hinged-hinged beams are prevented from vibrating at both ends because they are mechanically bound to a structure or machine bearings, vibration amplitudes at x = 0 and x = L are zero. Hence, $\sin(\lambda_j \cdot L) = 0$. This condition is accomplished if $\lambda_j \cdot L = 0, \pi, 2 \cdot \pi, 3 \cdot \pi \ldots j \cdot \pi$. Therefore:

Formula 3.43 Spatial frequencies formula for hinged-hinged beams

$$\lambda_j = \frac{j\cdot\pi}{L} = \frac{2\cdot\pi}{L_j}$$

From the Euler-Bernoulli equation, it is derived the natural frequency formula.

Formula 3.44 Natural frequencies formula for hinged-hinged beams

$$\omega_{nj} = \frac{(j\cdot\pi)^2}{L^2}\cdot\sqrt[2]{\frac{E\cdot I_z}{\rho\cdot A}}$$

The product $\rho \cdot A$ is the beam mass per unit length, usually designated with μ; it is expressed in kg mass/m. The coefficient $(j \cdot \pi)^2$ is only valid for hinged-hinged (also called pinned-pinned or simply supported) beams. Therefore, to extend this formula to other support types, coefficient $(\pi \cdot j)^2$ is replaced by a general coefficient α_j, which depends on the vibration mode j and the beam's support type. Therefore, Formula 3.44 of natural frequencies is converted in the following expression.

Formula 3.45 Beam natural frequency for any support type

$$\omega_{nj} = \frac{\alpha_j}{L^2} \cdot \sqrt[2]{\frac{E \cdot I_z}{\rho \cdot A}}$$

The values of the α_j coefficient[12] are listed in Table 3.2 for vibration modes 1 to 5. See in Figure 3.12, the graphical representation of coefficient α_j for six beam support types[13].

It is possible to simplify the natural frequency formula for uniform cylindrical beams, also valid for shafts, by substituting in Formula 3.45, the term $\sqrt[2]{E/\rho}$ by the stress propagation velocity c and $\sqrt[2]{I_z/A}$ by the D/4 ratio. The resultant expression is:

Formula 3.46 Natural frequency of a uniform circular beam or shaft

$$\omega_{nj} = \frac{\alpha_j}{4} \cdot c \cdot \frac{D}{L^2}$$

TABLE 3.1
Analogies Between λ and ω

λ	Is Analogous to Frequency	ω
L_j	is analogous to the period	T
$\frac{2 \cdot \pi}{L_j}$	is analogous to	$\frac{2 \cdot \pi}{T}$

TABLE 3.2
Values of α_j Coefficient for Several Types of Support and Vibration Mode

Values of α_j

support type at x = 0 and x= L	Mode 1	Mode 2	Mode 3	Mode 4	Mode 5
Hinged – free or chopper's wing	0,00	15,42	50,00	104,00	178,00
Clamped-free or cantilever	3,52	22,00	61,70	120,90	200,00
Hinged-hinged or simply supported	9,87	39,48	88,83	157,91	246,74
Clamped–hinged	15,42	49,97	104,00	178,28	272,02
Clamped-clamped and free-free	22,40	61,70	120,90	199,90	298,60

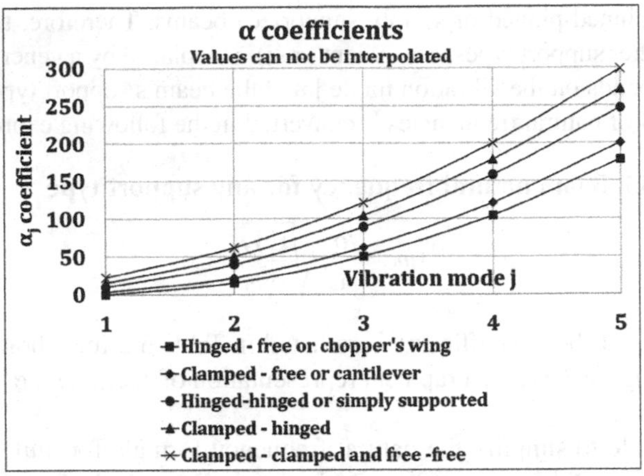

FIGURE 3.12 α_j coefficients of the natural frequencies' Formula 3.45.

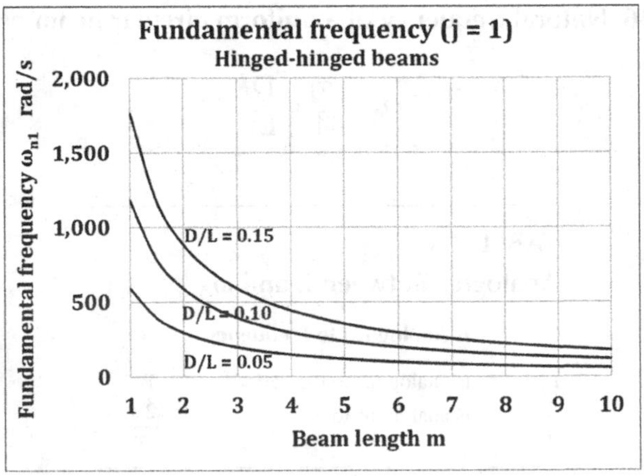

FIGURE 3.13 Natural frequencies for j = 1 versus beam's D/L ratio and length. c = 4,773 m/s.

This formula demonstrates that the natural frequency in a circular beam is proportional to the wave velocity and the beam diameter.

Figure 3.13 shows the natural frequency curves of a hinged-hinged beam as a function of the beam's length for three different D/L ratios. These curves reveal that the lower the beam diameter/length ratio, the lower the natural frequency. In other words: thin and long beams have lesser natural frequencies than gross and short beams.

Valid only for hinged-hinged beams

The beam's natural frequencies strongly depend on the beam's slenderness. The D/L ratio is used as a slenderness measurement. Beam's typical D/L ratio is

0.10 (not higher than 0.15). For longer beams, the natural frequency tends asymptotically to zero.

In summary: a beam, or a shaft, has infinite natural frequencies. These frequencies depend on the beam's support type and its slenderness. Vibration amplitude is different from one beam point to another because it has a sinusoidal distribution along the beam. The higher the beam slenderness, the lower the natural frequency. A freely vibrating beam shape, given by the function $\delta_j(t)$, depends on vibration mode j and the support type.

3.4.5 Practical Assessment. Turbogenerator Set Frequencies

Calculate the shaft first three resonance velocities in the machine shaft of a 60 Hz power plant. Assume that the shaft is equivalent to a hinged-hinged beam. The following table contains the shaft technical specifications.

Data:

Young's modulus E = 3.04E+10 kg/m^2
Steel density = 801 kg mass /m^3
Shaft length = 3.0 m
Shaft uniform diameter D = 0.45 m

Calculations:

Wave velocity:
c = (2.04E + 10/801)$^{1/2}$ = 5, 047 m/s
α_j coefficients values (from Table 3.1):
$\alpha_1 = 9.9$ $\alpha_2 = 39.5$ $\alpha_3 = 88.8$

Natural frequencies:

$$\omega_{n1} = \frac{\alpha_1}{4}\cdot c\cdot\frac{D}{L^2} = \frac{9.9}{4}\times 5{,}047\times\frac{0.45}{3^2} = 624.6\,\text{rad/s}$$

$$\omega_{n2} = \frac{\alpha_2}{4}\cdot c\cdot\frac{D}{L^2} = \frac{39.5}{4}\times 5{,}047\times\frac{0.45}{3^2} = 2{,}492.0\,\text{rad/s}$$

$$\omega_{n3} = \frac{\alpha_3}{4}\cdot c\cdot\frac{D}{L^2} = \frac{88.8}{4}\times 5{,}047\times\frac{0.45}{3^2} = 5{,}602.1\,\text{rad/s}$$

Resonance velocities:

$N_1 = 624.6 \times 30/\pi = 5{,}965$ RPM
$N_2 = 2{,}495.0 \times 30/\pi = 23{,}826$ RPM
$N_3 = 5{,}602.1 \times 30/\pi = 53{,}496$ RPM

The conclusion is that as a 60 Hz generator rotates at 3,600 RPM, the machine is not shaken during the startup because the minimum resonance velocity is 5,965 RPM.

3.5 THE WAVE EQUATION

Propagation of a physical magnitude like sound, electromagnetism, heat, mechanical stress, deflection and other properties in a solid, liquid or gaseous medium is explained by a mathematical expression known as the wave equation. This equation demonstrates how torsional and shear waves propagate energy from one location to another through a solid medium. In the case of a machine shaft, this medium is steel.

Figure 3.14 shows a shaft driven by a machine. It is assumed that the system is frictionless and that the shaft inertia is included in the load inertia. The torsional deflection is indicated as angle ϑ. It is zero at the load side (x = 0) and maximum at the shaft coupling extreme (x = L). The deflection angle linearly increases between both extremes, from the load to the turbine or motor coupling. See the right Figure 3.14. Load inertia and shaft rigidity torques oppose the applied torque, so the torque equilibrium is expressed:

Formula 3.47 Applied torque T_a = load and shaft inertial reactions T_L + shaft rigidity reaction T_R

$$T_a = T_L + T_R$$

The proportionality coefficient between the reaction torque T_R and the deflection angle is the shaft stiffness coefficient k_t, equal to $G \cdot I_p/L$. The formula of both reactions T_R and T_L are:

Formula 3.48 Shaft reaction torques as a function of ϑ

$$T_R = \frac{G \cdot I_p}{L} \cdot \vartheta \quad T_L = J \cdot \ddot{\vartheta}$$

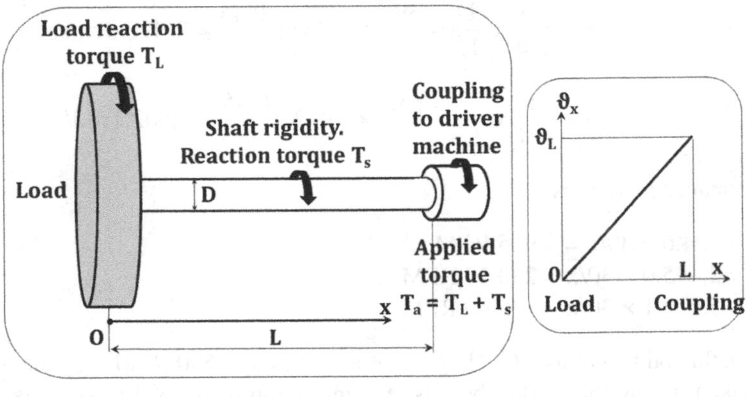

FIGURE 3.14 Torsional system. Applied and reaction torques.

In these formulas, G is the shear modulus, and I_p is the polar moment of inertia. The inertial reaction T_L was discussed in Chapter 1. These formulas demonstrate that at constant RPM, the rigidity reaction equals the applied torque to the load. Revisit the graphical interpretation of angle ϑ in Figure 2.2. The equation of motion is readily written with the above concepts.

Formula 3.49 Equation of motion of torsional vibration

$$T_a(t) = J \cdot \ddot{\vartheta}(t) + \frac{G \cdot I_p}{L} \cdot \vartheta(t)$$

This equation represents the equilibrium between the applied torque and the inertia and rigidity components' resistant torque. The system does not contain a damping reaction because it is conservative.

3.5.1 Derivation of the Wave Equation

Torsional vibration formulas derive from a second-order partial differential equation called the wave equation. Partial derivatives are used because the results are expressed as a function of time and the spatial coordinate identifying the machine's studied point. This equation is used in many science fields, including vibration engineering. Therefore, in this text, it is applied to rotating machines to calculate their natural frequencies.

The integration of the wave equation allows deriving a formula to calculate the shaft-rotor natural frequencies. The wave equation assumes that beams or shafts are continuous systems with distributed mechanical properties.

Assume an elemental shaft of length dx, as shown in Figure 3.15. As torsion is a linear phenomenon along the shaft, it is $\vartheta/L = d\vartheta/dx$ at any shaft section.

Hence, Formula 3.38 of T_R is written as follows.

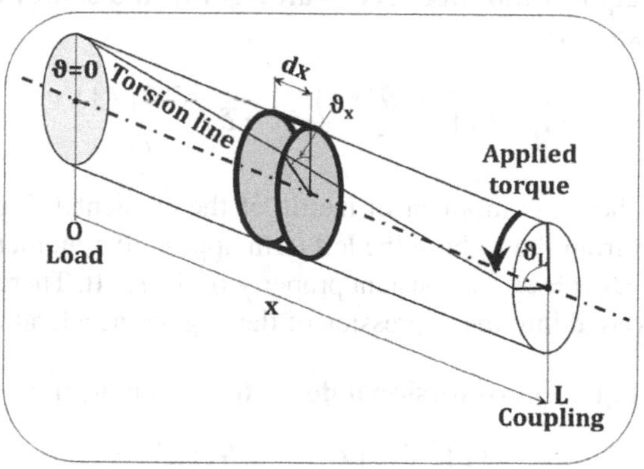

FIGURE 3.15 Elemental shaft and torsion angle ϑ.

Formula 3.50 Rigidity reaction torque as a function of dϑ/dx

$$T_s = G \cdot I_p \cdot \frac{d\vartheta}{dx}$$

Where subscript s stands for the shaft.

The shaft angular deflection at any section of the shaft depends on the time and distance to the applied torque. This double dependency obliges the use of partial derivatives in the mathematical model of torsional vibration. If the shaft is running at a constant speed, the inertial shaft reaction torque is zero. Therefore, in that case, the applied torque is only equal to the shaft rigidity torque reaction, and so no torsional vibration exists. The following formula gives the shaft torque change δT_s along the shaft's elemental length δx.

Formula 3.51 Torque change at the elemental shaft

$$\delta T_s = \frac{dT_s}{dx} \cdot \delta x$$

Replacing T_s of Formula 3.50 in Formula 3.51, the torque change in the elemental shaft is:

Formula 3.52 Rigidity torque change in the elemental shaft

$$\delta T_s = G \cdot I_p \cdot \frac{\partial^2 \vartheta(x,t)}{\partial x^2} \cdot \delta x$$

According to Newton's second law, the deflection angle is accelerating due to the applied torque overcoming the shaft rigidity reaction. Therefore, the rigidity torque of Formula 3.52 is equal to the moment of inertia times the angular acceleration, as shown in Formula 3.53.

Formula 3.53 Equilibrium between shaft rigidity and shaft inertial reactions, in |kg · m|

$$\delta T_s = G \cdot I_p \cdot \frac{\partial^2 \vartheta(x,t)}{\partial x^2} \cdot \delta x = \delta J \cdot \frac{\partial^2 \vartheta(x,t)}{\partial t^2}$$

Where δJ is the mass moment of inertia of the elemental shaft[14], moving δJ of Formula 3.53 from the right to the left term appears the moment of inertia per unit length $\mu_J = \delta J/\delta x = J/L$, a constant property of the shaft. Therefore, the above equation is converted into the expression of the angular acceleration.

Formula 3.54 Equation of torsional deflection acceleration

$$\frac{G \cdot I_p}{\mu_J} \cdot \frac{\partial^2 \vartheta(x,t)}{\partial x^2} = \frac{\partial^2 \vartheta(x,t)}{\partial t^2}$$

This expression is the equation of the angular acceleration of the elemental shaft produced by the shaft rigidity reaction. As the μ_J/I_p ratio[15] is equal to the shaft density, this equation is expressed.

Formula 3.55 Wave equation of torsional deflections, in |1/s²|

$$\frac{G}{\rho} \cdot \frac{\partial^2 \vartheta(x,t)}{\partial x^2} = \frac{\partial^2 \vartheta(x,t)}{\partial t^2}$$

This equation says that as ϑ is a function of coordinate x and time t, the oscillation behaves as a travelling wave along the shaft. It is demonstrated below that a standing wave for each harmonic determines the amplitude of the travelling waves. The G/ρ ratio's square root is the torsional T-wave velocity along the shaft, constant for every specific type of steel. As noted previously, the typical value of this velocity is 3,100 m/s in inox steel and is noted with the letter c (do not confuse with longitudinal waves velocity, 67% higher). Therefore, the standard form of the torsional wave equation is.

Formula 3.56 The standard form of the wave equation in |1/m²|

$$\frac{\partial^2 \vartheta(x,t)}{\partial x^2} = \frac{1}{c^2} \cdot \frac{\partial^2 \vartheta(x,t)}{\partial t^2}$$

Therefore, the spatial distribution of torsional vibration amplitudes along the shaft (first term of 3.56) is proportional to vibration acceleration (second term of 3.56). The proportionality coefficient is the inverse of the square of the torsional wave velocity.

3.5.2 Solution to the Wave Equation

Identical to the solution to the Euler-Bernoulli equation, the solution to the wave equation is formed by the product of two functions: $\vartheta(x,t) = \vartheta(x) \cdot \vartheta(t)$. These two functions are the solution to two differential equations named temporal and spatial differential equations. Both are derived as explained in paragraphs a) and b).

a) The temporal equation is formed with the following formula of vibration acceleration seen in Chapter 1: $d^2\vartheta(t)/dt^2 = -\omega_n^2 \cdot \vartheta(t)$. This term is moved to the left, so forming the temporal Formula 3.47. The solution to this equation is $\vartheta(t)$.
b) The spatial equation is formed by replacing, in Formula 3.56, the acceleration by $-\omega_n^2 \cdot \vartheta(x)$ and moving this product to the left term. The solution to this equation is $\vartheta(x)$.

So, the wave equation solution is the solution's product of the two differential equations. See Formula 3.47.

Formula 3.57 Differential equations of torsional vibration

$$\frac{d^2\vartheta(x)}{dx^2}+\frac{\omega_n^2}{c^2}\cdot\vartheta(x)=0\quad\frac{d^2\vartheta(t)}{dt^2}+\omega_n^2\cdot\vartheta(t)=0$$

Spatial differential equation Temporal differential equation
The above equations are valid for one vibration mode each.

3.5.2.1 Solution to the Spatial Equation

From the differential equations theory, the solution to: y"(x) + a · y(x) = 0 is:

$$y(x)=C1\cdot\cos\left(\sqrt[2]{a}\cdot x\right)+C2\cdot\sin\left(\sqrt[2]{a}\cdot x\right)$$

Therefore, the following formula is the general solution to the spatial equation for vibration mode j.

Formula 3.58 General solution to the spatial differential equation

$$\vartheta_j(x)=B_{1j}\cdot\cos\left(\frac{\omega_{nj}}{c}\cdot x\right)+B_{2j}\cdot\left(\sin\frac{\omega_{nj}}{c}\cdot x\right)$$

Where ω_{nj}/c is the spatial frequency λ_j of the stationary wave, whose amplitude is δ(x), and ω_{nj} is the natural frequency of the j harmonic. Initial conditions at the shaft load extreme (see Figure 3.15) are: [x = 0, $\vartheta_j(0) = 0$]; therefore, $B_{1j} = 0$.

The function ϑ(x) describes the shaft shape when the shaft vibrates at one of its natural frequencies. Therefore, the δ(x) ordinates are the vibration amplitude at any coordinate x for a specific vibration mode. As the δ(x) waves are standing, the amplitudes do not change in time. The spatial frequency λ_j is derived from the following physical condition: at x > L, no reaction torque exists. Therefore, at x = L it is dϑ/dx = 0. By applying this boundary condition, the natural frequencies formula is returned, as demonstrated below.

Formula 3.59 Boundary condition at x = L

$$\frac{d\vartheta(x)}{dx}=B_j\cdot\lambda_j\cdot\cos\left(\lambda_j\cdot L\right)=0$$

This condition is fulfilled if $\cos(\lambda_j \cdot L) = 0$, which means that the angle $\lambda_j \cdot L$ is equal to π/2 or a whole odd number multiple of π/2 of the following series: [3, 5, 7, 9…]. Accordingly, the spatial frequency formula is:

Formula 3.60 Spatial frequency of torsional waves

$$\lambda_j = \left(j - \frac{1}{2}\right)\cdot\frac{\pi}{L}$$

Thus, the solution to the spatial differential equation or amplitudes function for vibration mode j is:

Formula 3.61 Spatial function of torsional vibration amplitudes

$$\vartheta_j(x) = B_j \cdot \sin(\lambda_j \cdot x)$$

See in Figure 3.16, the representation of this function for vibration modes 1, 3 and 4.

Figure 3.16 an example of spatial waves for vibration modes 1, 3 and 4. In this figure, the abscissas x units are in % of the total shaft length L. Vibration mode 4 has a wavelength of 57% of the shaft length. It means that shaft sections at x = 0% and x_a = 57% do not vibrate with mode 4. Therefore, there are 1/0.57 = 1.75 cycles of vibration of mode 4 in the shaft length. Instead, at x = 57%, harmonic 3 has an amplitude of 0.58°.

3.5.2.2 Solution to the Temporal Equation

The general solution to this differential equation for vibration mode j is similar to the spatial equation (see footnote of the last section).

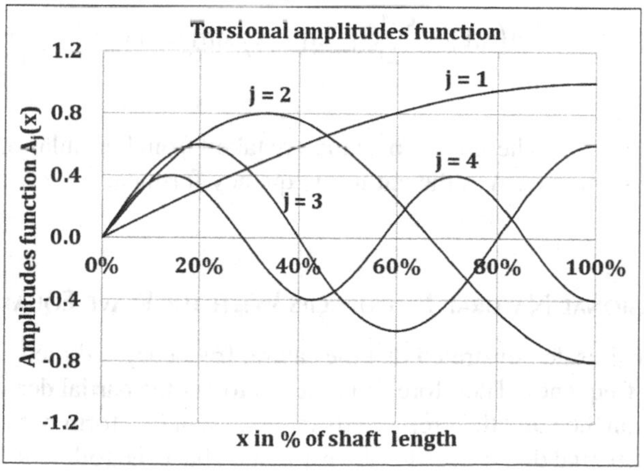

FIGURE 3.16 Amplitude stationary waves along the shaft for c = 3,100 m/s.

Formula 3.62 General solution to the temporal differential equation

$$C_j(t) = A_{1j} \cdot \cos(\omega_{nj} \cdot t) + A_{2j} \cdot \sin(\omega_{nj} \cdot t)$$

For initial conditions of [$t = 0, \vartheta_j(0) = 0$]; it is $A_{1j} = 0$; therefore, the solution to the temporal equation is:

Formula 3.63 Temporal function of torsional vibrations

$$\vartheta_j(t) = A_j \cdot \sin(\omega_{nj} \cdot t)$$

If the initial vibration velocity $\dot{\vartheta}_0$ is prefixed, the derivative of $\vartheta_j(t)$ respect to time is equal to $\dot{\vartheta}_0$ at $t = 0$. As $\dot{\vartheta}(t) = A_j \cdot \omega_{nj} \cdot \cos(\omega_{nj} \cdot t)$, at $t = 0$ this derivative is: $\dot{\vartheta}_0 = A_j \cdot \omega_{nj}$. Therefore, the temporal solution to the wave equation is:

Formula 3.64 Temporal solution of the wave equation for torsional vibration

$$\vartheta_j(t) = \frac{\dot{\vartheta}_0}{\omega_{nj}} \cdot \sin(\omega_{nj} \cdot t)$$

3.5.2.3 General solution of the wave equation

The general formula of the torsional vibration of mode j is the product of Formulas 3.61 and 3.64. The complete solution is the sum of all j mode solutions from 1 to infinite.

Formula 3.65 General solution to the wave equation for torsional vibrations

$$\vartheta(x,t) = \sum_{j=1}^{\infty} C_j \cdot \sin(\lambda_j \cdot x) \cdot \sin(\omega_{nj} \cdot t)$$

Where $C_j = A_j \cdot B_j$. These two constants are taken from Formulas 3.51 and 3.54. The next sub-section derives the natural frequency formula.

3.5.3 Torsional Natural Frequencies With the Wave Equation

In this section, it is demonstrated that the natural frequency is directly proportional to the spatial frequency. Therefore, it is needed to get the partial derivatives of the general solution 3.65 and then replace them in the standard form of the wave equation 3.56. The partial derivatives for the harmonic j have the following expressions.

Formula 3.66 Partial derivatives of ϑ(x,t)

$$\frac{\partial \vartheta(x,t)}{\partial x} = C_j \cdot \lambda_j \cdot \cos(\lambda_j \cdot x) \cdot \sin(\omega_j \cdot t) \rightarrow \frac{\partial^2 \vartheta(x,t)}{\partial x^2} = -C_j \cdot \lambda_j^2 \cdot \sin(\lambda_j \cdot x) \cdot \sin(\omega_j \cdot t)$$

$$\frac{\partial \vartheta(x,t)}{\partial t} = C_j \cdot \sin(\lambda_j \cdot x) \cdot \omega_j \cdot \cos(\omega_j \cdot t) \rightarrow \frac{\partial^2 \vartheta(x,t)}{\partial t^2} = -C_j \cdot \sin(\lambda_j \cdot x) \cdot \omega_{nj}^2 \cdot \sin(\omega_j \cdot t)$$

The relation between spatial and natural frequencies is obtained by replacing the second partial derivatives of Formulas 3.66 in the wave equation 3.56.

Formula 3.67 Natural frequency versus spatial frequency

$$\omega_{nj} = c \cdot \lambda_j$$

Thus, the natural frequency is obtained by multiplying Formula 3.60 by the wave velocity c.

Formula 3.68 Natural frequency of torsional vibration mode j

$$\omega_{nj} = \left(j - \frac{1}{2} \right) \cdot \frac{\pi}{L} \cdot c$$

Formula 3.68 is an important mathematical expression of continuous systems theory because it readily calculates a shaft's natural frequencies. Based on the above formula, the fundamental harmonic frequency (j = 1) is given by the following expression:

Formula 3.69 Fundamental frequency (j = 1) of torsional waves

$$\omega_{n1} = \frac{\pi}{2} \cdot \frac{c}{L}$$

Therefore, the harmonic frequencies are directly proportional to the first harmonic frequency ω_{n1}. Their formulas are: $\omega_{n2} = 3 \cdot \omega_{n1}$, $\omega_{n3} = 5 \cdot \omega_{n1}$, $\omega_{n4} = 7 \cdot \omega_{n1}$, $\omega_{n5} = 9 \cdot \omega_{n1}$, etc. See in the chart in Figure 3.17 the frequency curves as a function of the shaft length for vibration modes 1 to 4. These curves show that the longer the shaft, the lower the shaft's natural frequencies. The SDOF frequency ($\omega_n = c/L$) curve has been included to compare natural frequencies between continuous and SDOF systems.

Therefore, Formula 3.69 demonstrates that continuous systems' natural fundamental frequency is $\pi/2$ higher than a discrete SDOF system, which means a 57% difference. For example, for the case of a 4 m shaft length, the SDOF natural frequency is 3,100/4 = 775 rad/s, while Formula 3.59 returns ω_{n1} = 1,217 rad/s that is 57% higher than the SDOF frequency.

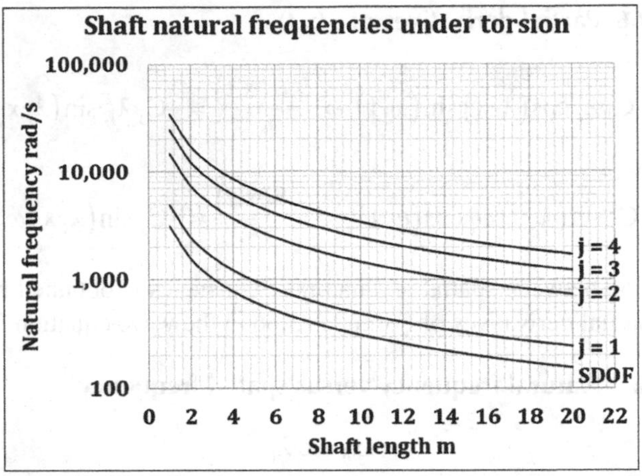

FIGURE 3.17 Natural frequency of the first four vibration modes, for $c_s = 3{,}100$ m/s.

As the typical stress propagation velocity of torsional waves in steel is 3,100 m/s, the first natural frequency may be estimated with the following rule of thumb expression.

Formula 3.70 Rule of thumb formula for the first three natural frequencies

$$\omega_{n1} = \frac{\pi \cdot c}{2 \cdot L} \cong \frac{5{,}000}{L} \quad \omega_{n2} = \frac{3 \cdot \pi \cdot c}{2 \cdot L} \cong \frac{15{,}000}{L} \quad \omega_{n3} = \frac{5 \cdot \pi \cdot c}{2 \cdot L} \cong \frac{25{,}000}{L}$$

Where ω_{nj} are in rad/s and L in m. The difference between adjacent natural frequencies is:

Formula 3.71 Difference between successive natural frequencies

$$\Delta\omega_{nj} = \omega_{nj} - \omega_{n\,j+1} = \frac{c}{L} \cdot \pi$$

Therefore, high-order harmonic frequencies are easily calculated by adding $c \cdot \pi / L$ to the previous natural frequency. The L/c ratio is the torsional stress propagation time from one shaft end to the other. For a 4 m shaft, the propagation time is 0.0013 seg, which is considered negligible.

3.5.4 Practical Assessment. Oil Drill Rig

As an example, assume an oil well-implemented built with a drill rig rotating at 620 RPM. The shaft length is 400 m. Determine if there is any risk of resonance during startups and the resonances velocities in RPM. Assume a propagation velocity of stress T waves of 3,150 m/s.

Shaft rotating frequency: $\omega_s = \pi \times 620/30 = 64.9$ rad/s
The shaft's natural frequencies are:

$$\text{Fundamental harmonic: } \omega_1 = (1-1/2)\times\pi\times3{,}150/400 = 12.4\,\text{rad/s}$$

$$\text{Second harmonic: } (2-1/2)\times\pi\times3{,}150/400 = 37.1\,\text{rad/s}$$

$$\text{Third harmonic: } (3-1/2)\times\pi\times3{,}150/400 = 61.9\,\text{rad/s}$$

$$\text{Fourth harmonic: } (4-1/2)\times\pi\times3{,}150/400 = 86.6\,\text{rad/s}$$

These results show that the shaft rotating frequency is between the third and the fourth harmonic. Thus, the shaft is on three occasions under resonance during startup before reaching its operating speed of 64.9 rad/s. Therefore, during the shaft startup process, transient resonance will happen at 12.4, 37.1 and 61.9 rad/s, which corresponds to 118, 354 and 591 RPM, respectively. As the shaft velocity is 620 RPM, the third harmonic proximity (4.9% speed difference) represents a risk. This closeness puts the shaft at permanent risk of resonance and in the aftermath of fatigue and breakage.

NOTES

1 Recommended books for Chapter 3: 1. Mechanical Vibrations by S. Graham Kelly. Schaum's Outline Series. McGraw Hill, 1996 – 2. Advanced Vibration Analysis. S. Graham Kelly. CRC Press. Taylor and Francis Group, 2007 – 3. Fundamentals of Vibration Engineering by Isidor Bykhovsky. MIR Publishers, Moscow. 1st published 1972 – 4. Mechanical Vibrations by J.P. Den Hartog. Dover Publications, Inc. New York, 1985 – 5. Mechanical Vibration by Haym Benaroya, Mark Nagurka and Seon Han. CRC Press. Taylor and Francis Group, fourth edition 2018.

2 Vibration mode refers to a vibration component of determined frequency. It must not be confused with fundamental vibration forms of Section 1.11.

3 Bibliography for this chapter:

1. Vibration Engineering by Isidor Bykhovsky. MIR Publishers, Moscow. 1st published 1972. Section 12.
2. Mechanical Vibration by Haym Benaroya, Mark Nagurka and Seon Han. CRC Press. Taylor and Francis Group, fourth edition 2018. Chapter 7.
3. Mechanical Vibrations by J.P. Den Hartog. Dover Publications, Inc. New York, 1985. Section 4.6.
4. 1. Mechanical Vibrations by S. Graham Kelly. Schaum's Outline Series. McGraw Hill, 1996 Chapter 7.

4 The sum of two 3×3 matrices is done with the following formula:

$$\begin{bmatrix} a_{11} & a_{12} & a_{13} \\ a_{21} & a_{22} & a_{23} \\ a_{31} & a_{32} & a_{33} \end{bmatrix} + \begin{bmatrix} b_{11} & b_{12} & b_{13} \\ b_{21} & b_{22} & b_{23} \\ b_{31} & b_{32} & b_{33} \end{bmatrix} = \begin{bmatrix} a_{11}+b_{11} & a_{12}+b_{12} & a_{13}+b_{13} \\ a_{21}+b_{21} & a_{22}+b_{22} & a_{23}+b_{23} \\ a_{31}+b_{31} & a_{32}+b_{32} & a_{33}+b_{33} \end{bmatrix}$$

5 Physical properties are constant numbers that characterize a system, such as rigidity coefficient, friction ratio, mass moment of inertia, natural frequencies, etc. State variables are the system's position, velocity, and acceleration.

6 For an application of this method to variable frequency drives see the paper Torsional Analysis of Variable Frequency Drives, by Fred Szenasi of Engineering Dynamics, Inc. San Antonio, Texas. On site: http://www.engdyn.com/images/uploads/60-torsional_analyses_of_variable_frequency_drives_frs.pdf

7 The sum of two 2×2 matrices is done with the following general formula:

$$\begin{bmatrix} a_{11} & a_{12} \\ a_{21} & a_{22} \end{bmatrix} + \begin{bmatrix} b_{11} & b_{12} \\ b_{21} & b_{22} \end{bmatrix} = \begin{bmatrix} a_{11}+b_{11} & a_{12}+b_{12} \\ a_{21}+b_{21} & a_{22}+b_{22} \end{bmatrix}$$

8 It's recommended to consult the site: https://en.wikipedia.org/wiki/Euler%E2%80%93Bernoulli_beam_theory

9 See a thorough analysis of the Euler-Bernoulli beams in. Advanced Vibration Analysis. S. Graham Kelly. CRC Press. Taylor and Francis Group, 2007 Section 2.11

1. Mechanical Vibrations by S. Graham Kelly. Schaum's Outline Series. McGraw Hill, 1996 – 2 – 3.– 4.– 5. Mechanical Vibration by Haym Benaroya, Mark Nagurka and Seon Han. CRC Press. Taylor and Francis Group, fourth edition 2018.

10 Fundamentals of Vibration Engineering by Isidor Bykhovsky. MIR Publishers, Moscow. 1st published 1972, pages 92 and 93.

11 See the book Mechanical Vibrations by J.P. Den Hartog. Dover Publications, Inc. New York, 1985 , Section 4.6.

12 Many other vibration texts also reproduce this table with minor differences.

13 Consult Mechanical Vibrations by J.P. Den Hartog. Dover Publications, Inc. New York, 1985, 4th edition, Appendix, Section V, table in page 432.

14 $\delta J = \delta_m \cdot D^2/8$

15 Considering the total shaft: $J = m \cdot D^2/8$; then, $\mu_J = J/L = m \cdot D^2/(8 \cdot L)$. As $I_p = \pi \cdot D^4/32$, it results $\mu_J = \rho \cdot I_p$.

Part II

Turbo Machines and Ship Vibrations

4 Critical Velocity of Turbomachines

4.1 INTRODUCTION TO THE CRITICAL VELOCITY

Rotor-shaft resonance is due to shaft eccentricity and deflection. These mechanical imperfections generate centrifugal forces that excite a vertical vibration known as lateral or transversal. Ergo, at a certain speed, the machine undergoes a strong shaking produced by the resonance of shaft, wheels, couplings, and other critical components; therefore, severe damages must be expected when a rotating machine is running at critical velocity, especially in the blade roots.

Resonance is one of the greatest machine destructive phenomenon. Even if the vibration is not violent, the impending accumulated fatigue of materials may generate catastrophic consequences to the machine. As 90% of structural cracks and fractures are due to fatigue produced by vibration, it is of utmost importance to monitor and systematically balance the rotor-shaft according to a sound maintenance schedule. See Chapter 10, an introduction to monitoring systems and standards.

Turbomachine velocities at which resonance peaks occur are called critical speeds. Of course, critical velocities within the machine's operating range are the most important to identify to prevent resonance in the design stage or the field. This prevention implies knowing the value of critical speeds from the beginning of the turbomachine's engineering. Procedures for calculating critical speeds have been intensively investigated, and the most important of them are exposed in this chapter.[2] Turbomachines, such as turbines, rotary compressors, and centrifugal pumps, have a low damping ratio value because their technical conception avoids friction in their parts to prevent energy losses during their operation. Therefore, turbomachines are studied as frictionless systems for critical velocity calculation purposes.

The critical velocity is equal to any rotor-shaft set's resonance frequencies (from now on, the rotor-shaft set will be named only rotor-shaft). Resonance causes noises and dangerous high amplitude vibrations. If the first critical velocity is lower than the operating speed, the resonance will happen during startup. However, in turbomachines, this process is extremely fast because it usually takes only a few milliseconds.

Are critical velocities always the same? The answer is no. For example, a too-low clearance of support bearings produces critical velocities higher than calculated. The aftermath of this low clearance is an undesirable rubbing between the shaft and the rotor that increases the critical speed. There also exist causes that produce lower critical velocities than calculated. For example, spherical bearings

DOI: 10.1201/9781003175230-6

allow some shaft deformation, which produces a less rigid shaft, and keyways also reduce the shaft rigidity. These low rigidities ease the creation of critical speeds lower than the calculated velocities at the design stage. The vibration theory demonstrates that the second critical speed is four times the fundamental. The third is nine times the fundamental, and the fourth is sixteen times the fundamental.

A perfect balanced rotor does not exist. Therefore, in practice, some imbalance is tolerated in rotors. Tolerances are set in the ISO Standard 1940-1[3]. This standard specifies the acceptable level of unbalancing moments according to the rotor type and the operating velocity.

4.1.1 Calculation and Measurement of the Resonant Frequency

A bump on a machine or structure, whose natural frequency must be known, is a practical test easy to do in the field by a skilled technician. At the machine's standstill, the operator knocks on its surface with a hammer to induce the machine's natural vibration. Simultaneously, an accelerometer installed on the machine records the vibration produced by the bump. This action shows the machine's natural frequency in the instrument display. According to Fourier's theory, the hammer bump is an impulse excitation that creates infinite harmonic waves. However, only a group of them have significant amplitudes and frequency. Among these, waves of natural frequency have a much higher amplitude than all the other waves. So, in a vibrometer display, natural frequencies are easily identified.

During the test, knocking is done at several places. Each knock will show the machine's natural frequency in its display. As massive machines have lower resonant frequencies than smaller machines, the bump test must create a low-frequency response. That is why large machines are impacted with rubber or wood and small machines with metal or rigid plastic.

Resonance is corrected by moving the resonant frequency away from the excitation frequency. The resonant frequency may be changed by increasing or decreasing the machine rigidity or adding or removing mass pieces from the machine. If the machine can operate at another velocity without affecting its performance, it is good to change its speed.

4.1.2 Type of Rotors

A rotor is classified according to the ratio between rated and resonant velocities. If the rotor speed is lower than 70% of the resonant frequency, the rotor is classified as rigid. Balancing of rotors at velocities higher than their rated speed holds for a range between 0 and 70% of the critical speed. This validity is because deflections produced by centrifugal forces are lower than at velocities higher than 70% of the critical velocity. Thus, rigid rotors have a smooth running at any velocity within that range. This type of rotor is less exposed to efforts derived from an unbalanced condition, and centrifugal deflection changes are minimal. These small deflections are due to a magnification factor lower than 1 at $u < 0.7$. Therefore, the shaft deflection of rigid rotors is similar to the deflection of the last balancing.

Rigid rotors have low inertia forces, and the bearing deflection absorbs most of the centrifugal force energy. Therefore, rigid rotors don't need frequent rebalancing when they run at velocities lower than the balancing speed. Fortunately, most turbomachines, like steam and gas turbines, centrifugal pumps and compressors, operate in the rigidity range.

If the rotor-rated velocity is higher than 70% of the critical speed, centrifugal forces are more significant than in the rigidity range; therefore, they are subject to changing deflections. Hence, this type of rotors is called flexible. The aftermath of higher velocities is that these forces more easily bend the shaft; therefore, these deflections may produce severe vibration. The shaft deflection is no longer equal to the deflection it had during the last balancing; then, the machine needs to be rebalanced. The most common machines with flexible rotors are high-velocity turbines (gas or steam), paper machine rollers and multistage pumps.

4.2 RAYLEIGH-RITZ METHOD

A uniformly loaded beam or shaft's natural frequency is obtained with the Rayleigh-Ritz method, valid only for frictionless systems. This method is based on the hypothesis that a frictionless bent shaft, freely gyrating (no moment is applied), has an angular speed equal to the shaft's natural frequency. In that case, the kinetic and potential energies are equal. Another important assumption of the Rayleigh-Ritz hypothesis is that, during rotation, the shaft shape is given by the elastic curve produced by static loads. Therefore, it does not consider shaft deformations produced by centrifugal forces. See in Figure 4.1 a rotating shaft according to this hypothesis. As the shaft is uniformly loaded, the maximum deflection is in the middle.

The kinetic energy is maximum when the maximum deflection section is in both positions M because the linear velocity is maximum at these points. It is zero in positions L or H, where velocity is zero because in these points, the velocity changes from the upward to the downward direction. Conversely, the potential energy is maximum in points L or H because, at these points, the kinetic energy

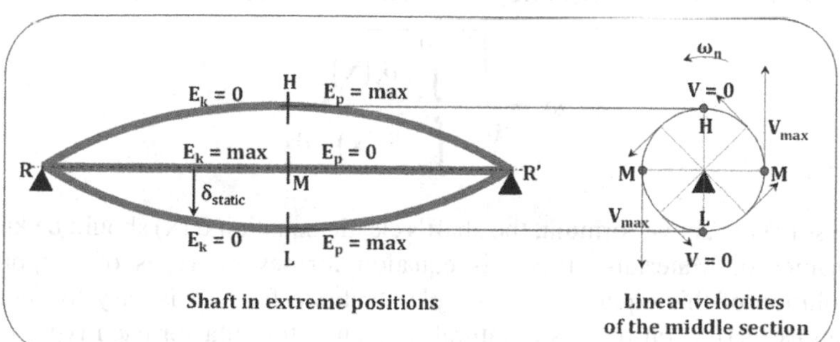

FIGURE 4.1 Shaft gyrating as per Rayleigh-Ritz hypothesis.

is zero, then all the shaft energy is potential. Hence, the potential energy is zero in the middle points M, where kinetic energy is maximum. As the total energy is constant, the maximum kinetic energy is equal to the maximum potential energy. From this equality, it is possible to clear the natural frequency, as demonstrated below.

The rotor-shaft is a continuous system; then, the kinetic and potential energies should be determined for a shaft sector of differential length and then integrate this expression between x = 0 and the shaft length L. The potential and kinetic energy of a differential shaft section are:

Formula 4.1 Potential and kinetic energy of a differential shaft section

$$dE_p = \frac{dm \cdot g \cdot \delta_s(x)}{2} \quad dE_k = \frac{1}{2} \cdot dm \cdot \left[\delta_s(x) \cdot \omega_n\right]^2$$

Where $\delta_s(x)$ is the static deflection curve, known as elastic in Mechanics of Materials. The shaft rigidity reaction is proportional to this deflection; therefore, the potential energy should be divided by 2. The elemental mass dm is substituted by ρ·A·dx, where ρ is the mass density, A is the shaft cross-section, and dx is the shaft's elemental length. Therefore, by introducing these physical properties in Formulas 4.1 and integrating between zero and the shaft length, the potential and kinetic energy formulas are obtained.

Formula 4.2 Potential and kinetic energy of the shaft

$$E_k = \rho \cdot A \cdot \omega_n^2 \cdot \int_{x=0}^{L} \frac{\delta_s(x)^2 \cdot dx}{2} \quad E_p = \rho \cdot A \cdot g \cdot \int_{x=0}^{L} \frac{\delta_s(x) \cdot dx}{2}$$

The equality of the potential and kinetic energy is the essence of the Rayleigh-Ritz method. Therefore, this equality produces an expression that allows clearing the following natural frequency formula.

Formula 4.3 Natural frequency of a shaft under uniform load

$$\omega_n = \sqrt[2]{g \cdot \frac{\int_{x=0}^{L} \delta_s(x) \cdot dx}{\int_{x=0}^{L} \delta_s(x)^2 \cdot dx}}$$

To solve the above formula, the shaft's elastic equation $\delta_s(x)$ should be known. Mechanics of Materials gives this equation for several types of support. As the solution of this equation is a simple algebraic form, it is easy to integrate. The above derivation returns a natural frequency formula for each type of support. These formulas are in Table 4.1, along with the elastic of static deflections formulas.

TABLE 4.1
Formulas of Shaft's Elastic Equations and Natural Frequencies Under Uniform Load

Type of Supports	Elastic Equation with Uniform Load Used in Formula 4.3 to Calculate Formulas of the Right Column. Symmetric Load	Natural Frequency
Simply supported or hinged - hinged	$\delta_s(x)=\frac{q}{24\cdot E\cdot I_z}\cdot\left(x^4-2\cdot L\cdot x^3+L^3\cdot x\right)$	$\omega_n=\sqrt[2]{97.5\cdot g\cdot\frac{E\cdot I_z}{q\cdot L^4}}$
Clamped-clamped	$\delta_s(x)=\frac{q}{24\cdot E\cdot I_z}\cdot\left(x^4-2\cdot L\cdot x^3+L^2\cdot x^2\right)$	$\omega_n=\sqrt[2]{504\cdot g\cdot\frac{E\cdot I_z}{q\cdot L^4}}$
Clamped-hinged	$\delta_s(x)=\frac{q}{48\cdot E\cdot I_z}\cdot\left(2\cdot x^4-3\cdot L\cdot x^3+L^3\cdot x\right)$	$\omega_n=\sqrt[2]{5\cdot g\cdot\frac{E\cdot I_z}{q\cdot L^4}}$
Clamped-free	$\delta_s(x)=\frac{q}{24\cdot E\cdot I_z}\cdot\left(x^4-4\cdot L\cdot x^3+6\cdot L^2\cdot x^2\right)$	$\omega_n=\sqrt[2]{12.5\cdot g\cdot\frac{E\cdot I_z}{q\cdot L^4}}$

4.2.1 Critical Velocity Versus Static Deflection

A more practical expression of natural frequency is to put it as a function of the maximum static deflection. The general formula of the maximum static deflection of uniformly loaded shafts is the following.

Formula 4.4 Shaft maximum static deflection under uniform load

$$\delta_{lq}=\beta_q\cdot\frac{q\cdot L^4}{E\cdot I_z}$$

Where β_q coefficient depends on the supports type. These coefficients are derived in the Mechanics of Materials and shown in Table 4.2. The β value is written with the subscript q to indicate that the load is uniform. Therefore, replacing in the natural frequency formulas of Table 4.2, the term $[q\cdot L^4/(E\cdot I_z)]$ by the term $[\beta_q/\delta_s]$ and introducing the corresponding β_q coefficient, formulas of the shaft's natural frequency and critical velocity are obtained. They are shown in the third and fourth columns of Table 4.2[4]. These formulas are only valid for static deflections expressed in mm. The result is in rad/s for the natural frequency and RPM for the critical velocity.

Table 4.2 shows that the type of support has a significant incidence in the natural frequency or critical velocity values. Another physical conclusion, derived from the coefficient's magnitude, is that, at the same deflection, uniformly loaded shafts have critical velocities higher than concentrated shafts because these have higher deflection.

4.2.2 A Practical Determination of Critical Velocity

Take, for example, be the following turbomachine, whose critical velocity is requested. Assume that the shaft and rotor weights are uniform loads, and the type of support is clamped-hinged. Use the Rayleigh-Ritz method.

TABLE 4.2
Shaft Natural Frequency and Critical Velocity Formulas

Type of Supports	Coefficients β_q	Natural Frequency	Critical Velocity
Simply supported or hinged - hinged	$\frac{5}{384}$	$\omega_n = \frac{111.6}{\sqrt[2]{\delta_l}}$	$N_c = \frac{1{,}066}{\sqrt[2]{\delta_l}}$
Clamped-clamped	$\frac{1}{384}$	$\omega_n = \frac{113.5}{\sqrt[2]{\delta_l}}$	$N_c = \frac{1{,}084}{\sqrt[2]{\delta_{ls}}}$
Clamped-hinged or clamped-simply	$\frac{1}{185}$	$\omega_n = \frac{16.2}{\sqrt[2]{\delta_l}}$	$N_c = \frac{155}{\sqrt[2]{\delta_l}}$
Clamped-free	$\frac{1}{8}$	$\omega_n = \frac{123.6}{\sqrt[2]{\delta_l}}$	$N_c = \frac{1{,}180}{\sqrt[2]{\delta_l}}$

Input data:
Shaft diameter: D = 0.4 m
Shaft length: L = 3 m
Young modulus: E = 196.1 GPa equivalent to: 2.0×10^{10} kg/m²
Rotor weight over shaft weight ratio: $W_w/W_s = 2.5$
Steel specific weight: 7,860 kg/m³
Calculation:
Shaft weight: $W_s = \pi \times 0.4^2/4 \times 3 \times 7{,}860 = 2{,}693$ kg
Rotor weight: $W_w = 2.5 \times 2{,}693 = 7{,}408$ kg
Total weight: $W = W_s + W_w = 10{,}371$ kg
Uniform load: $q = 9, 425.5/3 = 3{,}457$ kg/m
Support coefficient: $\beta_q = 1/185$
Area moment of inertia: $I_z = \pi \times 0.4^4/64 = 0.00126\ m^4$
Maximum static deflection: $\delta_s = \beta_q \times qL^4/E \cdot I_z = 0.0102$ mm
Natural frequency: $\omega_n = 155/\delta_s^{1/2} \equiv 1{,}535$ RPM

4.2.3 Stepped Shafts

The critical velocity formula of stepped shafts is based on the Rayleigh-Ritz hypothesis, discussed in Section 4.2. This method's application is especially important because real shafts are, in general, stepped and not uniform. This model, proposed by Rayleigh-Ritz, is ideally done by splitting the shaft into sectors of different diameters each, as suggested in Figure 4.2. Therefore, every sector of constant diameter supports one concentrated load, formed by the shaft sector weight plus the wheel's weights of that sector. In Figure 4.3, loads per sector are designated W_i, and the support reactions are R_a and R_b ($R_a+R_b = W_1+W_2+W_3$).

As discussed in Section 4.2, every piece of the shaft has maximum kinetic energy equal to its maximum potential energy. Summing up all these energies and equating the total kinetic energy to the total potential energy as done with Formulas 4.2, the following expression returns.

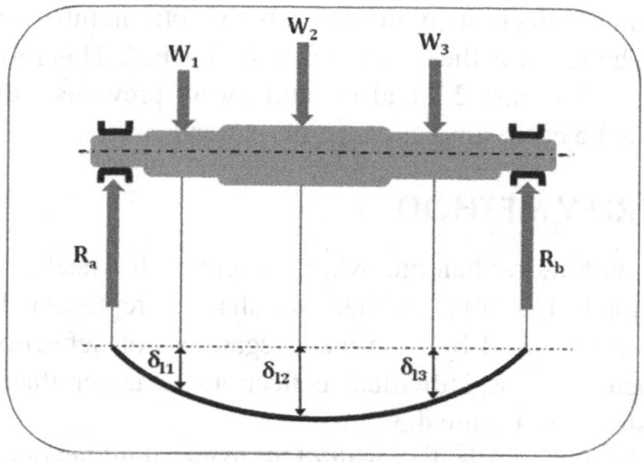

FIGURE 4.2 Variable diameter shaft with several concentrated loads.

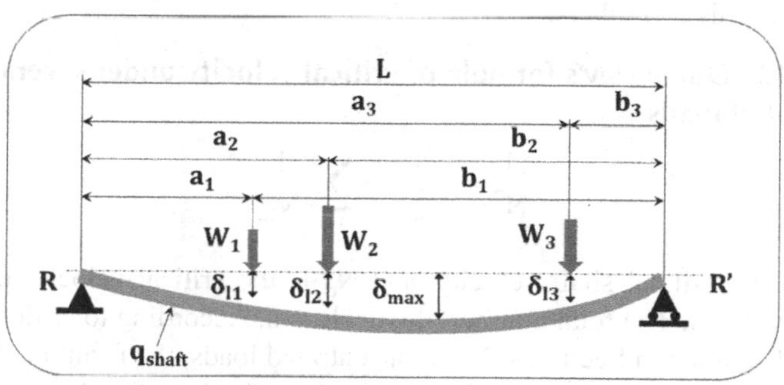

FIGURE 4.3 Simply supported shaft with combined concentrated and distributed loads.

Formula 4.5 Kinetic and potential energies during rotation under asymmetric loads

$$\sum_{i=1}^{n}\frac{1}{2}\cdot m_i\cdot(\delta_i\cdot\omega_n)^2=\sum_{i=1}^{n}\frac{1}{2}\cdot m_i\cdot g\cdot\delta_i$$

Where δ_i are the maximum deflections under each wheel, these values are read in the elastic produced by the rotor–shaft weight. This elastic is obtained by the Mohr method or, preferably, with specialized software. Therefore, clearing ω_n and converting it into RPM, the critical velocity of a stepped shaft is given by the following formula.

Formula 4.6 Critical velocity of a stepped shaft

$$N_c=946\cdot\sqrt[2]{\frac{\sum_{i=1}^{n}W_i\cdot\delta_i}{\sum_{i=1}^{n}W_i\cdot\delta_i^2}}$$

W_i is in kg, and deflections δ_i in mm. n is the total number of shaft sections between steps. The result is the critical velocity in RPM. This formula is the discrete expression of formula 2.7 and is valid for all previously discussed cases. This formula may be easily implemented in a spreadsheet.

4.3 DUNKERLEY METHOD

Turbomachines with more than one wheel, where each wheel or the coupling is a concentrated load on a constant diameter shaft, is represented in Figure 4.3. Deflections δ_l are produced by each load, regardless of deflections imposed by other loads. Each of these individual deflections is lesser than the maximum deflection δ_{max} shown in Figure 4.3.

The elastic of Figure 4.3 is the result of all loads simultaneously acting on the shaft or beam. That is why δ_l arrows do not reach the elastic curve in the above figure. Dunkerley has proposed an empirical formula to calculate the critical speed of a system. This formula is:

Formula 4.7 Dunkerley's formula of critical velocity under several concentrated loads

$$\frac{1}{N_c^2} = \frac{1}{N_{shaft}^2} + \sum_{i=1}^{n} \frac{1}{N_{ci}^2}$$

N_{shaft} is the critical shaft velocity, and N_{ci} is the critical velocity due to the weight of wheel i. The total number of wheels is n. According to critical velocity formulas discussed in Section 4.2 for concentrated loads, the formula of the critical velocity produced by loads W_i is given by formula $Nci = 946 / \sqrt[2]{\delta_i}$, where δ_i is the deflection under the load W_i^5. The critical shaft velocity is $Nshaft = K / \sqrt[2]{\delta_{shaft}}$. K is the numerator of formulas given in the fourth column of Table 4.2. N_{ci} and N_{shaft} are replaced in Dunkerley's Formula 4.7, and the critical velocity of the rotor-shaft N_c is cleared from the resultant expression.

Formula 4.8 Critical speed of a rotor-shaft with several concentrated loads

$$N_c = \frac{1}{\sqrt[2]{\frac{1}{N_{shaft}^2} + \sum_{i=0}^{n} \frac{1}{N_i^2}}} = \frac{1}{\sqrt[2]{\frac{\delta_{shaft}}{K_{shaft}^2} + \frac{\sum_{i=1}^{n} \delta_{li}}{894,916}}}$$

4.3.1 TURBOMACHINES WITH MORE THAN ONE WHEEL

As an alternative, there exists a good rule of thumb to calculate the critical velocity of turbomachines with stepped shaft and more than one wheel, based on the shaft maximum deflection.

Formula 4.9 Approximate formula of N_c for a stepped shaft with multiple wheels

$$N_c = \frac{1,012}{\sqrt[2]{\delta_{max}}}$$

The maximum deflection is estimated with the sum of all maximum deflections produced by each load. This calculation is an acceptable approach to simply supported shafts. However, with other types of supports, the difference may become unacceptable.

Formula 4.10 Maximum deflection under several non-symmetrical loads

$$\delta_{max} = \sum_{i=1}^{n} \delta_{max\,i}$$

The formula to calculate each maximum deflection depends on the β coefficient of the type of support.

Formula 4.11 Maximum shaft deflection

$$\delta_{max} = \beta_{max}(a, b, L) \cdot \frac{W \cdot L^3}{E \cdot I_z}$$

Coefficients β_{max} are a function of distances a and b, the shaft length L and the type of support. Formulas to calculate β_{max} and the coordinate x_{max}, where the maximum deflection is produced, are in Table 4.3.

TABLE 4.3
β_w Coefficients to Calculate Maximum Deflection Due to Non-Symmetrical Loads

Type of Support	β_{max} Formulas	Coordinate x of δ_{max}
Simply supported	$\beta_{max} = \frac{a^2 \cdot b}{9 \cdot L^4} \cdot (L+b) \cdot \sqrt{\frac{L+b}{3 \cdot a}}$	$x_{max} = \sqrt{\frac{L+b}{3 \cdot a}}$ a > b. If not, exchange a & b
Clamped–simply $\frac{a}{L} \le 0.414$	$\beta_{max1} = \frac{b^2 \cdot x_{max1}}{L^3} \cdot \left[\frac{a}{L} - \frac{2}{3} \cdot \left(1 + \frac{a}{2 \cdot L}\right) \cdot \frac{x_{max1}^2}{L^2}\right]$	$x_{max1} = \frac{b \cdot (L+a)}{L \cdot \left[1 + \frac{b}{2 \cdot L^2} \cdot (L+a)\right]}$
Clamped–simply $\frac{a}{L} \ge 0.414$	$\beta_{max2} = \frac{a \cdot x_{max2}^2}{L^3} \cdot \left[\left(1 - \frac{a^2}{L^2}\right) - \left(1 - \frac{a^2}{3 \cdot L^2}\right) \cdot \frac{x_{max2}}{L}\right]$	$x_{max2} = L \cdot \sqrt{\frac{a}{a + 2 \cdot L}}$
Clamped–clamped a > b	$\beta_{max} = \frac{2 \cdot a^3 \cdot b^2}{3 \cdot L^5} \cdot \left(\frac{L}{L + 2 \cdot a}\right)^2$	a > b If a < b exchange a and b
Clamped and free (cantilever beam)	$\beta_w = \frac{1}{3}$	

It is frequent to see shafts with free extremes and simply supported or with three or four bearings. These cases are solved with more sophisticated methods using specialized software. Some of them are described in old versions of the Engineer's Manual by Hütte.

4.4 CRITICAL VELOCITY ASSESSMENT. EXAMPLE

The assessment of the critical velocity of a four wheels turbine is requested. The shaft has a uniform diameter, with clamped-hinged supports.

Technical specifications
Shaft diameter D = 0.2 m
Shaft length L = 3 m
Young's modulus $E = 1.97 \times 10^{10}$ |kg/m²|
The wheels' weight and location on the shaft are given in the following table.

Wheel #	W_i |kg|	a_i |m|	b_i |m|
A	500	0.50	2.50
B	500	0.60	2.40
C	600	0.80	2.20
D	600	1.20	1.80

The above table contains the design data usually accessible to the field or design engineers

Calculation procedure
The calculation procedure has the following sequence.

1. Shaft deflection
2. Deflections produced by the wheels
3. Critical velocity by Dunkerley's formula

- **Shaft deflection**

  Shaft weight W_s.

$$W_s = \pi \cdot D^2/4 \cdot L \times \text{steel specific weight}\left(7{,}850\,\text{kg/m}^3\right) = 740|\text{kg}|$$

  Shaft weight per unit length q.

$$q = W/L = 247|\text{kg/m}|$$

  Moment of inertia I_z of the shaft cross-section.

$$I_z = \pi \cdot D^4/64 = 7.85 \times 10^{-5}|\text{m}^4|$$

  Shaft's maximum deflection due to its weight.

$$\delta_{\text{max shaft}} = 5/384 \cdot q_s \cdot L_s^{\,4}/(E \cdot I_z) \times 1{,}000 = 0.17|\text{mm}|$$

- **Deflection produced by the wheels**
  The distance from loads to the left bearing is a_i and to the right bearing is b_i. Values of a_i and b_i are data read in drawings or measured at the site. In this case, each wheel is an asymmetric load.
  Deflection formulas.

$$\beta i = ai2 \cdot bi3 \cdot (4 \cdot a_i + 3b_i)/(12 \cdot L6)$$

$$\delta_i = \beta_i \cdot W_i \cdot L_s^{3}/(E \cdot I_z) \times 1{,}000 |mm|$$

It is remarked that δ_i is not the maximum deflection produced by each wheel. It is just the deflection under the concentrated load W_i. The results of these two formulas are in the following table.

Wheel	β_{wi}	δ_{ii} mm at x = a
Wheel A	0.0027	0.024
Wheel B	0.0127	0.111
Wheel C	0.0192	0.201
Wheel D	0.0107	0.112

- **Critical velocities**
  The critical velocity formulas.
  Wheels: $Nwi = 946 / \sqrt[2]{\delta_i} |RPM|$
  Shaft: $Nshaft = 1{,}066 / \sqrt[2]{\delta_{shaft}} |RPM|$
  Dunkerley's formula: $N_c = 1/(1/N_{shaft}^{2} + 1/N_{wA}^{2+} 1/N_{wB}^{2+} 1/N_{wC}^{2+} 1/N_{wD}^{2})$
  The results of these formulas are in the following table.

Turbine part	RPM	rad/s
Wheel A	3,992	418
Wheel B	3,465	363
Wheel C	2,588	271
Wheel D	2,109	221
Shaft	2,599	272
N_c Dunkerley formula	1,223	128

Figure 4.4 shows the elastic curves due to the shaft and wheels' weight of this example. It must be noted that maximum deflections do not coincide with loads' position and are produced at different x coordinates. The x coordinate of the maximum deflection of each curve of Figure 4.4 is: $x_{max_i} = \dfrac{2 \cdot a_i \cdot L}{L + 2 \cdot a_i}$.

Usually, power plants have 3,000 or 3,600 RPM turbines. It means that the turbine of this case undergoes resonance during startup as velocity passes through 1,223 RPM. The critical frequency inversely depends on the number of wheels. Therefore, as the number of wheels decreases, the critical frequency increases.

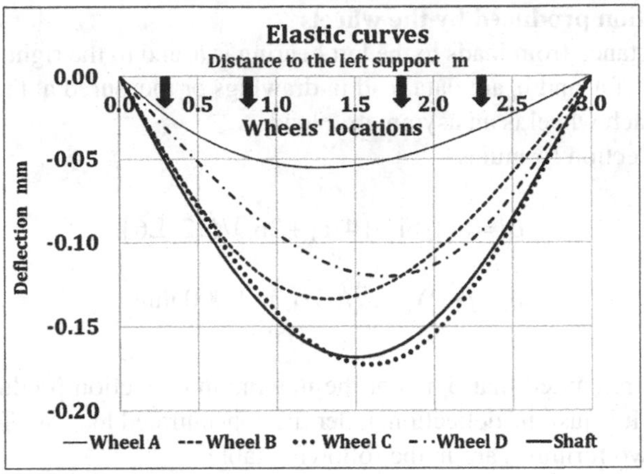

FIGURE 4.4 Elastic curves due to shaft and wheels weight.

4.5 ROTOR BALANCING

4.5.1 Conceptual Introduction to Balancing

Asymmetric mass distribution produces rotor unbalancing, one of the most common vibration sources in turbomachines. Therefore, rotor-shafts should be periodically balanced to prevent harm to the machine. Rotors have circular grooves or pre-bored holes on the rotor circumference, where neutralizing weights are added for static or dynamic balancing purposes.

Centrifugal forces are applied to each elemental mass of a rotor-shaft. The integration of all these elemental forces returns one resultant force plus a moment that tends to deviate the principal inertial axis from the geometric axis. If these axes do not coincide, the rotor-shaft assembly is subject to vibration. If the shaft eccentricity is zero, the resultant force may be null, but not the moment. Therefore, even with non-zero eccentricity, a moment arises. As zero eccentricity is almost impossible to get, the moment produced by centrifugal forces always exists in rotor-shafts. In summary: a statically balanced rotor is not forcefully dynamically balanced. Conversely, a dynamically balanced rotor is for sure statically balanced.

A common cause of unbalancing is improper rotor heating; therefore, in this case, some plastic deformation is likely to arise, which produces a dynamic unbalancing generated by the centrifugal forces' moment. Similarly, blade erosion, corrosion, or blade rupture or any mechanical work made on the rotor may cause a dynamic unbalancing. See Figure 4.5, which shows a graphical description of an ideal balanced and an unbalanced rotor.

The most important rotor balancing operations and their goals are:

- Static balancing purports to cancel the eccentricity.
- Dynamic balancing intends to cancel the moment created by centrifugal forces.

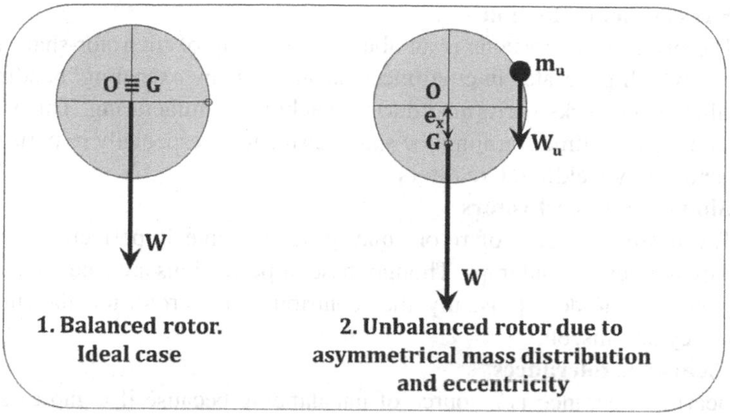

FIGURE 4.5 Balanced and unbalanced rotor schemes.

Sketch 1 of Figure 4.5 shows the ideal case, where the gyration axis center and gravity center are coincident. It does not happen in sketch 2, where there is a mass asymmetry due to an unbalancing mass m_u located on the rotor surface and the gravity center is not on the gyration radius. A common cause of unbalancing is a spurious protruding piece on the rotor combined with eccentricity. Under this condition, the rotor is unbalanced, and therefore, the turbomachine is subject to vibration.

Balancing machines detect unbalanced masses. These machines revolve the rotor on spring-mounted bearings; therefore, the rotor wobbles while the machine measures its motion amplitude and phase to calculate the unbalanced masses' location. After that, it must be chosen locations where balancing weights must be added. These weights are inserted in the rotor grooves or pre-bored holes.

4.5.2 Causes of an Unbalanced Rotor

Manufacturing imperfections in the rotor, wheels, blades and blades' shrouds and non-homogenous materials generate asymmetries that produce moments, which in many cases are significantly high, especially in the low-pressure stage steam turbines, which are characterized for having long blades. Other specific causes of unbalancing are briefly described in the following paragraphs.

- **Deposit buildup**
  The buildup of any spurious deposits is usually uneven, and; therefore, they produce an undesired unbalancing and vibration. Correct preventive maintenance helps mitigate these problems; nonetheless, the rotor may need to be removed for performing a new balancing in many cases.
- **Eccentricity**
  Eccentricity exists when the geometric centerline does not coincide with the rotating centerline. Even the manufacturer cannot remove the eccentricity. Nonetheless modern tools can reduce it to a neglectable value.

- **Stress relief distortion**
  The origin of distortions is an abnormal shaping of the rotor-shaft assembly, which generates internal mechanical stress by extruding, bending and other metalworks performed during machine manufacturing. This stress is relieved over time, forming the shape distortions, especially in parts manufactured by welding, like rotors.
- **Blowholes in cast rotors**
  The casting process of rotors may produce some imperfections such as blown holes or sand traps. Though these imperfections are small and almost impossible to detect visually, they contribute to the rotor unbalancing even if they are unseen.
- **Clearance tolerances**
  Bearing clearance is a source of unbalancing because if a gap exists, the shaft is displaced to one side of the geometric shaft axis. The shaft displacement mismatches the shaft's geometric and rotating axis.
- **Corrosion or wear**
  Corrosion and wear are seldom uniform; therefore, their uneven distribution produces a dynamical unbalance.
- **Keyways**
  If the rotor was initially balanced and keyways were installed on the shaft after the rotor balance, the keyways weight produces imbalance.
- **Thermal distortion**
  Minor manufacturing imperfections and uneven heating generate thermal distortions. This distortion formation is due to elevated operating temperatures; therefore, in some cases, the rotor must be balanced at its nominal operating temperatures.

Conclusion: many causes make a rotor unbalanced. If defects are minor, easily visualized and accessible, they are often fixed on site. Otherwise, it is recommended to remove the rotor and send it to a specialized workshop to perform a new balancing.

4.5.3 Static Balancing

A rotor is statically balanced when its mass is evenly distributed around its inertia axis. In this case, there is no resultant of centrifugal forces because the gravity center is on the inertia axis. In an ideally rigid shaft at a standstill, with zero eccentricity and all masses symmetric distributed around the inertia axis, the rotor remains at any angular position. Its angular stability is indifferent. As eccentricity is zero, the rotor weight does not produce any moment.

The static balancing essence is to create with a balancing mass a centrifugal force in opposition to the shaft weight. The rotor is mounted on steel blades (see Figure 4.6), after which the shaft gravity center is in its lowest position. The balancing mass m_b is added in opposition to the gravity center G. Therefore, as the centrifugal force F_c acting on the mass should be equal to the rotor weight W, the

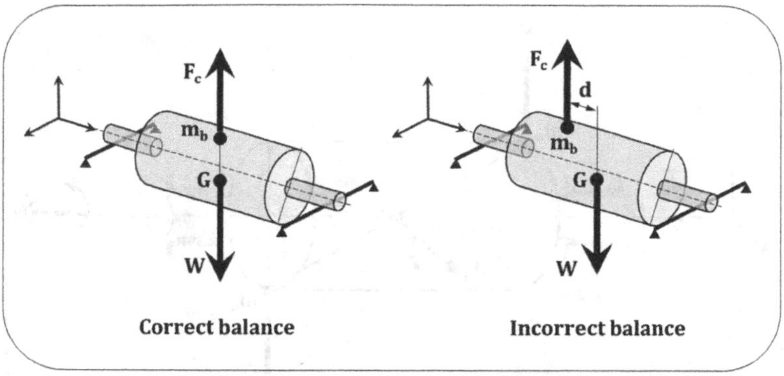

FIGURE 4.6 Static balancing.

mass m_b must be equal to $W/(e_x \cdot \omega^2)$. See the left scheme of Figure 4.6. However, if the gravity center G and the balancing mass m_b are not in the same perpendicular plane, an undesired couple arises, equal to the shaft weight W times the distance d. This moment is the aftermath of a dynamic unbalancing. See the right scheme in Figure 4.6.

Several methods carry out a static balancing acceptable for low-speed rotors of limited wheel length. If this length is small (short rotors), the distance d of Figure 4.6 may be insignificant. However, static balancing is usually insufficient in high-speed long rotors; therefore, they must be dynamically balanced.

Proceeding by successive additions or removals of masses (trial and error method) in a wheel freely mounted on steel blades, W.J. Kearton[6] reports that it is possible to reduce the eccentricity to 0.01 mm.

4.5.4 Dynamic Balancing

4.5.4.1 Dynamically Unbalanced Rotor

A dynamically unbalanced rotor is shown in Figure 4.7. Two spurious protruding masses produce this unbalancing. Forces are indicated with solid lines, and moments are shown with dashed lines. In this figure, m_u are two equal spurious masses, and M_u is the unbalancing moment. In this case, the line m_u-m_u passes through the rotor gravity center G. At standstill; the rotor is statically balanced because equal masses are symmetrically located with respect to the gravity center G. However, when the rotor gyrates, two centrifugal forces arise, producing the unbalancing moment M_u, which tends to revolve the rotor in a clockwise direction, normal to the acting torque T_q that drives the rotor. As both centrifugal forces are equal and the eccentricity is zero, there is no resultant force. Still, there is a moment that revolves the rotor in the centrifugal forces' plane.

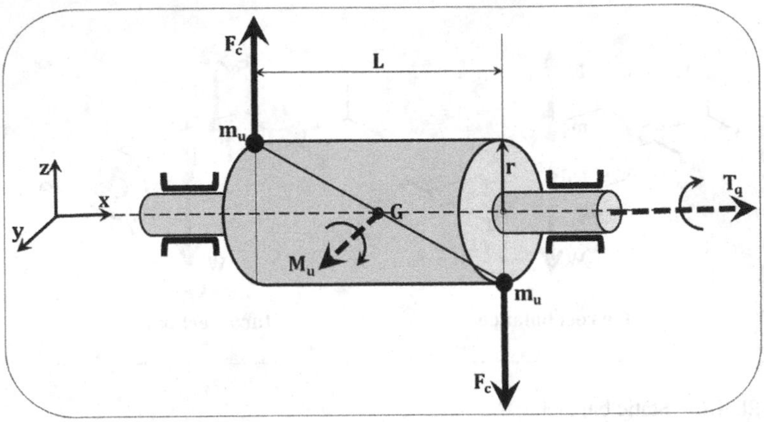

FIGURE 4.7 Dynamically unbalanced rotor.

The formula to calculate the unbalancing moment M_u[7] is the following.

Formula 4.12 Unbalancing moment

$$\bar{M}_u = \bar{F}_c \cdot L = \left(m_u \cdot \bar{r} \cdot \omega^2\right) \cdot L$$

As the rotor is revolving, the vector $\bar{M}_u$ is normal to the rotating axis and is also revolving because its tail is fixed to the inertia axis. Therefore, the torque T_q and the unbalancing moment M_u vectors are always at 90°.

Figure 4.8 shows the vectors that represent a dynamically unbalanced rotor. The centrifugal forces F_c and gyration axis are in the same plane (see shaded area); therefore, the unbalancing moment vector M_u is perpendicular to the forces' plane. The vector sum of torque T_q and unbalancing moment vector M_u is vector T_c, whose tail is fixed to point G. This vector forms the angle λ with the rotating axis.

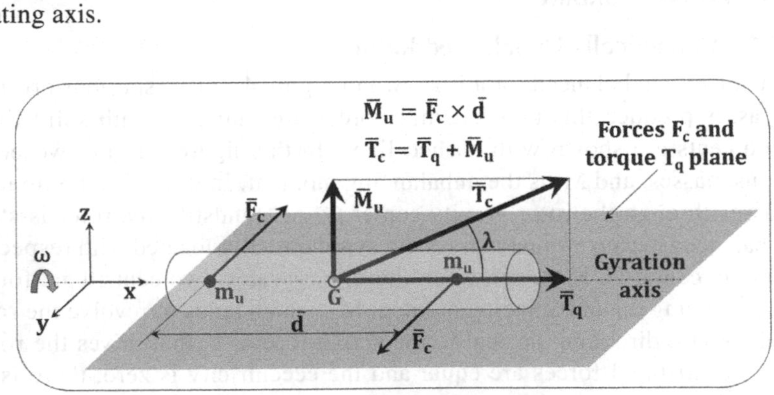

FIGURE 4.8 Dynamical imbalance. Combined torque formation.

Other important characteristics of these forces and torques are:

a) The torque T_q direction coincides with the gyration axis.
b) The combined torque T_c rotates around the gyration axis with its tail resting on the gravity center G, at the angular velocity ω. Therefore, the rotor is subject to a changing direction torque T_c, producing undesirable shaking.

Formulas to calculate vectors M_u and T_c are shown in Figure 4.8. As T_q is higher than M_u, the angle λ is small. Nonetheless, the resultant torque T_c makes the rotor wobble harmonically. This angular change of the combined torque direction shakes the turbomachine, which means that the rotor undergoes a complex vibration.

4.5.4.2 Balancing Masses Calculation

The dynamical balancing is the addition of two balancing masses, one on each rotor extreme. Centrifugal forces of these masses must be equal to the shaft weight's centrifugal force and the moment with respect to the section where the gravity center is located. The balancing masses are calculated using formulas derived from the free body diagram of Figure 4.9. Conditions of a dynamically balanced shaft are:

- Centrifugal forces acting on the balancing masses m_{b1} and m_{b2} are in equilibrium with the centrifugal force due to the shaft eccentricity F_{cs}.
- Moments produced by the balancing masses must cancel each other.

These conditions are expressed with the following formulas:

Formula 4.13 Shaft dynamical balancing conditions for the above figure

$$F_{cs} = F_{c1} + F_{c2}$$

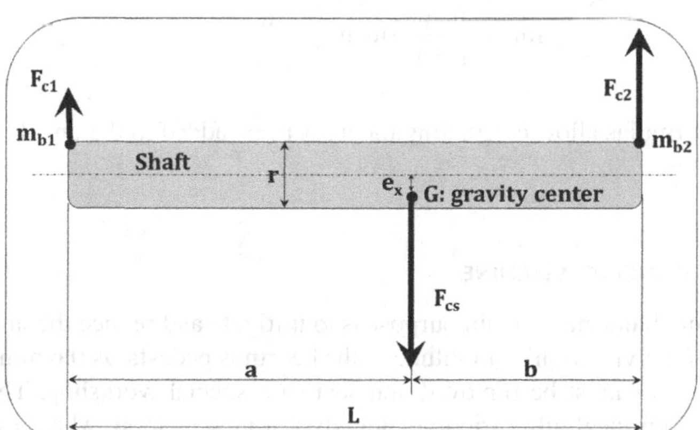

FIGURE 4.9 Free body diagram of a dynamically balanced shaft.

$$M_{b1} + M_{b2} = 0$$

Where:

F_{cs} is the centrifugal force produced by the shaft eccentricity e_x.

F_{c1} and F_{c2} are the centrifugal forces acting on the balancing masses m_{b1} and m_{b2}.

M_{b1} and M_{b2} are the moments with respect to the right cross-section, where the gravity center is. Centrifugal forces create both moments.

Figure 4.9 represents a dynamically balanced rotor that explains Formulas 4.14, 4.15 and 4.16 to calculate the balancing masses.

The total balancing mass value is derived from the centrifugal forces' equilibrium (the first equation of Formulas 4.13).

Formula 4.14 Total balancing mass

$$m_{b1} + m_{b2} = \frac{e_x}{r} \cdot m_s$$

Where r is the shaft radius, and m_s is the shaft mass. It means that with a shaft of 302 kg mass, 0.4 m diameter and 0.1 mm eccentricity, the total balancing mass is 0.15 kg mass. Therefore, the total weight of balancing masses is 1.48 kg. The exact location of each mass is derived from the balancing measurements. From the moments' equilibrium, given by $M_{b1} + M_{b2} = 0$, it results:

Formula 4.15 Moments' equilibrium of balancing masses

$$mb1 \cdot b - mb2 \cdot a = 0$$

Therefore, formulas of balancing masses are:

Formula 4.16 Balancing masses in dynamical balancing

$$m_{b1} = \frac{b}{L} \cdot \frac{e_x}{r} \cdot m_s \; m_{b2} = \frac{a}{L} \cdot \frac{e_x}{r} \cdot m_s$$

These formulas allow calculating the mass to be added to the unbalanced rotor on both sides.

4.5.5 Balancing Machine

The dynamic balancing's main purpose is to mitigate and reduce the unbalancing moments if the vibration's amplitude at the bearings pedestal is 0.6 mm or lower. If not, the rotor must be removed and sent to a special workshop. The rotor is suspended from negligible friction supports in the workshop, where even minor amplitudes are detected and registered using electronic instrumentation. As commented before, balancing is made by adding balancing weights at pre-machined tee-slots around the rotor circumference or inserting screws in pre-bored screwed holes.

Modern balancing machines are sophisticated. Thus the trial-and-error procedure used in the past is no longer necessary. The rotor is suspended in a manner that balancing is performed on two planes. However, steam turbine LP rotors are usually balanced in three planes due to their large dimensions. Balance on-site is not recommended if a precise balancing is required because measurements are affected by shaft misalignments, bush-bearing rigidity, and other mechanical causes.

Multiplane balancing velocity depends on the critical rotor speed. However, 400 RPM may be considered a typical testing velocity. LP rotors of steam turbines, which have significant long blades, are balanced at velocities higher than the usual balancing speed because their critical velocities are higher than this value. Therefore, these rotors are sometimes balanced in vacuum chambers to prevent the blades from overheating due to windage. Site conditions are simulated in these chambers as close as possible to reality.

NOTES

1 Recommended books for Chapter 4: 1. Handbook of Rotordynamics by Fredric F. Ehrich. McGraw Hill, Inc, 1992 – 2. Gas Turbine Engineering Handbook by Meherwan Boyce. Elsevier, 4th edition – 3. Gas Turbine Theory by Herb Saravanamutto, Gordon Rogers, Henry Cohen, Paul Straznicky. Person, Prentice Hall, 6th edition, 2009 – 4. Marks' Standard Handbook for Mechanical Engineers by Eugene A. Avallone and Theodore Baumeister, Mc Graw Hill, 10th edition

2 See a description of other methods to calculate natural frequencies in the Wiley Online Library.
https://onlinelibrary.wiley.com/doi/pdf/10.1002/9781119038122.app1

3 It is recommended to see the document Balance Quality Requirements of Rigid Rotors, published by IRD Balancing company in the site:
http://www.irdbalancing.com/assets/balance_quality_requirements_of_rigid_rotors.pdf,

4 Reminder: $N_c = 30/\pi \cdot \omega_n$

5 See the book Strength of Materials by Stephen Timoshenko. CBS Publishers & Distributors 3rd edition, Reprint 2002. Chapter V Deflection of Beams

6 See book:. 7th Steam Turbines by Heinz P. Bloch and Murai P. Singh. McGraw Hill 2nd edition, 2009.

7 In ISO standard 1940-1 this moment is designed U and expressed in g·mm

5 Lateral Vibration of Turbomachines

5.1 INTRODUCTION TO LATERAL VIBRATION

Figure 5.1 shows a shaft bent by three applied forces: the shaft weight $W_{s;}$ the rotor wheels' weight $W_{w;}$ and the centrifugal force F_c produced by the shaft deflection and eccentricity times the square of the angular velocity. The rotor-shaft weight is not a distributed force along its axis because of the existing gap between wheels. For example, a two wheels rotor has two concentrated loads at the points where they are located. Nonetheless, for simplicity purposes, a multiple-wheel rotor may be modelled as a uniform load. The wheels' weight cannot be disregarded because, in general, their weight is higher than the shaft weight.

The centrifugal force is distributed along the shaft because it acts on each elemental shaft's slice. However, it may be modelled as a concentrated load at the middle of the shaft to simplify the centrifugal deflection's mathematical formulas. The centrifugal force's gyration radius equals the eccentricity e_x of the shaft gravity center plus the bending deflection ($\delta_s + \delta_w$) produced by the rotor-shaft weight plus the deflection δ_c produced by the centrifugal force.

Nomenclature of this chapter:

W_s: shaft's weight. Produces a uniform load
W_w: wheels' weight. Produces concentrated or distributed load
F_c: centrifugal force. Due to RPM, deflections, eccentricity, and loads mass
δ_s: static deflection due to the shaft weight = W_s/k_q
δ_w: static deflection due to the wheels' weight
δ_b: static shaft bending deflection due to rotor-shaft weight. $\delta_b = \delta_s + \delta_w$
δ_c: deflection produced by the centrifugal force
e_x: eccentricity of the gravity center with respect to the geometric shaft axis
r_g: gyration radius of the centrifugal force $r_g = e_x + \delta_s + \delta_w + \delta_c$
a_v: vibration amplitude

Applying the superposition principle, each force produces a deflection regardless of the other forces' deflection. δ_s and δ_w are calculated using the Mechanics of Materials formulas for beam's uniform and concentrated loads. Shaft bending always happens because a perfectly rigid shaft does not exist (Young modulus is never infinite!).

Figure 5.2 shows the deflections geometry due to the rotor-shaft weight and the centrifugal force. Their geometry is conceptually described based on this figure, and their formulas are derived after these conceptual definitions.

DOI: 10.1201/9781003175230-7

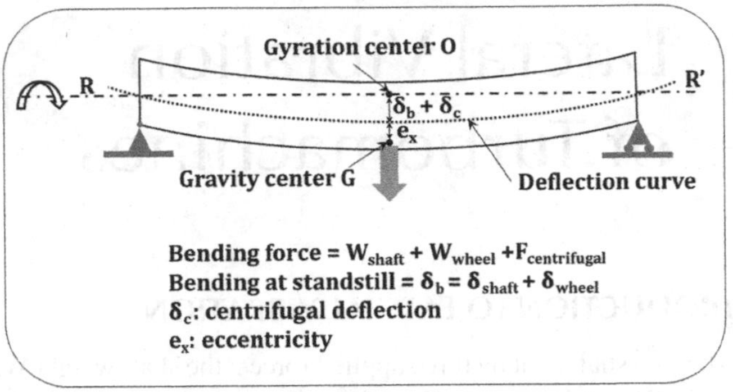

FIGURE 5.1 Misaligned shaft bent by its load and centrifugal force.

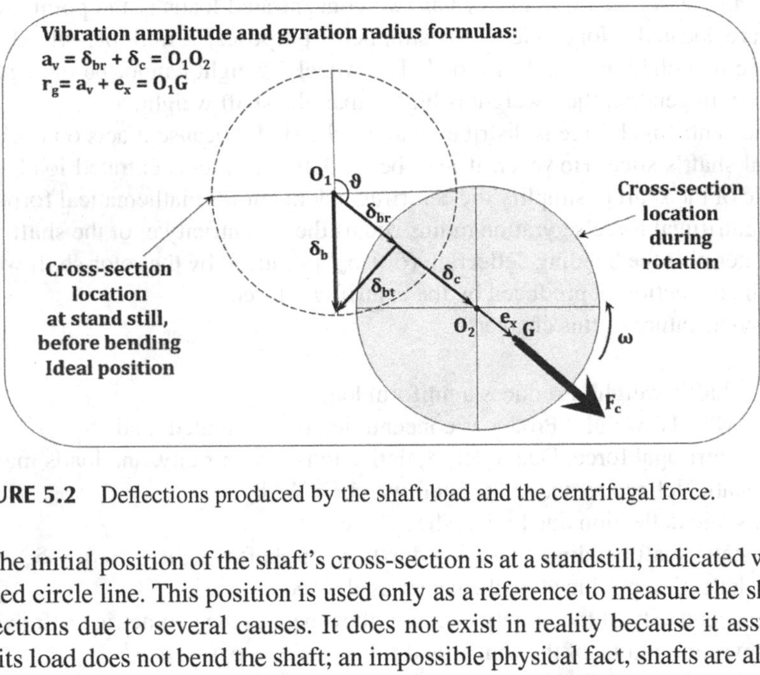

FIGURE 5.2 Deflections produced by the shaft load and the centrifugal force.

The initial position of the shaft's cross-section is at a standstill, indicated with a dashed circle line. This position is used only as a reference to measure the shaft's deflections due to several causes. It does not exist in reality because it assumes that its load does not bend the shaft; an impossible physical fact, shafts are always bent at a standstill due to their weight and rotor's load. In Figure 5.2, the grey circle with solid line is the cross-section's real position at the angle ϑ of the centrifugal force during rotation. The gyration center is the point O_1 in the shaft geometrical axis RR' shown in Figure 5.1.

At a standstill, the shaft and wheels' weight displace the cross-section downward a vertical bending deflection δ_b. (this cross-section's position is not shown in the scheme of Figure 5.2). The bending deflection δ_b is constant during gyration regardless of the angular position ϑ and the shaft velocity ω. It is represented by a vertical vector pointing downward. Its contribution to the gyration radius is its radial component $\delta_{br}(\vartheta)$ along the centrifugal force direction.

The shaft eccentricity e_x is the distance between the gravity center and the geometric axis RR' of the shaft. Its magnitude is small, but in any case, it is not neglectable. For example, consider a shaft eccentricity of 0.1 mm of a 10 tonne rotor-shaft weight, rotating at 3,600 RPM. If it is assumed that the weight is concentrated in the middle of the shaft, the centrifugal force formula returns a value of 14.5 tonnes. This value means that the shaft is enduring a centrifugal force 45% higher than its weight. This centrifugal force produces the deflection δ_c, whose formula is derived in Section 5.2.

If a vibrometer is applied on the shaft, it measures the vibration amplitude a_v as the displacement of the shaft's cross-section with respect to the ideal section at a standstill. That is the distance between the O_1 center and the O_2 center. This distance is formed by the radial component δ_{br} of δ_b plus the deflection δ_c produced by the centrifugal force, applied on the gravity center G.

The gyration radius of the centrifugal force is the distance O_1G. This length must be used to calculate the centrifugal force. As the eccentricity is the internal location of the gravity center, it affects the centrifugal force but not the vibration amplitude a_v. As the gyration radius depends on the vibration amplitude, its magnitude changes as the shaft revolves. Given that the centrifugal force magnitude ($F_c = m \cdot \omega^2 \cdot r_g$) depends on a changing gyration radius, it produces a changing centrifugal deflection δ_c.

The angle ϑ identifies the cross-section angular position. The $\omega \cdot t$ product can replace this angle. After this replacement, the following formulas demonstrate that the described deflections are a harmonic motion representing the transversal or lateral vibration.

5.2 LATERAL VIBRATION FORMULAS

By convention, downward forces and deflections are positive. Therefore, $\delta_{br}(\vartheta)$ must be positive for $90° > \vartheta < 270°$ (second and fourth quadrants); then, the formula of the radial component of δ_b, vibration amplitude a_v and gyration radius are:

Formula 5.1 Deflections, vibration amplitude and gyration radius

$$\delta_{br}(\vartheta) = -\delta_b \, . \cos\vartheta$$

$$\mathrm{av}(\vartheta) = \delta_{br}(\vartheta) + \delta c(\vartheta)$$

$$\mathrm{rg}(\vartheta) = \mathrm{av}(\vartheta) + \mathrm{ex}$$

As $\vartheta = \omega \cdot t$, Formulas 5.1 are converted into expressions of time used to produce the oscillograms in Figure 5.3.

Formula 5.2 Lateral vibration as a function of ϑ and t

$$\delta_{br}(\omega \cdot t) = -\delta_b \cdot \cos(\omega \cdot t)$$

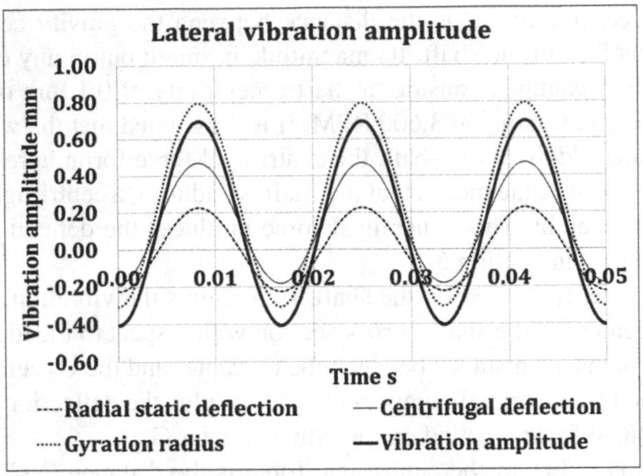

FIGURE 5.3 Lateral vibration produced by a bent shaft.

$$\mathrm{av}(\omega\cdot t)=\delta_{br}(\omega\cdot t)+\delta c(\omega\cdot t)$$

$$\mathrm{rg}(\omega\cdot t)=\mathrm{av}(\omega\cdot t)+\mathrm{ex}$$

In summary: eccentricity, radial bending deflection, centrifugal deflection and centrifugal force are all aligned and rotating around the center O_1 at the shaft angular velocity ω. The radial bending deflection δ_{br} and the centrifugal deflection δ_c sum is the vibration amplitude a_v. The vibration amplitude a_v and eccentricity is the centrifugal force's gyration radius. The eccentricity e_x is given by the manufacturer or measured at any specialized laboratory[2]. The formula to calculate the centrifugal deflection $\delta_c(\vartheta)$ is discussed in the next section.

5.3 CENTRIFUGAL DEFLECTION

The formula that expresses the equilibrium between the centrifugal force and the rigidity and friction reactions is the following:

Formula 5.3 Equilibrium of radial bending forces

$$\text{Centrifugal force} = \text{Shaft rigidity reaction} + \text{Friction reaction}$$

$$m\cdot\left[e_x+\delta_{br}(\omega\cdot t)+\delta_c(\omega\cdot t)\right]\cdot\omega^2=(k+j\cdot f\cdot\omega)\cdot\delta_c(\omega\cdot t)$$

The term between brackets is the gyration radius of Formula 5.2, m is the rotor-shaft mass, and ω is the shaft angular velocity in rad/s. Dividing this expression by m, the (k/m) ratio shows up and is replaced by the square of the rotor-shaft's natural frequency. By clearing δ_c from Formula 5.3, the shaft centrifugal deflection is given by:

Formula 5.4 Centrifugal deflection as a function of ω.

$$\delta_c(\omega\cdot t)=\frac{m\cdot\omega^2}{\sqrt[2]{\left(k-m\cdot\omega^2\right)^2+(f\cdot\omega)^2}}\cdot\left[e_x+\delta_{br}(\omega\cdot t)\right]$$

The term under the square root is known as mechanical impedance. It is the shaft's physical property opposing the vibratory motion. The above formula reveals that the model is a second-order system. Therefore, after some algebra and considering that: $fc = 2 \cdot \sqrt[2]{k \cdot m}, \zeta = f/fc, u = \omega/\omega n$, and $\omega_n^2 = k/m$, Formula 5.4 can be written in dimensionless notation:

Formula 5.5 Centrifugal deflection as a function of the frequency ratio u

$$\delta_c(u,t) = \frac{u^2}{\sqrt[2]{\left[1-u^2\right]^2 + 4 \cdot \zeta^2 \cdot u^2}} \cdot \left[e_x + \delta_{br}(\omega \cdot t)\right] = Q_c(u) \cdot \left[e_x + \delta_{br}(\omega \cdot t)\right]$$

Where Q_c is the centrifugal magnification factor, which at a standstill (u = 0) is zero, this factor must not be confused with the magnification factor of the static deflection defined in Chapter 1. See in Figure 5.4 the $Q_c(u)$ curves.

The magnification factor is equal to the vibration amplitude over the eccentricity plus the radial bending deflection. This magnification factor is maximum at $u = u_r$, though, due to the damping ratio's low values, like in turbomachines, it is generally accepted that the maximum is at u = 1. See formulas 1.74. At $\zeta = 0$, the amplitude vibration is theoretically infinite. However, all real machines have some friction that prevents the rotor from vibrating with infinite amplitude. Although resonance amplitudes do not reach an infinite value, they are always a threat to the machine's mechanical integrity.

Again, Figure 5.4 must not be confused with the family curves of Figure 1.32. In this case, the magnification factor is zero at zero frequency. At infinite frequency, the Q_c value tends to 1 because the mass cannot follow high-frequency excitation.

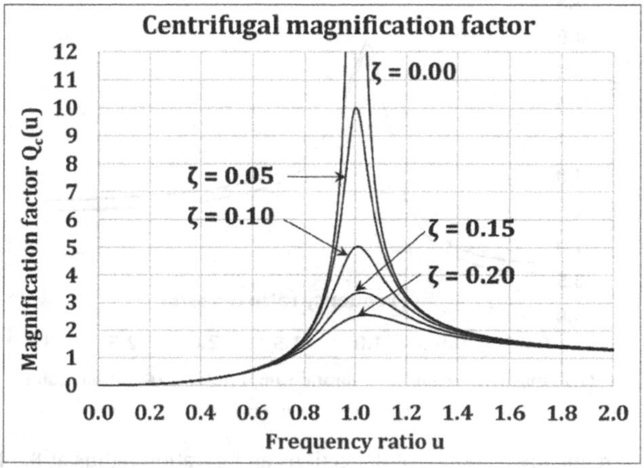

FIGURE 5.4 Centrifugal magnification factor curves.

5.4 GYRATION RADIUS FREQUENCY RESPONSE

Based on Formula 5.5, the centrifugal gyration radius formula may be expressed as a function of the centrifugal magnification factor.

Formula 5.6 Gyration radius

$$r_g(\omega) = \left[1 + Q_c(\omega)\right] \cdot \left[e_x + \delta_{br}(\vartheta)\right]$$

The maximum deflection is produced at $\vartheta = 180°$, then, as at that position is $\delta_{br} = \delta_b$, the gyration radius at that angular position is:

Formula 5.7 Frequency response of the gyration radius at $\vartheta = 180°$

$$r_g(\omega) = \left[1 + Q_c(\omega)\right] \cdot \left(e_x + \delta_b\right)$$

Figure 5.5 represents the frequency response of deflections and gyration radius at $\vartheta = 180°$ for a conservative system. The difference between the gyration radius curve and the total bending deflection curve is the eccentricity. The difference between the vibration amplitude and the centrifugal deflection is the radial bending deflection.

At infinite frequency, the magnification factor tends to 1; therefore, the gyration radius tends to $2 \cdot (e_x + \delta_{br})$. See Formula 5.7. The maximum gyration radius happens at $\vartheta = 180°$. The chart in Figure 5.5 illustrates the case of a turbomachine whose shaft has a natural frequency of 217 rad/s (2,067 RPM). As the rated turbine velocity is 188.5 rad/s (1,800 RPM), the turbomachine does not pass over the resonance during startups; therefore, the rotor is rigid. See Section 4.1.2.

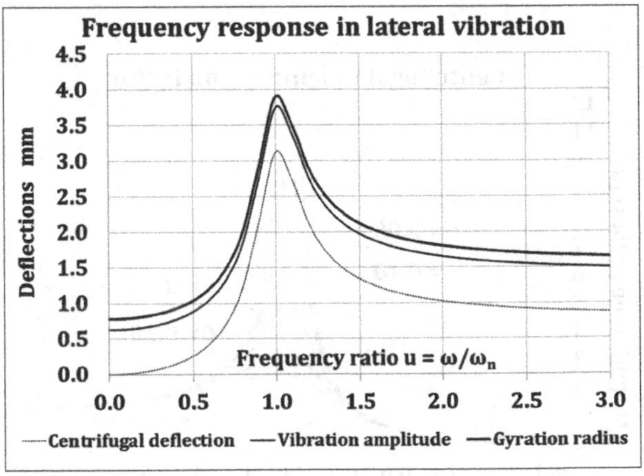

FIGURE 5.5 Frequency response of deflections and gyration radius at $\vartheta = 180°$.

5.4.1 Deflections and Gyration Radius at Singular Angles ϑ

As was seen before, deflections, vibration amplitude, and gyration radius depending on the shaft cross-section's angular location. Table 5.1 displays formulas to calculate them at ϑ singular angles of 0°, 90°, 270° and 360°.

The maximum deflections and gyration radius are produced at $\vartheta = 180°$, where the shaft load is arithmetically added to the centrifugal force because both are pointing downward. This angular location is the most stringent position for the shaft resistance. The minimum values are at $\vartheta = 0°$ because the centrifugal force and the rotor-shaft weight are in opposition. There are two singular angles during one revolution where the centrifugal deflection is equal and opposite to the bending deflection. At these two singular points, the vibration amplitude a_v is zero. Angles ϑ_{01} and ϑ_{02} formula is derived by equating the vibration amplitude formula to zero.

Formula 5.8 Equation of singular angles ϑ_{01} and ϑ_{02} at $a_v = 0$

$$\left[1+Q_c(\omega)\right]\cdot\left(c_x-\delta_b\cdot\cos\vartheta_0\right)=0$$

Solving this equation, the singular angles at which the gyration radius and the centrifugal deflection are zero at:

Formula 5.9 Singular angles for $r_g = \delta_c = 0$

$$\vartheta_{01}=\cos^{-1}\left(\frac{e_x}{\delta_b}\right)\vartheta_{02}=360°-\vartheta_{01}$$

The polar plot of Figure 5.6 represents the graphical interpretation of these two angles.

If the eccentricity e_x is higher than δ_b, no zero can exist, and both curves have positive deflection higher than zero. The minimum gyration radius is not at $\vartheta = 0°$ but ϑ_{01} and ϑ_{02}. If the eccentricity is zero; then, $\vartheta_{01} = 90°$ and $\vartheta_{02} = 270°$.

Figures 5.7 and 5.8 graphically show the gyration radius formation at the singular angles of 0° and 180°.

TABLE 5.1
Singular Vibration Amplitudes Formulas

Angle ϑ	Radial Bending δ_{br}	Centrifugal Deflection δ_c	Vibration Amplitude a_v	Gyration Radius r_g
0°	$\delta_{br} = -\delta_b$	$\delta_c = Q_c \cdot (e_x - \delta_b)$	$av = -\delta b + Q_c \cdot (e_x - \delta_b)$	$r_g = (1 + Q_c) \cdot (e_x - \delta_b)$
90° and 270°	$\delta_{br} = 0$	$\delta c = Q_c \cdot ex$	$av = Q_c \cdot ex$	$r_g = (1 + Q_c) \cdot e_x$
180°	$\delta_{br} = \delta_b$	$\delta_c = Q_c \cdot (e_x + \delta_b)$	$av = \delta b + Q_c \cdot (e_x + \delta_b)$	$r_g = (1 + Q_c) \cdot (e_x + \delta_b)$

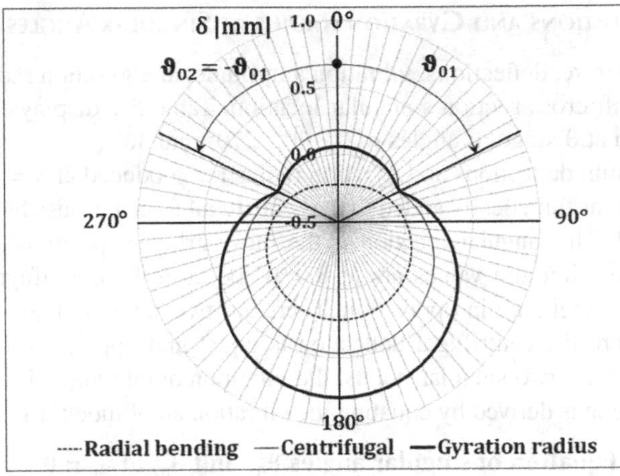

FIGURE 5.6 Polar plot of shaft deflections.

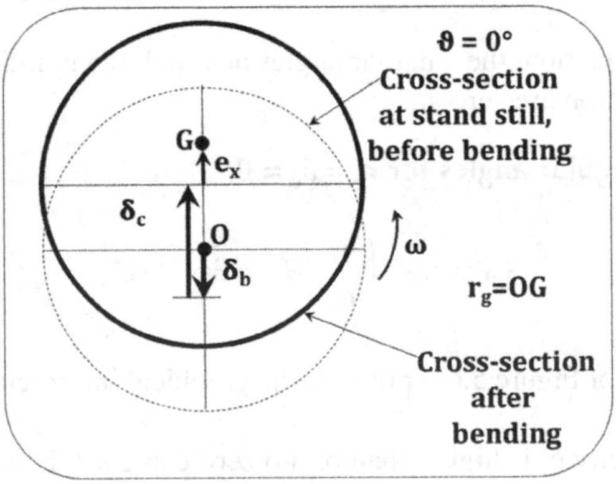

FIGURE 5.7 Gyration radius. Centrifugal force pointing downward.

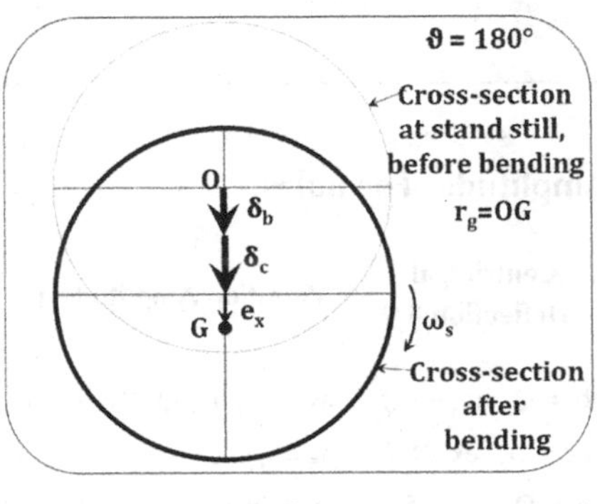

FIGURE 5.8 Gyration radius. Centrifugal force pointing upward.

Figure 5.7 shows the cross-section displaced upward because the centrifugal force is pointing in that direction. The net deflection (vibration amplitude) is the difference between the centrifugal and the bending deflections: ($\delta_c - \delta_b$), and the gyration radius is the net deflection plus the eccentricity. Figure 5.8 shows the maximum vibration amplitude because both the bending and centrifugal deflections are pointing downward. The gyration radius is maximum for the same reason.

5.5 NATURAL FREQUENCY VERSUS DEFLECTION

If a concentrated vertical load bends a shaft, it will experience lateral vibrations when gyrating. Therefore, the shaft acts as a vibrating system at the damped frequency ω_d. However, as rotating machines have a significantly low damping ratio, the shaft's damped frequency, resonant frequency, and natural frequency have similar values. Therefore, in the following explanation, the term natural frequency may be thought of as resonant frequency or critical velocity if the frequency is expressed in RPM.

The natural frequency is derived with the following simple manipulation of Formula 1.9: 1) Replace the coefficient k_w by the shaft weight to the static deflection ratio W/δ_0. 2) Replace the shaft mass by the ratio W/g. These simple replacements produced Formulas 5.10, proving that the natural frequency is a function of the shaft or beam's deflection.

Formula 5.10 Shaft natural frequency versus deflection in m

$$\omega_n = \sqrt[2]{\frac{g}{\delta_0}} = \frac{3.13}{\sqrt[2]{\delta_0}}$$

The static deflection δ_0 is the deflection at the load application point, expressed in meters and ω_n in rad/s. If the deflection is expressed in mm, the above formula is converted into the following:

Formula 5.11 Beam natural frequency versus deflection in mm

$$\omega_n = \frac{99.05}{\sqrt[2]{\delta_0}}$$

The chart in Figure 5.9 of natural frequency versus the static beam deflection is created based on the previous formula.

In the field, it is customary to refer to RPM and not in rad/s. Therefore, Formula 3.2 may be converted into the following expression:

Formula 5.12 Critical velocity in RPM. Concentrated load

$$N_c = \frac{30}{\pi} \bullet \omega_n = \frac{946}{\sqrt[2]{\delta_0}}$$

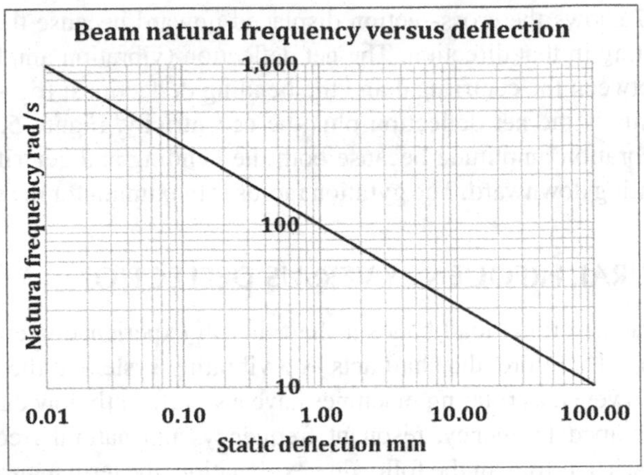

FIGURE 5.9 Natural frequency versus shaft static deflection.

Where N_c is the critical velocity in RPM, and the static deflection δ_0 is in mm. Conversely, if RPM's critical velocity is known, it is possible to estimate the shaft deflection with $894{,}538/N_c^2$. Formula 3.3 indicates that critical velocity and natural frequency are equivalent concepts. It is common in the technical literature to write the previous formula with δ_0 expressed in cm, which results in the following approximate expression: $Nc \simeq 300 / \sqrt[2]{\delta_0}$.

5.5.1 Correction by the Rotor Mass

The shaft's natural frequency given by the above formulas must be corrected due to the rotor mass's addition to obtaining the rotor-mass natural frequency. If in Formula 1.9 the mass includes the rotor mass, then it is seen as:

Formula 5.13 Rotor-shaft natural frequency

$$\omega_{n\ rotor-shaft} = \sqrt[2]{\frac{k_w}{m_{shaft} + m_{rotor}}}$$

Dividing the above formula by the shaft's natural frequency, it returns an expression of the shaft-rotor natural frequency as a function of the shaft's natural frequency and the rotor to shaft mass ratio.

Formula 5.14 Rotor-shaft's natural frequency versus the shaft's natural frequency

$$\omega_{n\ rotor-shaft} = \frac{\omega_{n\ shaft}}{\sqrt[2]{1 + m_{rotor} / m_{shaft}}}$$

Where $\omega_{n\ shaft}$ is given by Formula 5.10. In large turbines, typical values of m_{rotor}/m_{shaft} are between 2 and 5; therefore, $\omega_{n\ rotor-shaft}$ is approximately in the 0.6 to 0.4 range of $\omega_{n\ shaft}$. Therefore, the rotor-shaft natural frequency is 40% to 60% lower than the shaft's natural frequency.

5.5.2 Calculation of Shaft Deflection

According to Mechanics of Materials, shaft deflection produced by a symmetrical concentrated load is calculated with the following formula:Formula 5.15Shaft deflection under concentrated load

$$\delta_0 = \beta_w \cdot \frac{W \cdot L^3}{E \cdot I_z}$$

δ_0 is the deflection under the concentrated load.

β_w is a coefficient that depends on the type of supports and load. Subscript w means concentrated load, to differentiate this coefficient from the uniformly loaded shaft, named k_q.

Reference Table 5.2. If the load is symmetrical (applied in the middle of the shaft), β_w values are read in column 2. If the load is non-symmetrical, then column 3 gives the β_w formulas, where a is the load's distance to the left support and b is the distance to the right support. In any case, it is $a + b = L$.

E is the Young's modulus for steel.

I_z is the inertia moment of the shaft cross-section $= \frac{\pi \cdot D^4}{64}$. Where D is the shaft diameter.

The E.I_z product is the shaft flexural rigidity. The higher this product, the lower the deflection produced by the load. The rigidity coefficient used in the mechanical systems' equation of motion is equal to the flexural rigidity divided by the cube of the shaft length and the support coefficient β_w. Therefore: $k_w = W/\delta_0 = E.\ I_z/(\beta_w \cdot L^3)$.

TABLE 5.2
Deflection's Formulas of Coefficients β_w

Type of Support	β_w Symmetrical Load	β_w Non-Symmetrical Load
Hinged-hinged	1/48	$\frac{a^2 \cdot b^2}{3 \cdot L^4}$
Clamped-hinged	7/768 ≅ 1/110	$\frac{a^2 \cdot b^3}{12 \cdot L^6} \cdot (4 \cdot a + 3 \cdot b)$
Clamped-clamped (built in-built in)	1/192	$\frac{a^3 \cdot b^3}{3 \cdot L^6}$
Clamped-free	1/3	1/3

W: is the concentrated load on the shaft, including its weight.

The site's static deflection measurement is only feasible in easy-to-access shafts or long transmission shafts, such as in ships. However, that is not the case for the enclosed part of shaft turbines. See Table 5.2.

5.6 NATURAL FREQUENCY VERSUS STRESS PROPAGATION VELOCITY

As was seen in Section 5.5, the natural frequency of shafts or beams with a concentrated load is calculated as a function of their static deflection. Nonetheless, the natural frequency may also be expressed as a function of the longitudinal propagation velocity of stress. This formula has some advantages because it allows discovering a natural frequency formula versus the stress propagation velocity that is easy to use.

For a uniform circular shaft of length L and diameter D, the rigidity coefficient and mass are:

$$k_w = \frac{1}{\beta_w} \cdot \frac{E \cdot I_z}{L^3} \quad m = \rho \cdot \frac{\pi \cdot D^2}{4} \cdot L$$

Where β_w is in Table 5.1. If the above formulas are replaced in Formula 1.9, the natural frequency appears expressed as a function of the propagation velocity c and the D/L^2 ratio:

Formula 5.16 Shaft natural frequency versus propagation velocity

$$\omega_n = \sqrt[2]{\frac{k_w}{m}} = \frac{1}{4 \cdot \sqrt[2]{\beta_w}} \cdot c \cdot \frac{D}{L^2}$$

For typical stainless steel, this velocity is of the order of 5,000 m/s. Therefore, as a rule of thumb, uniform circular shafts' natural frequency is calculated with the following formula:

Formula 5.17 Approximate natural frequency of uniform circular shafts

$$\omega_n \cong \frac{1{,}250}{\sqrt[2]{\beta_w}} \cdot \frac{D}{L^2}$$

Figures 5.10 and 5.11 show the shaft's natural frequency as a function of length and diameter. Curves of these figures are valid for E = 196 GPa and ρ = 801 kg mass/m^3. Figure 5.11 shows the natural frequency curves of a 300 mm diameter shaft for several types of support. It indicates that the highest frequency belongs to the clamped-clamped support and the lowest to the clamped free support.

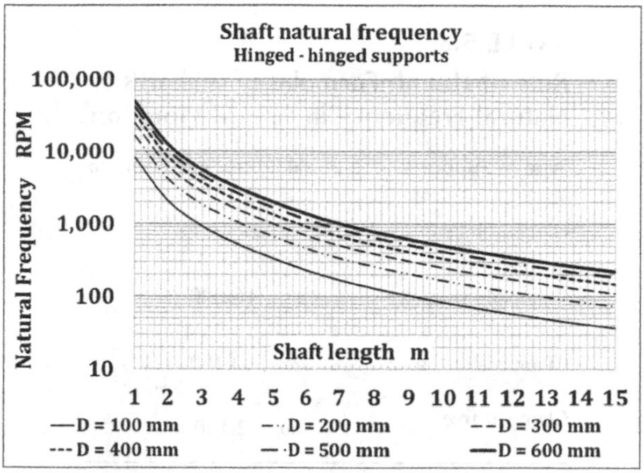

FIGURE 5.10 Shaft natural frequency.

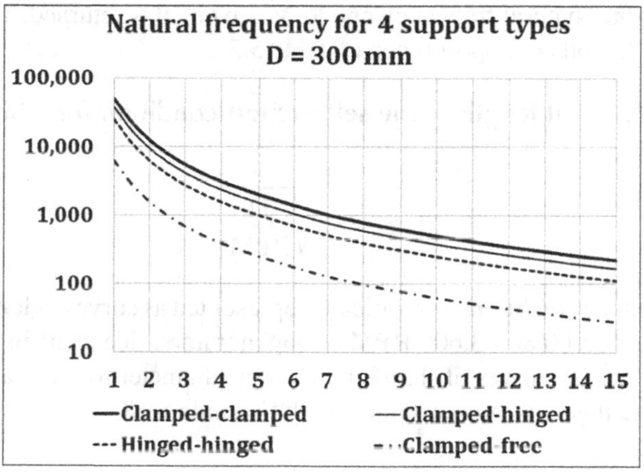

FIGURE 5.11 Natural frequency for different supports.

Formulas 5.16 and 5.17 prove that long shafts vibrate at a lower frequency than short shafts of the same diameter. They also express that the shaft's natural frequency is directly proportional to the diameter; then, the higher the diameter, the higher the natural frequency. Introducing the typical stress wave velocity in stainless steel shafts (c = 5,000 m/s) and the coefficients β_w of each support type, the rule of thumb of Formula 5.17 is converted into the expressions exhibited in Table 5.3:

5.6.1 Shaft Lateral Resonance in Power Plants

At a specific pair of shaft length and diameter values, the shaft natural frequency is equal to the shaft RPM; therefore, the shaft is subject to lateral vibration resonance

TABLE 5.3
Rule of Thumb Formulas of a Shaft's Natural Frequency in Lateral Vibration

Type of Supports	Natural Frequency Formula
Hinged-hinged	$\omega_n = 8{,}660 \times \frac{D}{L^2}$
Clamped-clamped	$\omega_n = 16{,}320 \times \frac{D}{L^2}$
Clamped-hinged	$\omega_n = 13{,}100 \times \frac{D}{L^2}$
Clamped-free	$\omega_n = 2{,}160 \times \frac{D}{L^2}$

under this condition. The formulas of Table 5.3 give this resonance combination. Clearing the shaft length from the formula of the hinged-hinged shaft, the length at which the shaft natural frequency and RPM are equal is returned. The same can be done with the other support types of Table 5.3.

Formula 5.18 Shaft length at the self-excited condition for a hinged-hinged shaft

$$L \cong 288 \times \sqrt[2]{\frac{D}{RPM}}$$

This resonance condition is graphically represented as curves of length L versus diameter D for 3,000 and 3,600 RPM turbogenerators. See chart in Figure 5.12. This chart shows the curves of shaft length versus diameter for the standard velocities of electrical generators, computed with Formula 5.18.

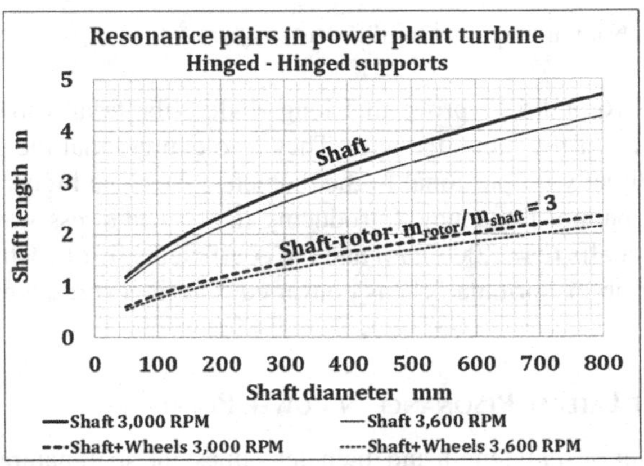

FIGURE 5.12 Self-excitation curves for turbogenerators.

Consider a turbogenerator set of 60 Hz; therefore, its rated velocity is 3,600 RPM. The shaft is assumed to have hinged-hinged supports. For example, if the shaft diameter is 700 mm and the shaft length is 4.0 m, the shaft's natural frequency coincides with its rotating velocity, producing a self-excited vibration.

The auto-excited resonance is avoided if the shaft length is different to 4.0 m. However, any pair of shaft length-diameter close to the curve of chart 5.12 may produce a significant vibration amplitude amplified due to the proximity to the resonance curve. The chart shows the resonance curves for a shaft-rotor assembly, assuming that the rotor mass is equivalent to three times the shaft mass. As the mass of the assembly is higher than the shaft, the natural frequencies are lower.

NOTES

1 Recommended books for Chapter 5: Mechanical Vibrations by S. Graham Kelly. Schaum's Outline Series. McGraw Hill, 1996 – 2. Advanced Vibration Analysis. S. Graham Kelly. CRC Press. Taylor and Francis Group, 2007 – 3. Fundamentals of Vibration Engineering by Isidor Bykhovsky. MIR Publishers, Moscow. 1st published 1972 – 4. Mechanical Vibrations by J.P. Den Hartog. Dover Publications, Inc. New York, 1985, 4th edition – 5. Mechanical Vibration by Haym Benaroya, Mark Nagurka and Seon Han. CRC Press. Taylor and Francis Group, fourth edition 2018.
6. Free Vibration Analysis of Beams and Shafts by Daniel Gorman. John Wiley & Sons. 1975 – 7. . Marks' Standard Handbook for Mechanical Engineers by Eugene A. Avallone and Theodore Baumeister, Mc Graw Hill, 10th edition

2 It is recommended to read some literature about eccentricity measurement. See an example in the Field Application Note of STI Vibration Monitoring Inc. in: https://www.stiweb.com/v/vspfiles/downloadables/appnotes/ecc.PDF

6 Vibratory Forces in Turbomachines

6.1 INTRODUCTION TO VIBRATORY FORCES

This chapter describes the most important forces inside a turbomachine[1] that produce or may produce vibration of vital parts of the machine. These forces are caused by unbalancing and the substance of work (gas, air, or steam) flowing through the blades. Mechanical imperfections, like shaft bending, are also examined. The main causes of these forces are summarized in the following list:

- Non-uniform flow force on blades.
- Pulsating torsional torque.
- Centrifugal forces on the rotor-shaft.
- Centrifugal forces on blades.
- Mass rotor unbalancing.

The flow impulse on the turbine blades has three intermittent components: tangential, axial, and radial. See Figure 6.1. These components are not steady forces but impulse forces whose duration is when the blade faces the nozzle flow; therefore, the impulse is extremely short. This tangential impulse on the blades produces an impulse torque on the rotor and shaft, whose inertia, friction, and rigidity react opposing the torque. The radial component is resisted by the shaft rigidity and the axial component by the thrust bearing. In the three cases, the applied force is intermittent at a frequency equal to the blades' quantity times the shaft RPM. The tangential impulse on the blades is resisted by the rotor-shaft inertia, friction and rigidity. This reaction creates a short transient of torsional vibration and the possibility of resonance.

As the shaft is bent, a radial force formed by the centrifugal force combined with the rotor-shaft weight is created. This force is radial and bends the shaft twice on each turn. This bending produces the lateral vibration discussed in Chapter 5. This force's impact on the pedestals and foundation is studied based on its horizontal and vertical components.

The radial force's horizontal component is applied to the bearings, producing a periodical overturning moment on pedestals. The vertical component is applied through the bearings and pedestals on the foundation. Therefore, pedestals and foundations are subject to periodic and significant forces while the turbomachine is running.

The flow impulse produces the blade's vibration because it is not a perfectly rigid piece. As there are no important friction forces, the blade's vibration does

DOI: 10.1201/9781003175230-8

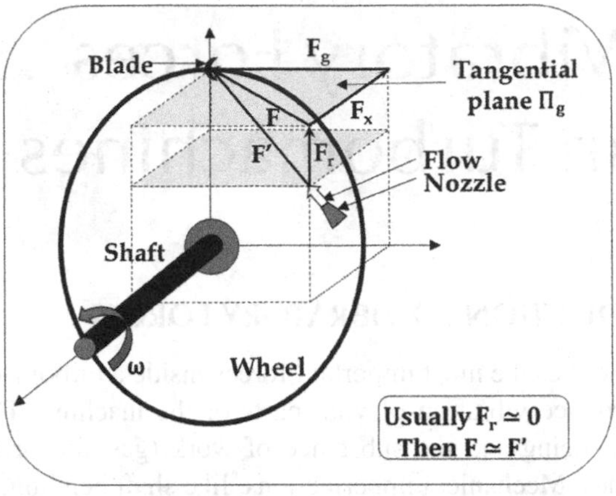

FIGURE 6.1 Schematics of applied forces on turbomachine blades.

not damp; therefore, the blades vibrate permanently while the turbine is running. The flow impulse is extremely short, and the blade receives as many impulses per shaft turn as nozzles are installed. Therefore, the higher the number of nozzles, the higher the excitation frequency. The aftermath of this vibration is the blades' fatigue and, eventually, their collapse risk.

All blades support the same centrifugal force that tends to pull out them from their roots. Because of the rotor symmetry, all centrifugal forces on blades are in equilibrium. These forces depend on the blade mass and the square of the shaft velocity. However, if one or more blades are broken, the rotor-shaft becomes unbalanced. In that case, the centrifugal forces are no longer in equilibrium, and the shaft loses the impulse of damaged blades and produces torsional vibration. Besides, the rotor masses' asymmetry creates harmonic moments that produce a complex vibration. The best solution to this problem is balancing the rotor in a specialized workshop.

6.2 FORCES ON BLADES AND BEARINGS

This section provides a short insight into the forces that make the turbomachine run and produce forces in addition to the wheels' driving forces. The total force acting on a blade is a centrifugal force formed by three components mentioned previously: tangential, axial and radial. See Figure 6.1. Tangential force F_g is tangent to the wheel; it is the force that drives the wheel. The curved flow path between blades generates the tangential force. See Figure 6.2. The axial force F_x is parallel to the shaft. It produces an axial thrust that pushes the rotor. It is never neglected because it exerts a significant effort, counteracted by a thrust bearing. In a reaction turbomachine, axial thrust is mitigated with a balancing system or pressure equalizing orifices drilled in the wheels. The radial force F_r is neglectable in axial turbomachines. Therefore, with little error, the force F applied to the blades, composed of the axial and tangential forces, is considered the blades' total

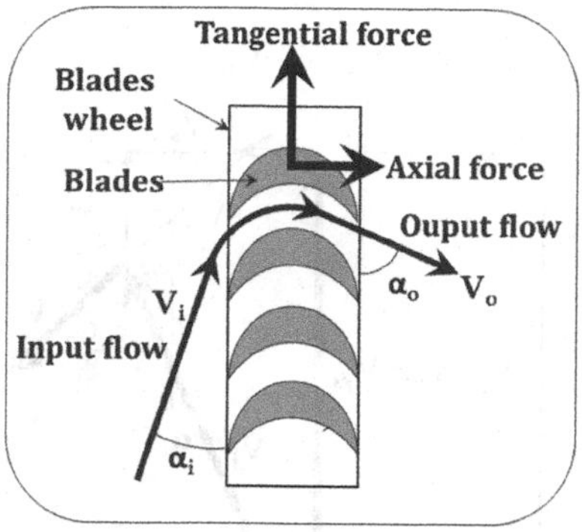

FIGURE 6.2 Graphical interpretation of Euler's formula.

force. The horizontal total force F and the tangential force F_g are in the plane Πg tangent to the wheel. The angle between the total and tangential forces is the flow input angle to the blades.

Euler's equation is the cornerstone of all types of turbomachines. It demonstrates that the tangential force on a blade is equal to the mass flow G_m times the flow tangential velocity V_g. For the same reason, the axial force equals the mass flow G_m times the flow axial velocity V_x. The tangential velocity and the axial velocity depend on the input flow velocity V_i and output flow velocity V_o and their angles α_i and α_o with respect to the wheel plane. The tangential and axial forces have the following formulas derived from the Euler equation: See in Figures 6.1 and 6.2 the graphical interpretation of this formula.

Formula 6.1 Flow forces on blades and thrust bearing based on Euler's equation

$$F_g = G_m \cdot V_g = G_m \cdot \left(V_i \cdot \cos\alpha_i + V_o \cos\alpha_o\right) F_x = G_m \cdot V_x = G_m \cdot \left(V_i \cdot \sin\alpha_i - V_o \sin\alpha_o\right)$$

Tangential force Axial thrust

The flow directions are designed so that the tangential forcc is maximized and the axial force is minimized. Therefore, according to Formula 6.1, the input angle α_i should be as small as possible to take advantage of the input flow kinetic energy. The typical value of α_i is 15°. In an impulse turbomachine, the output flow represents kinetic energy lost to the environment. Therefore, minimizing this energy is essential to optimizing turbomachine efficiency. The minimum output kinetic energy is produced by an output angle equal to 90°. Then the turbomachine efficiency is maximal at that angle. As turbomachines have different velocity regimes, it is not always possible to accomplish this condition.

Formula 6.1 is applied to calculate V_g and V_x. Velocity components V_g and V_x are graphically calculated, as indicated in Figure 6.3. However, a better option for turbomachines with more than one wheel is a spreadsheet using trigonometric

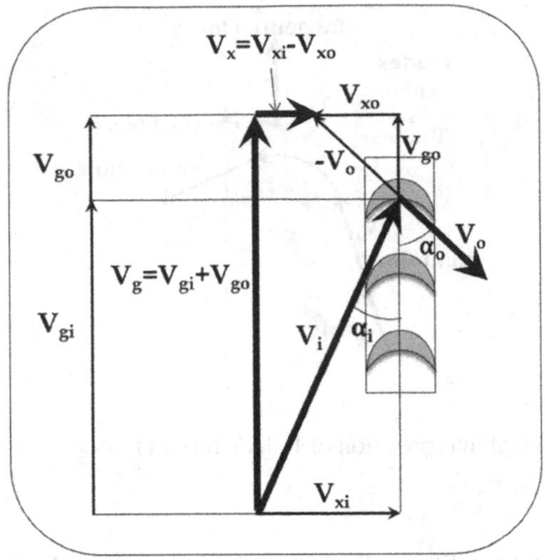

FIGURE 6.3 Flow velocities diagram for $\alpha_o < 90°$. Positive case.

formulas or specialized software. Vector velocities V_g and V_x formation is visualized in Figure 6.3 along with the input and output flow velocities V_i and V_o.

From Euler Formulas 6.1, the tangential and axial velocities are derived:

Formula 6.2 Tangential and axial velocity

$$V_g = V_i \cdot \cos\alpha_i + V_o \cos\alpha_o$$

$$V_x = V_i \cdot \sin\alpha_i - V_o \sin\alpha_o$$

These formulas explain the vector diagrams 6.3 and 6.4. In Figure 6.3, the tangential component of V_0 is V_{go} that is added to the wheel tangential velocity V_{gi}. Because the tangential velocity is enhanced, this condition is named the positive case. However, if α_o is higher than 90°, V_{go} is 180° reversed; therefore, it is subtracted from velocity V_{gi}, and the result is a tangential velocity reduction. As the tangential velocity is reduced, the wheel driving force, F_g, is also reduced. Hence, because of this reduction, this case is called the negative case. See Figure 6.4.

Formula 6.1 presents an essential physical conclusion of the Euler equation: the flow's forces do not depend on the substance's nature of the flow. This conclusion means that it does not matter if the flow is water (hydraulic turbomachine), gas (gas turbomachine), steam (steam turbomachine) or air (compressors or fans). Therefore, in all turbomachines, forces are only dependent on flow mass and flow velocities regardless of the flow's substance nature. The tangential and axial velocities must be interpreted as forces whose proportionality coefficient is the mass flow G_m.

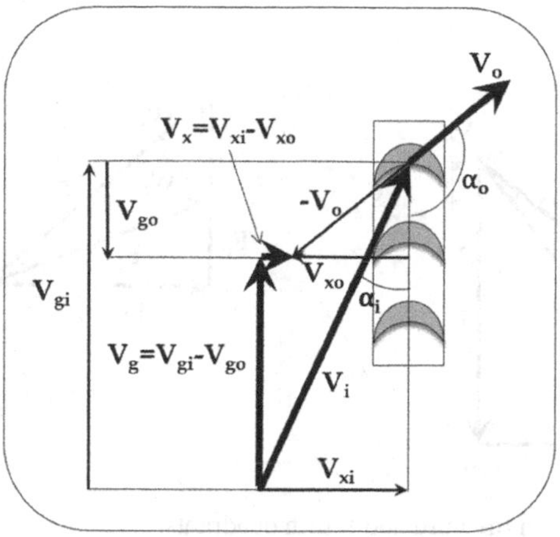

FIGURE 6.4 Flow velocities diagram for $\alpha_o > 90°$. Negative case.

6.3 RADIAL VIBRATORY FORCES

Radial forces act on the rotor-shaft during gyration are a) the shaft rotor weight that is always pointing downward and b) the centrifugal force that is a rotating radial force that is not constant during one rotation. These two forces are designated F_t in Figures 6.5 and 6.6, where subscript t stands for the total force. As these forces are applied to the rotor-shaft's gravity center, the centrifugal and the total force are spinning around the shaft's geometric axis, generating lateral vibration. The total force has two components: the vertical and the horizontal forces designated F_h and F_v in Figures 6.5 and 6.6.

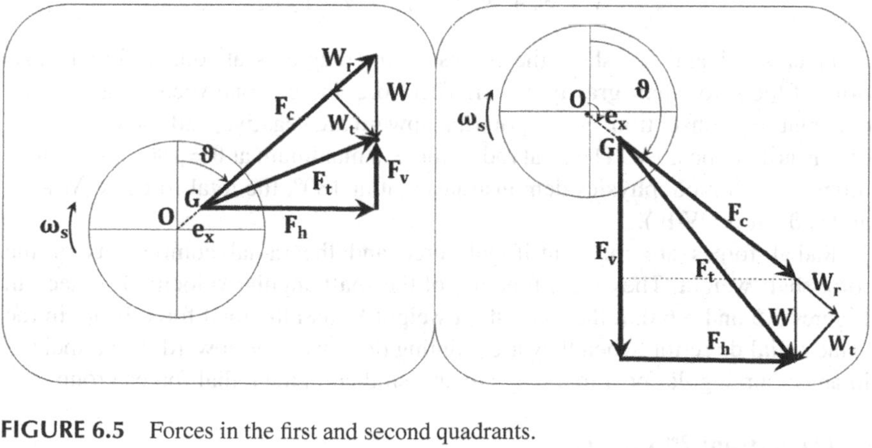

FIGURE 6.5 Forces in the first and second quadrants.

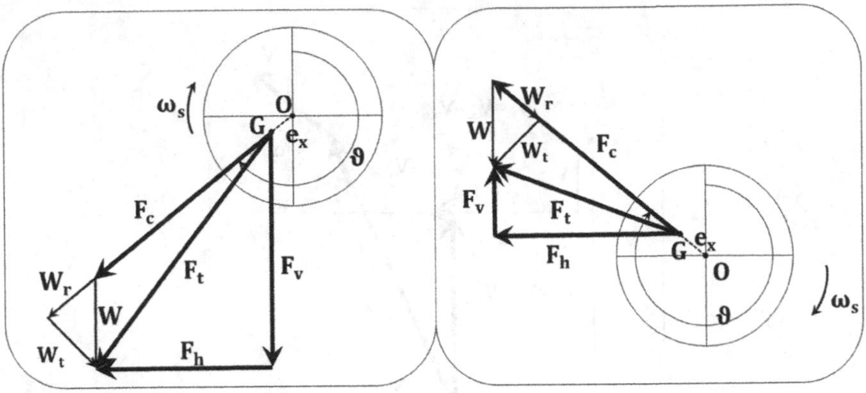

FIGURE 6.6 Forces in the third and fourth quadrants.

The forces represented in these Figures are:

W: rotor-shaft weight. It is applied to the gravity center and is a vertical force pointing downward

W_r: radial rotor-shaft weight component along the centrifugal force direction

W_t: transversal rotor-shaft weight component perpendicular to the centrifugal force direction

F_c: centrifugal force in a gyratory radial direction

F_v: vertical component of the total force

F_h: horizontal component of the total force

F_t: total force

The vector formulas that allow calculating the total force are the following:

Formula 6.3 Vector formulas of forces shown in Figures 6.5 and 6.6

$$\overline{W} = \overline{W}_r + \overline{W}_t \quad \overline{F}_t = \overline{F}_c + \overline{W} = \overline{F}_v + \overline{F}_h$$

Figures 6.5 and 6.6 show the forces vector diagrams at four different positions of the rotor-shaft gravity center. Therefore, there is one vector diagram per quadrant. By convention, forces pointing upward are negative, and forces pointing downward are positive. The total radial force is maximum at $\vartheta = 180°$ and a minimum at 0°. Simple physics demonstrates that at 180°, the total force is ($W+F_c$), and at 0°, it is ($W-F_c$).

Radial forces are the centrifugal force and the radial components of the rotor-shaft weight. They are a function of the shaft angular velocity. It is seen in Figures 6.5 and 6.6 that the rotor-shaft weight W and the total force F_t are in the same radial direction when they are pointing downward or upward. Nevertheless, in any other angular position, they are also studied in the radial forces group.

1st quadrant 2nd quadrant
3rd quadrant 4th quadrant

6.3.1 Assessment of Radial Vibratory Forces

The centrifugal force F_c is a function of angular speed ω and the gravity center angle ϑ or, what is the same, the (ω·t) product. Therefore, after this replacement, all acting forces are a function of time, proving in formulas that they are vibratory. The total force F_t and the shaft-weight components W_r and W_t are also a function of ω and ϑ or the angle ω·t. Formulas to calculate these forces are derived from simple trigonometric relations of Figures 6.5 and 6.6.

Formula 6.4 Radial forces model

$$F_c(\omega \cdot t) = m \cdot r_g(\omega \cdot t) \cdot \omega^2$$

$$W_r(t) = -W \cdot \cos\vartheta = -W \cdot \cos(\omega \cdot t)$$

$$W_t(t) = W \cdot \sin\vartheta = W \cdot \sin(\omega \cdot t)$$

$$F_{total} = \sqrt[2]{(F_c + W_r)^2 + W_t^2}$$

Where:

m: rotor-shaft mass

$rg(\omega)$: gyration radius $= [1 + Q_c(\omega)] \cdot (e_x - \delta_b \cdot \cos\vartheta)$

ω : shaft angular velocity $= \vartheta / t$

ϑ: the angular position of the gravity center

Replacing ϑ by (ω·t) as suggested in Formula 6.4, they are converted into time functions. These functions are useful when studying the vibratory forces applied to the turbomachine and its foundation. Figure 6.7 shows an example of vibratory forces. The centrifugal force's intensity is higher than the radial component of the rotor-shaft weight. This figure demonstrates that the shaft is subject to a

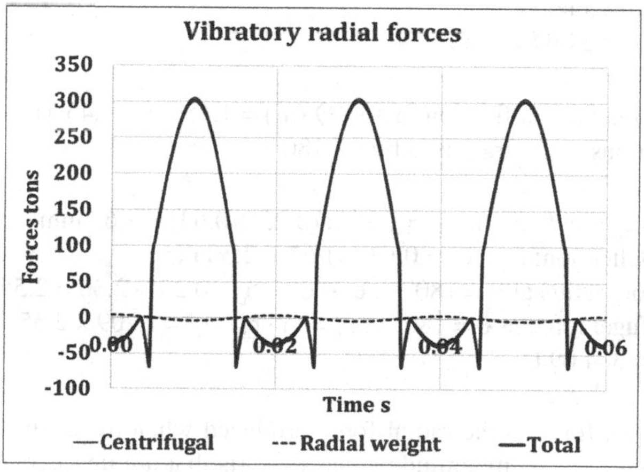

FIGURE 6.7 Radial vibratory forces.

continuous "reciprocating knocking" of radial forces, which produces the deflections studied in Chapter 5.

6.3.2 Technical Scenario and Assessment Request

Consider a 3,000 RPM turbomachine with a shaft diameter and length of 0.4 m and 4 m, respectively. At a standstill, the shaft static deflection δ_b is 0.63 mm. The shaft eccentricity is 0.25 mm. The shaft weight is 3.95 tonnes, and the rotor weight is 7.90 tonnes. The load is symmetrically concentrated, and the type of support is hinged-hinged ($\beta_w = 1/48$). The system is conservative.

1) Calculate the radial forces at $\vartheta = 180°$
2) Plot radial vibratory forces against time
3) Calculate the total force if balancing reduces the eccentricity to zero.

- **Calculation:**
  Shaft velocity in rad/s: $\omega_s = \pi/30 \times 3{,}000 = 314$ rad/s
  Rotor-shaft mass: m = (3.95 + 7.90)/9.81 × 1,000 = 1,209 kg mass
  Natural frequency. Use the formula derived from the stress velocity propagation of Chapter 1.

  Total weight to shaft weight ratio: $(W_s + W_w)/W_s = (3.95 + 7.90)/3.95 = 2$
  Where W_s is the shaft weight, and W_w is the rotor's weight

$$\omega_n \cong \frac{1{,}250}{\sqrt[2]{\beta_w}} \cdot \frac{D}{L^2} = 1{,}250/\left(1/48\right)1/2/2 \times 0.4/42 = 217\,\text{rad/s}$$

  Equivalent to $30/\pi \times 217 = 2{,}067$ RPM
  Natural frequency ratio:
  $u_n = 3{,}000/2{,}067 = 1.45$

  Centrifugal magnification factor: $Q_c(u_n) = 1.45^2/(1 - 1.45^2) = 1.90$
  Deflections and gyration radius at 180°:

  Centrifugal deflection: $\delta_c = 1.90 \times (0.25 + 0.63) = 1.67$ mm
  Total deflection: $\delta_b + \delta_c = 0.63 + 1.67 = 2.30$ mm
  Gyration radius at $\vartheta = 180°$: $e_x + \delta_b + \delta_c = 0.25 + 2.30 = 2.55$ mm
  Centrifugal force at $\vartheta = 180°$: $F_c = m \cdot r_g \cdot \omega_s^2 = 1{,}209 \times 2.55/1{,}000 \times 314^2/1{,}000 = 304.69$ t.

This centrifugal force is the radial force produced when the centrifugal force is pointing downward. Its magnitude creates efforts that are unacceptable. The turbine cannot operate under this condition. Nonetheless this case is exposed to illustrate the risks created by the vibratory forces.

Rotor-shaft weight: $W = 3.95 + 7.90 = 11.85$ t
Total force at $\vartheta = 180°$: $F_t = 304.69 + 11.85 = 316.54$ t

Figure 6.7 shows the oscillogram of vibratory radial forces of this example.

The vibration represented by Figure 6.7 is mainly due to centrifugal forces. The radial component of the shaft-rotor assembly is small compared to the centrifugal forces; therefore, its influence is barely noted.

- **Total force for $e_x = 0.0$ mm**
  Centrifugal deflection: $\delta_c = 1.90 \times 0.63 = 1.20$ mm
  Gyration radius at $\vartheta = 180°$: $\delta_b + \delta_c = 0.63 + 1, 20 = 1.83$ mm
  Centrifugal force at $\vartheta = 180°$: $F_c = 1{,}611 \times 1.83/1{,}000 \times 314^2/1{,}000 = 218.01$ t
  Total force at $\vartheta = 180°$: $F_t = 218.0 + 11.85 = 229.86$ t
  Total force reduction with $e_x = 0.0$ mm: $(229.86/316.54 - 1) \times 100 = -27.4\%$

By reducing eccentricity to zero the vertical force still has an unacceptable magnitude. Therefore, it is recommended to review the magnification factor calculation and introduce changes in the mechanical design that reduce the gyration radius to an acceptable value.

The turbomachine velocity is 314.0/217.0 × 100 = 144.7% of the natural frequency. In general, it is recommended to set the machine velocity outside the 70% to 130% range of the resonant frequency (2,067 RPM). The critical range of velocities is 1,446.9 to 2,687.1 RPM. Therefore, the rated velocity (3,000 RPM) is not in the critical range. However, it is close to its upper limit. Additionally, during startup, the machine will be subject to resonance because the rated velocity is higher than the resonant velocity. This resonance is a problem adding to the excessive vertical force seen before.

6.4 VERTICAL AND HORIZONTAL VIBRATORY FORCES

Vector diagrams of Figures 6.5 and 6.6 show that vertical and horizontal forces are easily calculated using trigonometric formulas:

Formula 6.5 Forces applied to the shaft, pedestals, and foundations

$$F_v = -F_c \cdot \cos\vartheta - W\left(\text{from } \vartheta > 270° \text{to } \vartheta < 90°\right)$$

$$F_v = -F_c \cdot \cos\vartheta + W\left(\text{from } \vartheta > 90° \text{to } \vartheta < 270°\right)$$

The minus sign of vertical forces obeys the convention previously mentioned that forces pointing downward are positive, and those pointing upward are negative.

In Figure 6.8, the total force's vertical component, corresponding to the last example, is represented against time in cartesian coordinates. It shows that the maximum compression effort is 316.8 Tn, which means 158.4 tonnes per pedestal if the turbomachine has two pedestals. As was said before, this force is unacceptable.

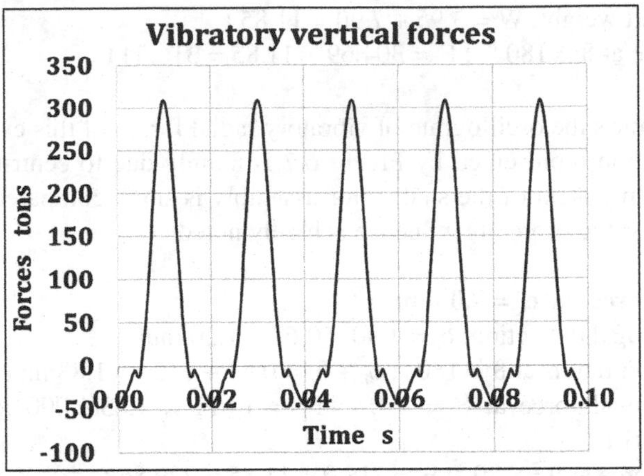

FIGURE 6.8 Vibratory vertical forces.

Otherwise, the load is distributed among all pedestals according to the wheels' load configuration. The vertical force represents a significant dynamic compression effort for the turbomachine pedestals and foundation. These forces are equivalent to a reciprocating knocking on the pedestal and foundation and must be considered at the turbomachine design stage, especially when designing the pedestals, journals, and foundations. Efforts transmitted to the pedestal are, in many cases, isolated by mounting the turbomachine on springs or rubber rings around the journals or bearings. There are many types of commercially available isolators of all sizes.

The ripple shown in Figure 6.8 is due to the radial component of the rotor-shaft weight. The maximum vertical force must be compared with the soil resistance to determine if it is strong enough to minimize the potential foundation settlement due to these forces. However, it is difficult to prevent the foundation from displacing vertically because the machine and foundation weights reduce the air to the soil's void volumes ratio.

6.4.1 Horizontal Vibratory Force

The horizontal component of the radial force may be positive or negative. The difference is irrelevant because it only says that positive horizontal forces are in the first and second quadrant, pointing from left to right. Conversely, negative horizontal forces are in the third and fourth quadrant, pointing from right to left. The horizontal force formula is:

Formula 6.6 Horizontal force component of F_t

$$F_h = F_c \cdot \sin\vartheta = F_c \cdot \sin(\omega \cdot t)$$

The horizontal force applies an overturning moment and force to the turbomachine pedestals and applies horizontal forces to both the foundation and the soil's foundation. The detailed formula of the centrifugal force is:

Formula 6.7 Extended form of the centrifugal force

$$F_c = m\cdot\left[1+Q_c(\omega)\right]\cdot\left[e_x-\delta_b\cdot\cos(\omega\cdot t)\right]$$

The bending deflection δ_b is moved out from the second term between brackets to derive Formula 6.8 to calculate the horizontal force. Therefore, the complete expression of the horizontal force is formed by a two-term product, shown in Formula 6.8. The first term between curly brackets is the centrifugal force $F_{c\pi 0}$ pointing downward at $\vartheta = 180°$ for $e_x = 0$. This force is ideal because zero eccentricity is very unlikely to happen. The factor $h(\omega\cdot t)$ converts the centrifugal force $F_{c\pi 0}$ into the horizontal force $F_h(\omega\cdot t)$.

Formula 6.8 Vibratory horizontal force

$$F_h(\omega\cdot t) = F_{c\pi 0} + h(\omega\cdot t) = \left\{m\cdot\left[1+Q_c(\omega)\right]\cdot\delta_b\cdot\omega^2\right\}\cdot\left\{\left[\frac{e_x}{\delta_b}-\cos(\omega\cdot t)\right]\cdot\sin(\omega\cdot t)\right\}$$

Therefore, Formula 6.8 can be split as follows:

Formula 6.9 Horizontal force versus time and frequency

$$F_{c\pi 0}(\omega) = m\cdot\left[1+Q_c(\omega)\right]\cdot\delta_b\cdot\omega^2$$

$$h(\omega\cdot t) = \left[\frac{e_x}{\delta_b}-\cos(\omega\cdot t)\right]\cdot\sin(\omega\cdot t)$$

$$F_h(\omega\cdot t) = h(\omega\cdot t)\cdot F_{c\pi 0}(\omega)$$

Formula 6.9 expresses that the centrifugal force $F_{c\pi 0}$ is constant if rotation velocity is unchanged; however, $h(\omega\cdot t)$ is a harmonic function indicating that the horizontal force is vibrating at frequency ω. At $t = 0$, this function is zero, and it is also zero at any time that $\sin(\omega\cdot t)$ is zero. The presence of the reciprocating radial component $\delta_{br}(\omega\cdot t)$ precludes the motion to follow a pure sinusoidal curve, as shown in Figure 6.9.

The above discussion concludes that the horizontal force alternately pushes the pedestals toward their left and right sides in each shaft turn. If abnormal operating conditions happen, such as a blade rupture, uneven heating, etc., the

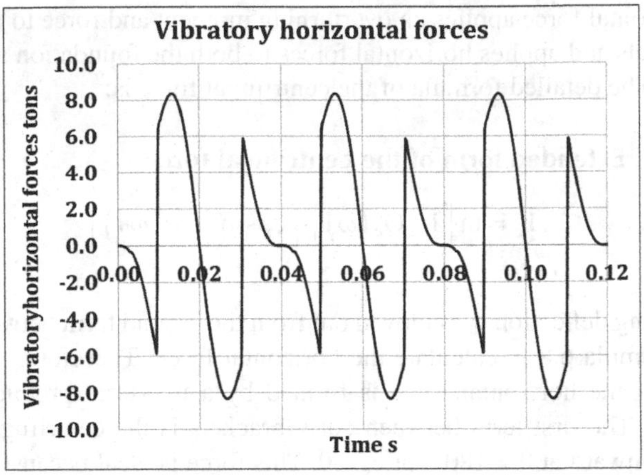

FIGURE 6.9 Oscillogram of h(ω·t) ratio and horizontal vibratory forces.

rotor unbalance increases. Therefore, pedestals and the turbomachine frame and foundation are subject to oscillating rollover moments that may harm them. Furthermore, as horizontal forces induce transversal vibration, pedestals' steel is exposed to fatigue and cracks. Therefore, it is recommended the periodic balancing of unbalanced rotors to reduce vibratory forces. This procedure extends the turbomachine lifetime.

6.4.1.1 Maximum Horizontal Force

The maximum horizontal force angle is obtained by the procedure of replacing in Formula 6.8 the (ω·t) product by ϑ, and then taking the derivative of this formula with respect to ϑ and equating the result to zero. At a constant velocity ω, this mathematical operation returns the following equation to calculate the maximum horizontal force angle:

Formula 6.10 Derivative of horizontal force formula

$$\frac{dF_h(\vartheta)}{d\vartheta} = F_{c\pi 0}(\omega)\cdot\left(\frac{e_x}{\delta_b}\cdot\cos\vartheta - \cos^2\vartheta + \sin^2\vartheta\right) = 0$$

Therefore, the ratio $h(\omega \cdot t) = F_h(\omega \cdot t)/F_{c\pi 0}(\omega)$ is maximum at the angle given by the following equation:

Formula 6.11 Equation to calculate the angle ϑ of maximum horizontal force

$$\frac{e_x}{\delta_b}\cdot\cos\vartheta - \cos^2\vartheta + \sin^2\vartheta = 0$$

This equation returns two angles at which the horizontal force is maximum. The first angle is:

Formula 6.12 Angle of maximum horizontal force

$$\vartheta_{\max 1}\left(\frac{e_x}{\delta_b}\right) = \arccos\left\{0.25 \times \left[\frac{e_x}{\delta_b} - \sqrt[2]{\left(\frac{e_x}{\delta_b}\right)^2 - 8}\right]\right\}$$

The second angle is $\vartheta_{hmax2} = 2 \cdot \pi - \vartheta_{hmax1}$. The absolute value of the horizontal force is the same for ϑ_{hmax1} and ϑ_{hmax2}. This means that these two forces are equal and have opposite directions. By replacing any of the angles ϑ_{hmax} in the formula of $h(\omega \cdot t)$, the maximum ratio $h(\omega \cdot t)$ is obtained. This maximum $h(\omega \cdot t)$ times $F_{c\pi 0}$ is the maximum horizontal force. The above formula demonstrates that the maximum h ratio only depends on the e_x/δ_b ratio. Therefore, the maximum horizontal force is calculated with the following formula:

Formula 6.13 Maximum horizontal force

$$h_{\max} = \left[\frac{e_x}{\delta_b} - \cos\vartheta_{hmax}\right] \cdot \sin\vartheta_{hmax}$$

$$F_{hmax} = h_{\max} \cdot F_{c\pi 0}$$

Alternatively, Figure 6.10 can be used to calculate ϑ_{hmax} and h_{max}. With the e_x/δ_b ratio, the ϑ_{hmax1} is given on the left vertical axis using the thin curve.

With the same e_x/δ_b ratio, the maximum $h(\omega \cdot t)$ is read on the right vertical axis using the thick curve. For example, for an e_x/δ_b ratio equal to 0.9, the maximum

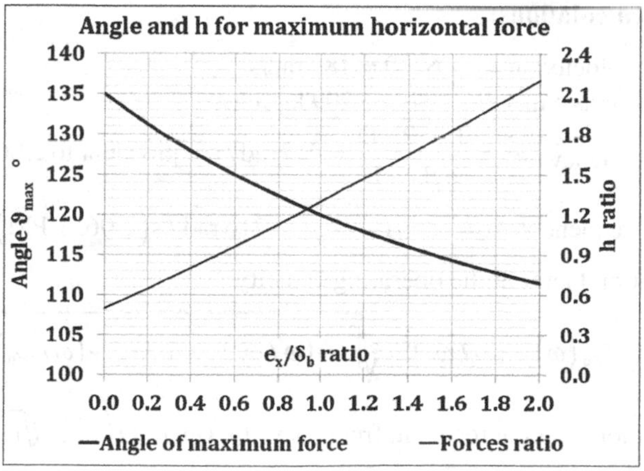

FIGURE 6.10 Angle of maximum horizontal force and maximum forces ratio.

angle is produced at 121.5°, and the maximum h ratio is 1.20. For zero eccentricity, the maximum horizontal force angle is 135°, and the maximum h ratio is 0.5. It means that the maximum horizontal force is equal to 50% of the centrifugal force $F_{c\pi0}$. For any e_x/δ_b ratio higher than 0.65, the horizontal force is higher than the centrifugal force $F_{c\pi0}$.

The h ratio curve is a right line ($R^2 = 0.999$); therefore, as a reasonable approach, the maximum h ratio can be calculated with: $h_{max} = 0.86 \times (e_x/\delta_b) + 0.46$.

Figure 6.10 indicates that the maximum horizontal forces are produced when the centrifugal force angle ϑ is between 90° and 270°, in the second and third quadrant.

6.4.2 Assessment of Vibratory Forces on Pedestals

The following example calculates the maximum vertical and horizontal forces on a turbomachine's pedestals at two different velocities: a) at rated velocity and b) at resonance velocity.

Technical specifications:

Application: power plant.
Generation frequency = 60 Hz
Shaft length L = 3.0 m
Shaft diameter D = 0.40 m
Steel modulus E = 190 GPa
Eccentricity $e_x = 0.15$ mm
Static bending deflection $\delta_b = 0.171$ mm
Shaft diameter D =0.4 m
Shaft length L = 4 m
Shaft and wheels weight W =11,853 kg. Assume this is a concentrated load
Shaft rated velocity N = 1,800 RPM
Friction ratio $\zeta = 0.05$

Previous calculations:

Operating velocity: $\omega = \pi \cdot N/30 = 189$ rad/s
Rotor-shaft mass: m = W/9.81 = 1,209 kg mass

Natural frequency: $\omega_n = \dfrac{1{,}250}{\sqrt[2]{\beta_w}} \cdot \dfrac{D}{L^2} = 217\,\text{rad/s, equivalent to}\,2{,}067\,\text{RPM}$

Resonant frequency: $\omega_r = \omega_n \cdot \sqrt[2]{1 - 2\cdot\zeta^2} = 216\,\text{rad/s}\,(2{,}062\,\text{RPM})$

Magnification factor at the operating velocity:

$$Q_c(\omega) = (\omega/\omega_n)^2 / \sqrt[2]{\left[1 - (\omega/\omega_n)^2\right]^2 + 4\cdot\zeta^2\cdot(\omega/\omega_n)^2} = 2.95$$

Magnification factor at resonant frequency: $Q_c(\omega_r) = 1/\left(2\cdot\zeta\cdot\sqrt[2]{1-\zeta^2}\right) = 10.01$

Resonance risk range: $0.7 \cdot \omega_r$ to $1.3 \cdot \omega_r$ that is: 151 to 281 rad/s. As $\omega = 189$ rad/s, the operating velocity is within the risk range.

Maximum vertical forces calculation ($\vartheta = 180°$):

At rated velocity: $F_{v\,max} = W + m \cdot \left[1 + Q_c(\omega_s)\right] \cdot (e_x + \delta_b) \cdot \omega_s^2 / 1000 = 144\,t$

At resonance velocity: $F_{v\,max} = W + m \cdot \left[1 + Q_c(\omega_r)\right] \cdot (e_x + \delta_b) \cdot \omega_r^2 / 1000 = 496\,t$

This formula is used to compute the curve $F_{vmax}(\omega)$ in Figure 11.12.

Maximum horizontal forces calculation

e_x/δ_b ratio = 0.239

The angle of maximum horizontal force at

$$\vartheta_{h\,max} = \arccos\left\{-\left[-\frac{e_x}{\delta_b} + \sqrt[2]{\left(\frac{e_x}{\delta_b}\right)^2 + 8}\right]/4\right\} = 2.29 \text{ rad or } 131.1°$$

Conversion factor: $h = \frac{F_{h\,max}}{F_{c\pi 0}} = \left(\frac{e_x}{\delta_b} - \cos\vartheta_{h\,max}\right) \cdot \sin\vartheta_{h\,max} = 0.675$

Centrifugal force for eccentricity $e_x = 0$: $F_{c\pi 0} = m \cdot [1 + Q_c(\omega)] \cdot \delta_b \cdot \omega^2/1{,}000 = 106.7$ t

Horizontal force at rated speed: $F_{h\,max} = F_{c\pi 0} \cdot h = 72.0$ t

Horizontal force at rated resonance speed:

$F_{h\,res} = m \cdot \left[1 + Q_c(\omega_r)\right] \cdot \delta_b \cdot \omega_r^2 \cdot h / 1{,}000 = 263.7\,t$

This formula is used to compute the curve $F_h(\omega)$ in Figure 6.11. As this formula is valid for any frequency ω, ω_r is replaced by ω.

The above forces are used to calculate the stress of bearings, pedestals, and foundation. Due to the magnitude of the force returned by the above calculation, it is anticipated that they are unacceptable, and some mitigating measures must be taken in the design stage. The force transmitted from the foundation to the

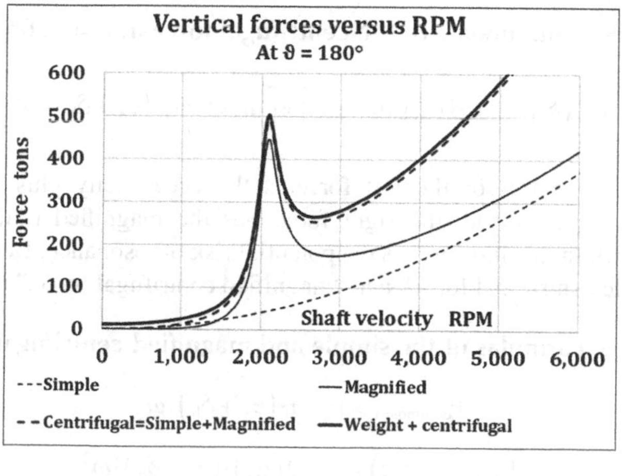

FIGURE 6.11 Vertical forces at 180° versus shaft speed.

soil is formed by the foundation, machine weight, and the vertical and horizontal dynamic forces. The soil's capacity must support all these forces.

6.5 FREQUENCY RESPONSE OF VIBRATORY FORCES

This section describes the study of forces versus frequency for turbomachines with an unbalanced rotor. The forces to be considered are dynamic because of the changing centrifugal forces applied during rotation. The main purpose of the frequency response study is to calculate the efforts produced under resonance conditions. Usually, the frequency response model is prepared for the worst-case scenario using the maximum vertical and horizontal forces.

During startup or under changing velocities, service, pedestals, and foundation are exposed to vibration forces of different frequencies and often to resonance. Therefore, it is important to know the amplitude of the forces versus frequency to assess the installation's mechanical integrity risks when the turbomachine velocity is changed or subject to speed transients. The vertical and horizontal forces Formulas 6.5 and 6.9 allow representing the forces as a function of the shaft speed and assess the riskiest shaft velocities.

6.5.1 FREQUENCY RESPONSE OF THE VERTICAL FORCE

The most stringent condition is when the centrifugal force is pointing downward, given by the following formula:

Formula 6.14 Maximum vertical force as a function of shaft velocity ω

$$F_{v\,max}(\omega) = W + F_{c\pi}(\omega) = W + m \cdot \left[1 + Q_c(\omega)\right] \cdot (e_x + \delta_b) \cdot \omega^2$$

It is possible to identify two centrifugal forces in this formula, expressed between brackets in the following expression:

Formula 6.15 Components of the centrifugal force at ϑ = 180°

$$F_{c\pi}(\omega) = \left[m \cdot (e_x + \delta_b) \cdot \omega^2\right] + \left[m \cdot Q(\omega) \cdot (e_x + \delta_b) \cdot \omega^2\right]$$

The gyration radius of the left force is the eccentricity plus the bending deflection ($e_x+\delta_b$). Instead, the right force has the magnified radius equal to $Q(\omega) \cdot (e_x + \delta_b)$. In this last force is the potential risk of resonance. Both forces are called "simple centrifugal force" and "magnified centrifugal force," respectively.

Formula 6.16 Formulas of the simple and magnified centrifugal force

$$F_{cv\,simple}(\omega) = m \cdot (e_x + \delta_b) \cdot \omega^2$$

$$F_{cv\,magnified}(\omega) = m \cdot Q(\omega_s) \cdot (e_x + \delta_b) \cdot \omega^2$$

The simple force parabolically increases with the shaft velocity and, theoretically, may reach an infinite value at an infinite frequency, which, of course, is impossible in practice. However, beyond the nominal speed value (1,800 RPM, for example), this force may seriously harm the turbomachine. For this reason, the turbomachine over-speed due to the sudden disconnection of a significant electrical load should always be avoided to prevent damages to bearings and pedestals. See the chart in Figure 6.11.

The magnified centrifugal force abnormally increases in two cases: close to the resonant frequency or extremely high speed. In Figure 6.11, the rated velocity (1,800 RPM) is lesser than the resonance speed. Therefore, this a rigid rotor.

The example shown in Figure 6.11 represents the turbomachine magnification factor derived in Section 6.5.2. Under the resonance conditions, the scenario changes abruptly. The maximum vertical force escalates to 500 tonnes at resonance. This effort is an abnormally high force that will produce severe damages to the turbomachine.

Machines should always have overspeed detectors and automatic shut-off. If the turbomachine is driving a generator, a sudden electrical load shedding may produce a serious accident. As soon as disconnection happens, the generator resisting torque disappears, and the turbomachine tends to accelerate. However, the over-speed protection system must prevent this type of damages by shutting off the fuel or steam valve.

6.5.2 Frequency Response of the Horizontal Force

The calculation of horizontal forces intensity due to changing shaft speed is done with Formula 6.17, converted into the following form:

Formula 6.17 Maximum horizontal force versus frequency

$$F_{h\,max}(\omega_s) = m \cdot \left[1 + Q_c(\omega)\right] \cdot \delta_b \cdot \omega^2 \cdot h = F_{c\pi 0}(\omega) \cdot h(\omega)$$

This study is based on the maximum horizontal force, which happens at $\vartheta_{h\,max}$. Just as was done for the vertical forces in Section 6.5.1, the horizontal force is decomposed into two forces: the simple and the magnified forces, as shown in the following expressions:

Formula 6.18 Formulas of the simple and magnified centrifugal horizontal forces

$$F_{ch\,simple}(\omega_s) = m \cdot \delta_b \cdot \omega^2 \cdot h(\omega)$$

$$F_{ch\,magnified}(\omega_s) = m \cdot Q(\omega) \cdot \delta_b \cdot \omega^2 \cdot h(\omega)$$

See in Figure 6.12 the frequency response curves of horizontal forces. They show a resonance peak and the quadratic growth for frequencies higher than the resonant frequency.

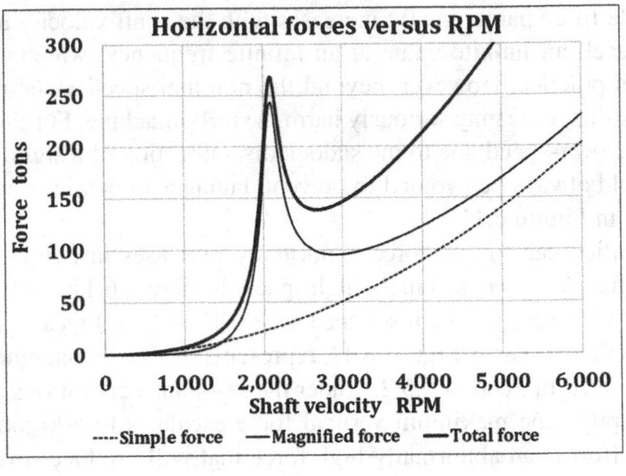

FIGURE 6.12 Horizontal forces at ϑ = 180°.

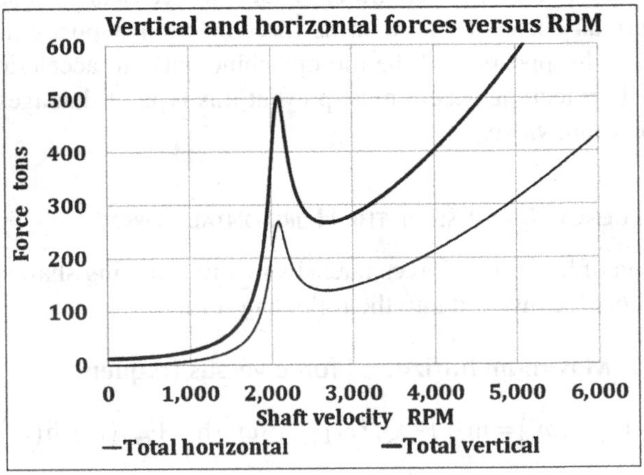

FIGURE 6.13 Comparison of vertical and horizontal forces.

Figure 6.13 shows a comparison between the frequency curves of the vertical and horizontal forces. Both present a resonance peak at the same frequency and a quadratic growth at higher frequencies.

Figures 6.12 and 6.13 show that this example's turbomachine cannot operate at post resonance velocities because forces are significantly high. Solving these problems is usually costly. The turbomachine's reinforcement to withstand dynamic forces will likely demand an unaffordable amount of money. Therefore, to mitigate these dynamic forces, they must be calculated in the design stage to create a solution before the turbomachine is manufactured.

6.6 BLADE SUBJECT TO IMPULSE FORCE

Blade or rotor vibration is one of the most significant problems for turbomachine mechanical integrity. In general, blade vibration is due to turbulent or non-uniform inside the turbomachine. This section refers to a simplified model of blade vibration, assuming that the flow impact is the cause of vibration every time the blade is exposed to the nozzle flow output[2]. It is not an exact model but helps to grasp this important aspect of a turbomachine[3].

Blades are always subject to centrifugal force during the rotor-shaft gyration. When the blade is in front of a nozzle, another force arises that is the flow impact. Centrifugal force tends to stretch the blade, which must not produce stress beyond the steel elastic limit. The flow impact produces a short-duration dynamic force. This impact is a source of vibration. However, as the root/blade structure is rigid enough, no plastic strain happens under normal operation. Another essential feature of blade configuration is that the force magnitude is proportional to the second power of the blade's tip radius (r_t) minus the second power of the root radius (r_r). This dependence anticipates significant centrifugal forces in long blades, such as those installed in the turbine's low-pressure zone.

The sketch of Figure 6.14 shows the centrifugal forces acting on the blades. Under ideal conditions, all centrifugal forces are balanced (see the left schematic); therefore, no transversal forces are generated.

Should one or more blades fail, an unbalanced centrifugal force arises, as shown on the right side of Figure 6.14. This force produces lateral vibration, and therefore, may inflict severe damages to the turbine. Because of centrifugal forces' efforts and vibration, blades are shrouded to prevent displacements during the operation. However, shrouding may fail and produce severe accidents because of an unbalanced rotor. As the normal operation is severely affected by the blade failure, the turbine should be stopped to fix it as soon as it is detected.

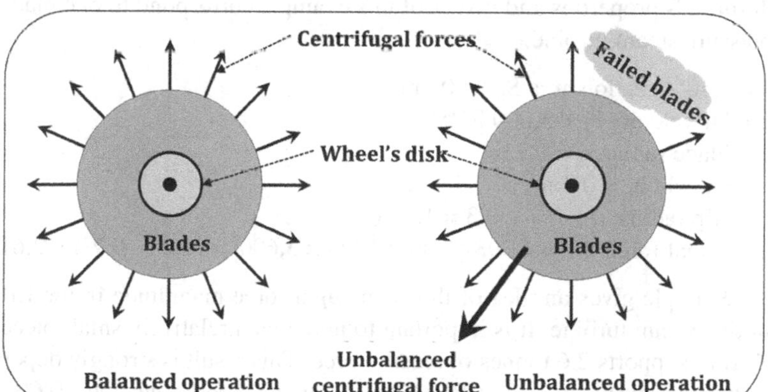

FIGURE 6.14 Balanced and unbalanced centrifugal forces.

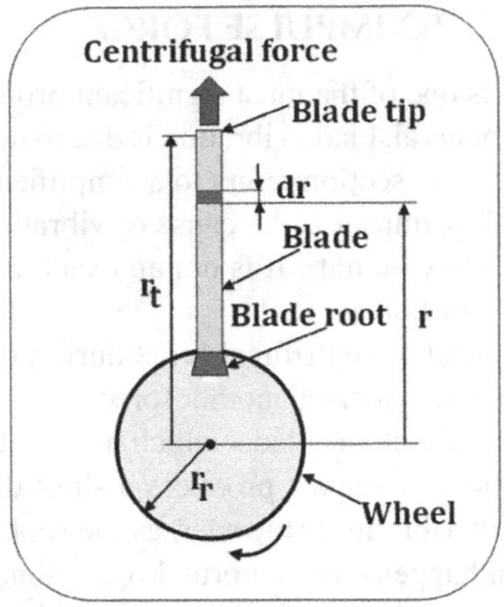

FIGURE 6.15 Scheme of a blade.

If one blade fails, the centrifugal force on the opposed blade is not compensated, as shown in Figure 6.14; therefore, the shaft endures a rotating force that tends to increase the shaft deflection. This centrifugal force is a significant vibratory force, as proved in the following example, which shakes the shaft at a frequency equal to the shaft RPM.

The centrifugal force on a steel blade is calculated with: $F_c = 4.398 \times S_b \cdot N^2 \cdot \left(r_t^2 - r_r^2\right)$. See in Figure 6.15 a sketch of a blade that explains r_t and r_r. S_b is the blade cross-section, and N is the shaft RPM.

6.6.1 Example of Centrifugal Force on a Blade

This formula's properties and values of this example correspond to one blade in a low-pressure steam turbine.

Blade cross-section area $S_b = 90$ mm^2
Angular velocity N = 3,600 RPM
Root blade radius $r_r = 0.7$ m
Blade height $h_b = 0.3$ m
Blade tip radius: $r_t = 0.7 + 0.3 = 1.0$ m
Centrifugal force: $Fc = 4.398 \times (90 \times 10^{-6}) \times 3{,}600^2 \times (1.0^2 - 0.7^2) = 2{,}616$ kg

This example gives an idea of the centrifugal force magnitude in the LP section of this steam turbine. It is important to note that a relatively small piece of a blade's root supports 2.6 tonnes of tensile force. This result is strongly dependent on the turbine speed. At 3,000 RPM, the centrifugal force will be 30.6% less, approximately 1.8 tonnes. The calculated centrifugal force produces a significant effort on the shaft, adding more deflection and increasing the centrifugal force due to this additional deflection. Due to a failed blade, the centrifugal force may reach values higher than 100 tonnes, making impossible the turbine operation.

6.6.2 Vibration Produced by the Flow Impact on Blades

Flow passing through blades is a fast vibration source due to its turbulence inside the blade and the gap between the nozzle, the blade, and the flow impact periodicity. The flow turbulence creates a blade vibration of low amplitude, which is harmful because it produces fatigue, the primary cause of blade collapse.

The blade mass and rigidity oppose the flow impact. Therefore, as the first approach to vibration is produced by this impact, the blade is modelled as a second-order system with no friction in this text. This system receives a periodic impulse force during rotation equal to as many impacts per turn as the nozzles' quantity installed around the wheel. Therefore, according to Chapter 1, the blade response is oscillatory with extremely small friction ($\zeta \cong 0$). The shrouding increases the blade's rigidity; hence the resonance is produced at high frequencies, and the shroud restrains the vibration amplitude. Nonetheless, the shroud does not prevent the blade.

The successive impulses may be considered a harmonic excitation of frequency equal to the nozzles' quantity times the wheel's angular velocity. Therefore, the frequency formula is:

Formula 6.19 Blade excitation frequency in rad/s

$$\omega = \frac{\pi}{30} \cdot \mathrm{N} \cdot \mathrm{z}$$

N is the wheel RPM, and z is the total quantity of nozzles around the wheel.

The impulse force and flow variation changes between the blades, make the blade vibrate. The blade is strongly built in the wheel hub, and a shroud tightly grabs its tips, then each blade is considered a clamped-clamped beam in this simplified model.

The flow force is an impulse of an extremely short duration calculated with Euler's Formula 6.1. However, due to its neglectable duration, it does not participate in the equation of motion. Theoretically, its value is zero after the impulse has happened at the instant $t = 0$. Therefore, the flow force must be multiplied by the unitary impulse function 1(t), which is infinite at $t = 0$ and is zero for any $t > 0$. The integral of 1(t) is equal to 1. As the equation of motion must be valid for $t > 0$, the blade's inertia and rigidity reactions are equal to zero, anticipating that the solution to the equation of motion is the blade's natural response. Therefore, as the first approach to a frictionless blade vibration, the following equation of motion may be considered:

Formula 6.20 Simplified form of a blade equation of motion

$$\mathrm{m_b} \cdot \ddot{\delta}(\mathrm{t}) + \mathrm{k_b} \cdot \delta(\mathrm{t}) = 0$$

In this formula, m_b is the mass blade, k_b is the blade rigidity and δ the blade's vibration amplitude. The solution is an undamped harmonic function, discussed in Chapter 1.

As the blade inevitably vibrates, it must be determined if the impulse force's frequency may produce the blade resonance at any of its natural frequencies. The methods used in Chapter 1 return the following SDOF model, which corresponds to a second-order system with zero damping ratio. As was seen in Chapter 1, the system has an initial static deflection $\delta_0 = F_f/k_b$:

Formula 6.21 SDOF model of a blade frequency response

$$\delta(\omega) = \frac{k_b \cdot \delta_0}{-m_b \cdot \omega^2 + k_b} = \frac{\omega_n^2 \cdot \delta_0}{\omega_n^2 - \omega^2}$$

In this SDOF system, and the natural frequency is $\sqrt[2]{k_b / m_b}$; therefore, the blade may experience resonance at its natural frequency, excited by the flow impulse force. However, the blade is a continuous system whose natural frequencies are higher than this formula. The application of the Euler-Bernoulli equation provides Formula 6.22 for the blade's natural frequencies:

Formula 6.22 Blade's natural frequencies

$$\omega_{nj} = \frac{\alpha_j}{L^2} \cdot \sqrt[2]{\frac{E \cdot I_z}{\rho \cdot A}}$$

The values of α_j for a clamped-clamped beam are in Table 3.2. This formula's result must be compared with the exciting frequency given by Formula 6.19 to determine resonance risks.

A great number of factors affect vibrations in turbine blades. Temperature variation from ambient to operating conditions with the associated changes in material properties can change the blade's natural frequencies up to 20%. The stiffening effect of centrifugal force can significantly alter blade resonant frequencies with small engine speed changes. Natural frequencies can also be altered by blade erosion and corrosion and deposits on blades, such as saltwater ingestion and soot in the turbine section.

6.6.3 Assessment of Blades Resonance Risk

To illustrate a turbine's blades' resonance risk assessment, assume that jets impacting the blades come from six nozzles. The blades section is rectangular. This shape is not realistic, but it is a simplification that helps to create the model. The blades have the following dimensions:

Height: L = 0.8 m
Width: b = 70 mm
Thickness: a = 60 mm
Nozzle quantity: z = 6
Shaft velocity: N = 3,000 RPM
Steel density: ρ = 800 kg mass/m^3

- **Calculation as a continuous system:**
  Excitation frequency:

$$\omega = \frac{\pi}{30} \cdot N \cdot z = \frac{\pi}{30} \cdot 3{,}000 \cdot 6 = 1{,}885\,\text{rad/s}$$

Young's modulus: 200 GPa, equivalent to 200 × 102 × 10^6 = 20,400 × 10^6 kg/m^2

Moment of inertia: $I_z = \frac{1}{12} \cdot b \cdot a^3 = \frac{1}{12} \times 0.050 \times 0.060^3 = 9 \times 10^{-7}\,m^4$

Blade cross section: A = a · b = 0.05 × 0.06 = 0.003 m^2

Blade rigidity: $E \cdot I_z = 20{,}400 \times 10^6 \times 9 \times 10^{-7} = 18{,}360$ kg · m^2

Natural frequencies:

$$\omega_{nj} = \frac{\alpha_j}{L^2} \cdot \sqrt[2]{\frac{E \cdot I_z}{\rho \cdot A}} \quad \omega_{nj} = \frac{\alpha_j}{0.8^2} \cdot \sqrt[2]{\frac{18{,}360}{800 \times 0.0042}} = \alpha_j \times 116$$

$$\omega_{n1} = 22.0 \times 116 = 2{,}541\ \text{rad/s}$$

The first natural frequency is close to the excitation frequency. The difference is 35%.

$$\omega_{n2} = 61.7 \times 116 = 7{,}157\ \text{rad/s}$$

The conclusion is that the first natural frequency may create risks of resonance.

- **Calculation as SDOF system**
  Clamped-clamped coefficient $\beta_w = 1/192$

Rigidity coefficient: $k_w = \frac{E \cdot I_z}{\beta_w \cdot L^3} = \frac{192 \times 18{,}360}{0.8^3} = 6.885 \times 10^6\,\text{kg/m}$

Blade mass: m = ρ · a · b · h = 800 × 0.06 × 0.07 × 0.8 = 2.688 kg mass

Natural frequency: $\omega_{n\,SDOF} = \sqrt[2]{\frac{k_w}{m}} = \sqrt[2]{\frac{6.885 \times 10^6}{2.688}} = 1{,}600\,\text{rad/s}$

This frequency is too close to the excitation frequency but is not as realistic as the frequency ω_{n1} calculated before.

6.7 ROTOR-SHAFT SUBJECT TO PULSATING TORQUE

Rotor-shaft torsional vibration can be studied on the basis of the Jeffcott rotor model[4]. This model is a simplified one-wheel rotor installed on a uniform rigid but massless shaft with eccentricity. The rotor, in its most simple expression, is a hinged-hinged beam. However, the model can be converted into a more sophisticated version assuming that the extremes are linked to the ground by spring-damper sets that allow a model with more degrees of freedom. Again, the model must be considered as a first approach to the rotor-shaft torsional vibration[5].

As the shaft has eccentricity, it is assumed that the centrifugal force is exciting the system; therefore, the equation of motion of this simplified model is:

Formula 6.23 Equation of motion due to intermittent torque

$$J_p \cdot \ddot{\vartheta} + f_t \cdot \dot{\vartheta} + k_t \cdot \vartheta = m \cdot e_x \cdot e^{j\cdot\omega\cdot t}$$

J_p is the rotor moment of inertia, f_t is the rotating friction coefficient, and k_t is the shaft torsional rigidity.

The solution to this equation of motion was seen in Section 1.9 for linear systems. The conclusions are the same as those of the cited section. Therefore, the natural frequency of the rotor shaft is Formula 1.11.

Another simplified model arises assuming that a non-uniform flow transmits a pulsating torque to the rotor-shaft, as seen in the last section. Flow forces on the blade are calculated with Formula 6.1. However, it is assumed that the applied flow force on the blade is intermittent because of the gap between blades. As a result of this intermittent force, the applied torque is also intermittent, resisted by the rotor-shaft inertia, damping and rigidity generating torsional vibrations. Considerations about the impulse force on blades were done in Section 6.6.

The torque produced by this intermittent force on the blade is derived from Formula 6.1. However, its influence on the driving torque is small and likely negligible.:

$$T_{qi} = G_m \cdot V_g \cdot r_w \cdot 1(t)$$

In these formulas, T_{qi} is the intermittent torque produced by the flow on the blades. r_w is the turbomachine's wheel radius. 1(t) is the unitary impulse, that is, zero for $t > 0$. Therefore, the equation of motion is:

Formula 6.24 Equation of motion for an impulse force excitation

$$J_p \cdot \ddot{\vartheta} + f_t \cdot \dot{\vartheta} + k_t \cdot \vartheta = 0$$

This vibration frequency is calculated with the same formula used in Section 6.6 for blade vibration but multiplying by the blade's quantity b and not by the nozzles' quantity z. Therefore, the shaft-rotor torsional vibration has a frequency b/z higher than the blade's vibration, making a significant difference.

Formula 6.25 Rotor-shaft excitation frequency in rad/s

$$\omega = \frac{\pi}{30} \cdot N \cdot b$$

Therefore, the intermittent torque makes the rotor shaft endure a torsional vibration. Theoretically, these pulses of damped natural vibration are produced at a constant interval, that is, the period of the frequency ω.

Formula 6.26 Period between pulses of torsional vibration

$$T = \frac{120 \cdot \pi}{b \cdot RPM}$$

The blade gaps are small; therefore, the pulses sequence may be considered a permanent frequency of low amplitude. Forces of this vibration are not always significant. They are known as passing vibration. Low integer multiples of the fundamental frequency may occasionally create problems additional to those generated by the flow dynamics that complicate the turbine vibration analysis.

NOTES

1 Recommended books for Chapter 6: 1. Handbook of Rotordynamics by Fredric F. Ehrich. McGraw Hill, Inc, 1992 – 2. Gas Turbine Engineering Handbook by Meherwan Boyce. Elsevier, 4th edition – 3. Gas Turbine Theory by Herb Saravanamutto, Gordon Rogers, Henry Cohen, Paul Straznicky. Person, Prentice Hall, 6th edition, 2009 – 3. Roark's formulas for Stress and Strain by Warren Young, Richard Budynas and Ali Sadegh. McGraw Hill 8th edition, 2012 – 4. Mechanical Vibration by Haym Benaroya, Mark Nagurka and Seon Han. CRC Press. Taylor and Francis Group, fourth edition 2018– 5. The Design of High-Efficiency Turbomachinery and Gas Turbines, by David Gordon and Theododios Korakianitis – 6. Strength of Materials by Stephen Timoshenko. CBS Publishers & Distributors 3rd edition, Reprint 2002.

2 Consult the book Gas Turbine Theory by Herb Saravanamutto, Gordon Rogers, Henry Cohen, Paul Straznicky. Person, Prentice Hall, 6th edition, 2009.

3 See University of Windsor Scholarship, Major Papers 7-17-1969 Damping factors in turbine blade vibration by Walter J. Pastorius. https://scholar.uwindsor.ca/cgi/viewcontent.cgi?article=7582&context=etd

4 Consult the book Handbook of Rotordynamics by Fredric F. Ehrich. McGraw Hill, Inc, 1992. Sections 1.2.1 and 1.2.2

5 See book Mechanical Vibrations by J.P. Den Hartog. Dover Publications, Inc. New York, 1985, 4th edition Section 6.1.

7 Ship's Oscillation and Vibration

7.1 INTRODUCTION TO SHIPS

A ship[1] has four fundamental parts:

1. The hull is the "ship's skin." It is formed by rectangular steel plates welded together. Hulls may be simple, like US submarines or double like Russian submarines and in some oil tankers. A double hull can better resist damages to the hull or the enemy's artillery shelling.
2. The keel and ship's ribs form the hull's supporting structure. The ribs are strongly inserted in the keel, forming a solid assembly like the spine and ribs in the human's body. This assembly supports the decks, which are horizontal plates with structural purposes to resist the ship's torsional efforts.
3. The propulsion system is formed by the main engine, usually a diesel motor, the torque transmission shafting, and the propeller. The rudder is not part of the propulsion system; its only purpose is the ship's steering. Furthermore, all ships have a power plant to supply electricity to the vessel, which in electrically propelled ships forms part of the propulsion plant.
4. The superstructure protrudes from the upper deck with different applications. One of them is the command bridge from where the ship is steered, and the ship commander or navigation officer issues orders to the rest of the crew. Another special type of superstructure is the container carrier transporting thousands of containers, forming a huge sail that tends to deviate the ship from its course due to wind impact. This effect produced the Suez channel blockade in 2021 by a carrier transporting 12,000 containers.

The supporting structure and the hull are studied as a beam. Therefore, the Mechanics of Materials' theory and formulas to calculate the stress and deformation produced by efforts apply to ships. In this study, the ship is named the ship-beam, and as it has no supports with restrictions, the free-free beam formulas are applicable.

The most important vibration sources are the propeller, the main engine, and a rough sea. These vibrations must be mitigated because of the steel fatigue, passengers' and crew's comfort, and specific technical requirements for warships and fishing ships. In a rough sea, the ship's pitch may violently slam the careen (submerged part of the ship) against the water or receive water on deck smashing the superstructure. This slamming produces the hull and structure vibration

DOI: 10.1201/9781003175230-9

and deformation. Artillery cannons' reactions strongly hit the hull girder system producing free vibration that propagates into its structure at any of its natural frequencies. When cannons are firing, the reaction forces may displace the ship several meters. A 14,000 warship of 185 m length is displaced about 3 meters.

7.2 SHIP'S PROPULSION SYSTEM

A ship's propulsion plant consists of a rotary machine (turbine or diesel engine), gearbox, shaft, and propeller. It is common to use more than one machine in parallel with another on the same gearbox. Two or more gearboxes form another layout to drive more than one propeller. Steam reciprocating machines were used in the late 19th and early 20th centuries. In 1894, the English engineer Charles Parsons invented the steam turbine. In 1897, during a naval parade in front of Queen Victoria, he presented a ship, the Turbinia, with 44 tonnes of displacement. The exhibition was a success, and the Royal Navy adopted the propulsion turbine on its ships. Some warships have gas turbines combined with diesel motors to accelerate the ship and navigate at a higher velocity during a battle. This turbine type is called combat turbine[2].

Until the 1960s, ships' propulsion plants had boilers and steam turbines, but the diesel engine began to be adopted in subsequent years for its higher efficiency. This change was boosted by the oil price increase of the 1970s. The last steamship was launched in 1984. Today, most ships are powered by diesel engines and sometimes by gas turbines. A marine propulsion diagram is in Figure 7.1[3].

The following components form a propulsion plant:

1. Main Engine
   Today, the diesel motor is the most used machine for its lower fuel consumption with respect to turbines. As diesel engines are reciprocating machines, they are a significant source of vibration.
2. Shaft
   The shaft drives the main engine torque to the propeller. Several short shafts, mechanically joined by flanges, make up the propulsion shaft. If this shaft length is significant, then torsional vibrations are large and have a high frequency.

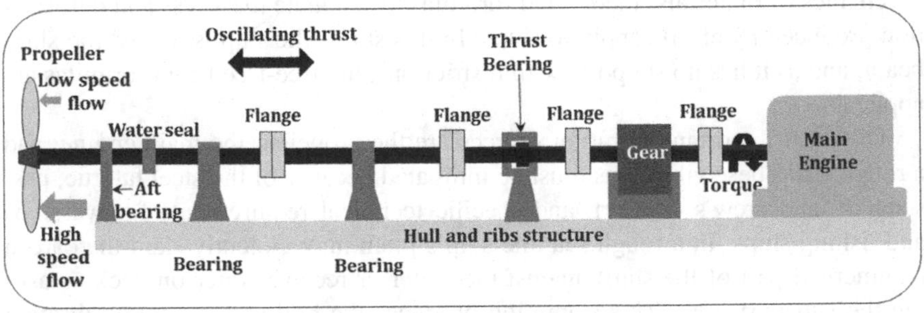

FIGURE 7.1 Scheme of a marine propulsion plant.

3. Gearbox
   The gearbox reduces the motor shaft speed to the specified propeller's rotational speed. On sailing, the propeller's RPM must be changed to adapt the ship's velocity to marine operations, such as navigate at cruiser velocity or dock the ship. They are also a vibration source, especially if teeth are worn, or the shaft is misaligned.
4. Bearings
   Their purposes are the shaft's support and alignment. The thrust bearing is a special bearing. This bearing supports the thrust created by the propeller and pushes the ship in its advancement direction. The thrust bearing is a strong structure designed to bear several tonnes of permanent and usually oscillating impulses. For example, a ship displacing 50,000 tonnes bears 460 tonnes of thrust when navigating at 30 knots (55.6 km/h).
5. Propeller
   The propeller generates a pressure difference between the inflow and outflow blades' faces. This pressure difference accelerates the water mass behind the ship and makes it move in the opposite direction. This flow is not uniform because of the different angles of attack that each blade has, and as a result, the thrust is a pulsating longitudinal force.

7.3 SHIP'S MOTIONS AND OSCILLATION

A vessel on the high sea is a free body floating in a medium of low viscosity. Therefore, it is subject to the forces of wind, currents, tides, waves, and other water movements typical of the sea's vastness. These phenomena take the ship out of balance, but various forces tend to restore it; therefore, they hardly succeed. Reactions that restore balance are intrinsic due to the ship's construction or created by controlled mechanisms that oppose the ship's oscillations. The ship moves angularly around three rotation axles, shown in Figure 7.2. Note in this figure that the arrows' size intends to represent each motion's importance on the ship's safety and comfort.

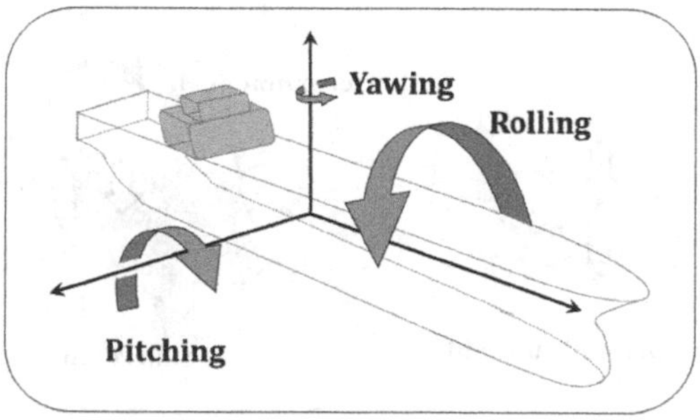

FIGURE 7.2 Ship's angular motions.

The ship's angular movements are:

a) Rolling motion around a longitudinal axis, the ship rotates around a longitudinal axis parallel to the X-axis, produced by transverse waves hitting any of the ship's sides. The roll angle may reach 40° or more under rough weather conditions. This roll angle is enlarged by the "free surfaces effect" of the storage tanks of liquids and even grains, whose displacements increase the rolling moment.
b) Pitching around a transversal axis. Long ships are prone to oscillate around a transversal axis, with pitch angles less than 5°. Instead, shorter ships reach 8°. The pitch frequency is 50% or less than the rolling frequency.
c) Yawing around a vertical axis. It is less harmful than the other two motions. This motion makes the ship deviate from the pre-set course due to the rudder's deflections and oscillations of the steering system.

7.3.1 Ship's Transversal Oscillation

According to Archimedes' principle, the ship receives an upward force equal to the careen's displaced water. In the left Figure 7.3, the careen is the shadowed area. This force is equal to the ship's weight. Therefore, two vertical forces act on a ship in equilibrium: the ship's weight and the buoyant force Δ. The buoyant force is applied to the careen's barycenter B, called buoyant center or careen center. See points B_1 and B_2 in Figure 7.3. The ship's weight is applied to the gravity center G. Both forces are always vertical regardless of the roll angle. If both centers are on the same vertical, the ship is perfectly upright. Therefore, a perfectly upright ship meets two conditions: the ship's weight is equal to the buoyant force, and the buoyant center and the gravity center are in the same vertical.

If these conditions are not met, a moment arises that makes the ship oscillate around the point M, called the metacentre. The left drawing of Figure 7.3 shows a ship perfectly upright. The right drawing displays the ship rolling. It is a harmonic motion from port to starboard and vice-versa. The barycenter position changes

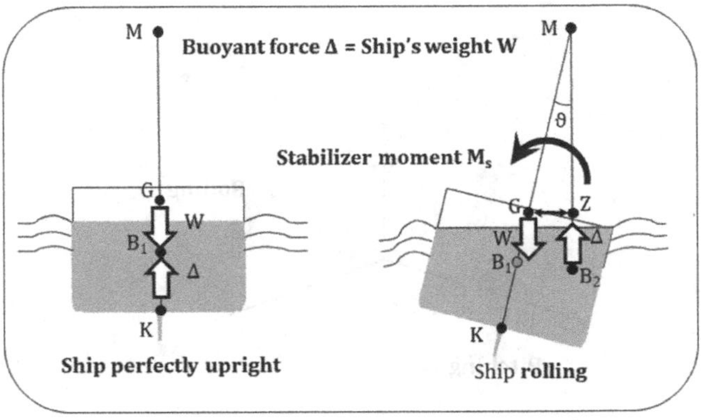

FIGURE 7.3 Ship's roll under stability conditions.

from B_1 to B_2 because the displaced volume of water has changed its shape. The vertical line that passes at B_2 intersects at the same metacentre point M. This is the point around which the ship oscillates. If the roll angle is moderate (<15°), the metacentre does not change place.

When the ship is rolling, the ship's weight and the buoyant force are no longer on the same vertical. Thus, they form a stabilizer moment M_s that tends to return the ship to the upright position. The buoyant force is now applied to point B_2. The vertical line of B_2 intersects the line of B_1G in the metacentre M. The location of this metacentre defines the ship's stability.

In Figure 7.3, there are three important points whose relative position defines whether a ship is in stable, unstable, or indifferent equilibrium. They are K, G and M. These letters stand for the keel, the gravity center, and the metacentre. The segment GM is called the metacentric height and is defined with the following formula:

Formula 7.1 Metacentric height

$$GM = KM - KGr_g = C_1 \cdot B.$$

The stabilizer moment is Δ·GZ, being GZ the righting lever. The formula to calculate this lever is GM·sin ϑ; therefore, the righting moment for this case is:

Formula 7.2 Ship's stabilizer moment

$$M_s = \Delta \cdot GZ = \Delta \cdot GM \cdot \sin \vartheta$$

Formula 7.2 is valid for ϑ angles lower than 15°. The replacement of sin ϑ by ϑ introduces an error of the order of 1.2% at 15°.

For roll angles higher than 15°, the righting lever is given by a curve of GZ versus the roll angle. The GZ curve is calculated for each ship under different conditions. Therefore, a valid curve or easy formula for all ships does not exist. As for rolling angles lower than 15°, it may be assumed that $\sin \vartheta \cong \vartheta$, Formula 7.2 may be converted in: $M_s = \Delta \cdot GM \cdot \vartheta$. Therefore, the ship is analogous to a rotating spring, whose rigidity coefficient is $\Delta \cdot GM$.

If the metacentre is between the buoyant center B and the gravity center G, the moment M_s is inverted; therefore, the ship is in unstable equilibrium that will produce the ship's collapse. In this case, the metacentric height is negative. If the metacentre coincides with the gravity center, the equilibrium is indifferent because the moment is zero. Hence, the ship's stability can be summarized as follows:

GM > 0 ship is in stable equilibrium
GM < 0 ship is in unstable equilibrium
GM = 0 ship is in indifferent equilibrium

In the stable equilibrium, the ship oscillates. In unstable equilibrium, it tends to collapse with the keel pointing upward. As an approach to the motion under stable conditions, the mechanical vibration equations are used to create a model that calculates the ship's natural frequency. The roll is an inevitable oscillation and an uncomfortable movement for the crew and passengers, although human beings usually tolerate a slow roll of small amplitude.

7.3.1.1 Roll's Natural Frequency

The natural frequency of the roll motion is derived under the following ideal conditions: a) no wind is pushing the superstructure, b) no waves or water currents are moving the ship, c) the hull's friction forces with water and air are negligible, d) the ship has rigidity reaction because GM is positive, and e) the ship is freely rolling due to some initial condition, like a force or moment of negligible duration. Under these conditions, the ship has inertial and rigidity reaction, then the equation of motion is: $J_s \cdot \ddot{\vartheta} + \Delta \cdot GM \cdot \vartheta = 0$, where J_s is the ship's mass moment of inertia with respect to the symmetry axis of the flotation area: $J_s = \int r^2 \cdot dm$. By analogy with the rotating SDOF systems, the equation of motion is:

Formula 7.3 Equation of motion of a rolling ship

$$J_s \cdot \ddot{\vartheta} + \Delta \cdot GM \cdot \vartheta = 0$$

By analogy with rotating SDOF systems discussed in Chapters 1 and 2, the ship's natural frequency may be calculated with the following formula.

Formula 7.3 Ship's natural frequency for the rolling motion

$$\omega_{n\,ship} = \sqrt[2]{\frac{\Delta \cdot GM}{J_s}}$$

The ship's polar mass moment of inertia J_s is calculated with the following formula:

Formula 7.4 Ship's mass polar moment of inertia

$$J_s = m_s \cdot r_g^{\,2}$$

The symbol r_g is the ship's gyration radius. There exists a direct relationship between the gyration radius and the ship's beam B (widest width at the waterline). This formula is derived from testing many warships and merchant ships; therefore, it is considered a good approach to the transversal radius of gyration.

Formula 7.5 Ship's gyration radius versus the ship's beam

$$r_g = C_1 \cdot B$$

C_1 is a coefficient that depends on the ship's type. Specialists have intensively discussed this coefficient. Many technical publications suggest a value between

0.38 and 0.44. The recommended paper in the footnote[4] has tables of C_1 for several types of ships. There are also some formulas to calculate C_1. The formula used by Bureau Veritas appears to have reasonable accuracy.

Formula 7.6 Bureau Veritas formula to calculate coefficient C_1 of Formula 7.7

$$C_1 = 0.289 \times \sqrt[2]{1 + 4 \cdot \left(\frac{KG}{B}\right)^2}$$

If Formulas 7.4 and 7.5 are replaced in Formula 7.3, the ship's natural frequency and the corresponding oscillation period are returned.

Formula 7.7 Ship's natural frequency at rolling and oscillation period

$$\omega_{ns} = 3.132 \times \frac{\sqrt[2]{GM}}{C_1 \cdot B} \quad T_{ns} = 2 \times \frac{C_1 \cdot B}{\sqrt[2]{GM}}$$

Formula 7.7 of the period T_{ns} reveals that wider ships have longer rolling periods. If GM and B are in meters, Formulas 7.7 return the natural frequency in rad/s and the period in seconds. The factor 3.132 is the square root of the gravity in m/s^2. The gravity appears in the derivation of Formula 7.7 because it is $g = \Delta/m_s$.

To apply these formulas, the metacentric height GM must be known. The metacentric height may be obtained from the careen's curves of attributes provided by the naval architect. These are a set of curves that show several careen's properties as a function of the ship's draught (distance between the waterline and the keel). One of these curves is the metacentric height GM.

If the careen's curves are not available, it is possible to force the ship to roll in calm sea and no wind and measure the oscillations period many times. In this case, adopting the most probable period, the metacentric height is cleared from the Formula 7.7 of T_{ns} to see the following expression:

Formula 7.8 Metacentric height versus measured oscillation period

$$GM_{test} = 4 \times \left(\frac{k}{T_{test}}\right)^2$$

Table 7.1 shows typical values of rolling periods without anti-roll devices installed.

This table may be used as a first idea of the roll period. Of course, it is recommended to calculate the roll period with the formulas mentioned above.

According to Formula 7.7, the roll period is inversely proportional to the metacentric height's square root. Therefore, there is a trade-off between the passengers' and crew's comfort and ship stability because a large metacentric height increases the uncomfortable roll frequency.

TABLE 7.1
Average Roll Period

Type of Ship	Typical Period |s|
Large passenger ship	23
Aircraft carrier	19
Medium size passenger ship	19
Cruiser warship	14
Large merchant ship	13
Destroyer	9

7.3.2 Ship's Longitudinal Oscillation

Ships can oscillate around a transversal axis, producing a motion called pitch. This oscillation study is similar to the transversal oscillation; therefore, pitch oscillation's mathematical model uses the same formulas. GM_l is the longitudinal metacentric height in the next formulas, and r_{gl} is the longitudinal radius of gyration. This radius can be estimated with the following formula:

Formula 7.9 Radius of longitudinal gyration

$$r_{gl} = 0.25 \times Lship$$

Therefore, natural frequency and pitch period are:

Formula 7.10 Ship's natural frequency when rolling

$$\omega_{nl} = 3.132 \times \frac{\sqrt[2]{GM_l}}{r_{gl}} \quad T_{nl} = 0.5 \times \frac{L_{ship}}{\sqrt[2]{GM_l}}$$

The longitudinal metacentric height is obtained from the ship's attribute curves, as seen earlier for transverse oscillations, or a longitudinal oscillation test is made up. The period of oscillations is measured in several tests. Therefore, the longitudinal metacentric height is calculated using the formula

Formula 7.11 Longitudinal metacentric height versus measured oscillation period

$$GM_{l\,test} = 0.25 \times \left(\frac{L_{ship}}{T_{test}} \right)^2$$

As the longitudinal radius of gyration is much higher than the transversal radius, the oscillation period is much longer than the transversal period. In general, the longitudinal period is 3 to 6 times the transversal period. Furthermore, longitudinal oscillations have a moderate amplitude, usually not higher than 8° in large ships. However, pitching can be dangerous if the ship's natural frequency is equal or close to the large waves' frequency hitting the ship's bow. This resonance's amplitudes have damaged a bow on many occasions because the ship bow slams against the water and vice-versa.

7.3.3 Ship's Equation of Motion

The solution to the equation of motion 7.3 reveals the rolling angle amplitude. By this equation by Js, the equation of motion is

Formula 7.12 Ship's equation of motion for rolling angles lower than 15°

$$\ddot{\vartheta}_s + \frac{\Delta \cdot GM}{J_s} \cdot \vartheta = 0$$

The coefficient of ϑ is the square of the natural frequency of Formula 7.3; therefore, the product $\Delta \cdot GM$ is the ship's rigidity to roll. It was seen in Chapter 1 that the general solution to the last equation of motion is:

$$\vartheta(t) = C_1 \cdot \cos(\omega_n \cdot t) + C_2 \cdot \sin(\omega_n \cdot t)$$

This formula is the solution to the equation of motion of a second-order frictionless system, rotating at its natural frequency. The integration constants C_1 and C_2 depend on the initial roll angle ϑ_0 and the initial angular velocity $\dot{\vartheta}_0$. Therefore, the solution to the ship's equation of motion is (see Formula 1.30):

Formula 7.13 Rolling angle oscillation formula

$$\vartheta(t) = \vartheta_a \cdot \cos(\omega_n \cdot t + \varphi)$$

Where amplitude and phase are calculated with the following formulas. These expressions need to know the ship's initial conditions ϑ_0 and $\dot{\vartheta}_0$.

Formula 7.14 Ship's oscillation when rolling

$$\vartheta_a = \sqrt[2]{\vartheta_0^2 + \left(\frac{\dot{\vartheta}_0}{\omega_n}\right)^2} \qquad \varphi = \tan^{-1}\left(\frac{\dot{\vartheta}_0}{\omega_n \cdot \vartheta_0}\right)$$

This oscillation is ideal because it does not consider the hull friction with water and the air's superstructure friction.

7.3.4 Absorption of Ship's Oscillations

There are several systems for absorbing the role of oscillations. A conceptual description of the most common systems used to mitigate the oscillations is given below.

7.3.4.1 Anti-Roll Tanks

These tanks were invented by Hermann Frahm (1867-1939) to stabilize ships sailing in a rough sea. Modern anti-roll tank systems are different from the original Frahm's design, but the physical principle is the same. In all cases, the water level inside the tank system oscillates at the sea waves frequency, the same as Frahm's mechanical absorber described in Chapter 9 ($\omega_{absorber} = \omega_{excitation}$). Today, there are sophisticated new designs of these tanks based on a mathematical model that considers all possible motions of a ship[5]. In this section, formulas and physical concepts discussed in Chapter 9 are used to model an anti-roll tank system with two degrees of freedom. The derivation assumes that the ship and tanks are SDOF systems whose equations of motion are coupled.

The equations of motion for the ship and anti-roll tanks, written as separate and independent machines (with no coupling), are:

Formula 7.15 Equations of motions of an uncoupled ship and anti-roll tanks

$$J_s \cdot \ddot{\vartheta}_s + f_s \cdot \dot{\vartheta}_s + k_s \cdot \vartheta_s = M \cdot \sin \omega \cdot t$$

$$0 = J_k \cdot \ddot{\vartheta}_k + f_k \cdot \dot{\vartheta}_k + k_k \cdot \vartheta_k$$

It is assumed that water waves are trains that induce a harmonic disturbance moment as described by this formula: $M \cdot \sin \omega \cdot t$. Subscript s stands for ship and subscript k for water inside the tanks. In this model, the wave frequency ω is assumed to be constant. The above equations presume that the ship and tank system are independent; however, when the anti-roll system is installed in the ship, a coupling arises, which converts the above equations into the following complex formulas.

Formula 7.16 Coupled equations of motion of the ship and anti-roll tanks set

$$-J_s \cdot \omega^2 \cdot \vartheta_s + k_s \cdot \vartheta_s + k_k \cdot (\vartheta_s - \vartheta_k) + j \cdot \omega \cdot f_s \cdot \vartheta_s + j \cdot \omega \cdot f_k \cdot (\vartheta_s - \vartheta_k) = M$$

$$-J_k \cdot \omega^2 \cdot \vartheta_k + k_k \cdot (\vartheta_k - \vartheta_s) + j \cdot f_k \cdot (\vartheta_k - \vartheta_s) = 0$$

The upper expression of the above formulas is the ship's equation of motion, and the other corresponds to the anti-roll tank system. The ship's equation's solution is important because it returns the roll motion damped by the anti-roll tanks. Formulas 7.16 may be simplified using the concept of mechanical impedance Z, the machine's component property opposed to the motion.

Formula 7.17 Equations of ϑ_s and ϑ_w for a ship with anti-roll tanks

$$Z_s \cdot \vartheta_s - Z_c \cdot \vartheta_w = M$$

$$-Z_c \cdot \vartheta_s + Z_k \cdot \vartheta_k = 0$$

Where:

Formula 7.18 Impedances of the ship and anti-roll tanks

$$Z_s(\omega) = -J_s \cdot \omega^2 + k_s + k_k + j \cdot \omega \cdot (f_s + f_k)$$

$$Z_c(\omega) = k_k + j \cdot \omega \cdot f_k$$

$$Z_k(\omega) = -J_k \cdot \omega^2 + k_k + j \cdot \omega \cdot f_k$$

Z_c is the coupling impedance between the ship and the water tanks, representing the tank's water friction and rigidity forces. The solution to Formulas 7.18 returns the ship's roll angle and the water's angle inside the anti-roll tanks as a function of the wave frequency ω.

Formula 7.19 Roll's amplitudes of ship and water of anti-roll tanks

$$\vartheta_s(\omega) = \frac{Z_k(\omega) \cdot M}{Z_s(\omega) \cdot Z_k(\omega) - Z_c^2}$$

$$\vartheta_k(\omega) = \frac{Z_t(\omega) \cdot M}{Z_s(\omega) \cdot Z_k(\omega) - Z_c^2}$$

The ship's roll attenuation a_r due to anti-roll tanks is:

Formula 7.20 Attenuation of the ship's roll at the sea waves frequency ω_w

$$a_{r\%}(\omega_w) = \left(1 - \frac{\vartheta_s}{\vartheta_{s0}}\right) \times 100 = \left(1 - \frac{Z_s(\omega_w) \cdot Z_k(\omega_w)}{Z_s(\omega_w) \cdot Z_k(\omega_w) - Z_t^2(\omega_w)}\right) \times 100$$

Where ϑ_{s0} is the ship roll angle with no anti-roll tanks in the ship.

The above formulas are used to display the ship and the tank's frequency response curves, where the best operating point is defined. See Figures 7.4 and 7.5 for frictionless ships and tanks. These figures have ordinates up to 120°, an impossible roll angle in any ship. Nonetheless, the ordinates have been enlarged to visualize better the ship and the tank system resonances.

The chart in Figure 7.4 displays the ship's roll angles (thick line) and the water inside the tanks (thin line). The abscissa axis is the water waves frequency, which corresponds to the exciting moment M. The dotted line is the initial roll angle, without the anti-roll tank system in operation. Any curve under this line indicates that at that frequency, the roll is attenuated. The thick line is the only one that exhibits a zero (BOP), meaning that the ship does not roll at that frequency. In this case, the frequency is 0.5 radians per second, which corresponds to 12.6 seconds between wave and wave. This frequency is the best operating point, where the natural frequency of the water oscillating inside the tanks is equal to the waves'

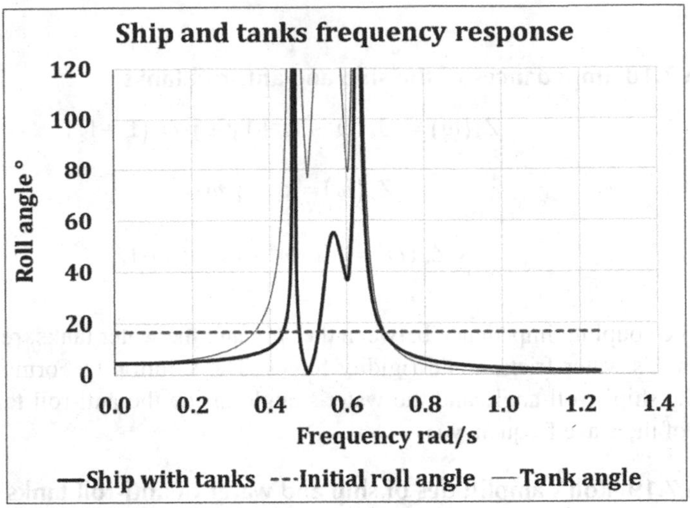

FIGURE 7.4 Frequency response curves of ship and tanks.

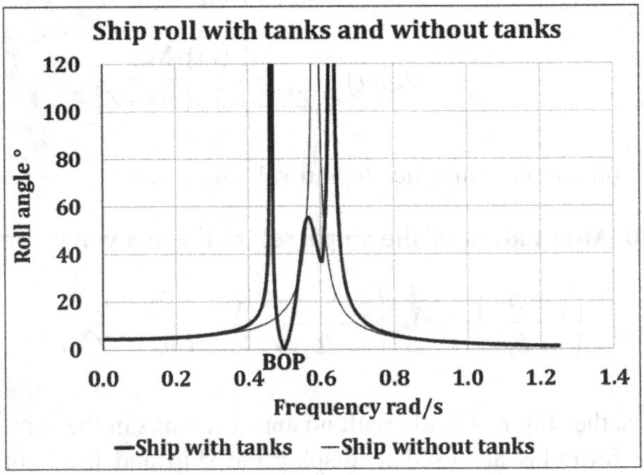

FIGURE 7.5 Ship's curves with and without anti-roll tanks.

exciting frequency. Both curves indicate that there are two resonant frequencies for the ship and two for the tank system.

Figure 7.5 superimposes the ship's curve of roll angles with a tank system (thick line) with the ship's curve without anti-roll tanks (thin line). Note that the ship has one resonant frequency different from the tanks' natural frequency in the last case. If the ship has no devices to prevent the rolling, its frequency response indicates it is always rolling. There is not a zero-rolling amplitude point in the thin curve.

The design procedure establishes that the tanks' natural frequency must be equal to the ship's natural frequency. This equality is obtained by adjusting the tank system's control devices. However, in a rough sea, this equality is rather difficult to meet.

The tanks are on the ship's sides. Usually, a pipe interconnects them. There is a flow regulating valve in this connecting pipe. This valve tunes the anti-roll system to the rolling frequency. Therefore, the moment created by the tanks is always opposed to the moment generated by the roll. In other designs, the tanks are independent one from the other. Most sophisticated systems inject air into the tanks to help to empty the tank. Some systems have an automatic feedback system because they measure the roll frequency with sensors to adjust the flow between tanks, using a controller as described in Chapter 11. The simplest anti-roll tank system does not need any external energy because the water flows from one tank to another by gravity. Transport companies for vehicles or containers very much appreciate this system because they use large roll-on roll-off or cargo ships, which require good stability to prevent accidents that might cause the merchandise to fall into the sea.

7.3.4.2 Bilge Keels and Stabilizer Fins

Undoubtedly, the oscillating sea waves do not always hit the ship with the same frequency; therefore, adding some devices, like the bilge keels or stabilizer fins, is advisable. See bilge keels in Figure 7.6. Bilge keels are passive systems that might be used in conjunction with other anti-roll systems, like fin stabilizers, shown in Figure 7.7. These fins are active systems because they require external energy to retract the fin or change its attack angle.

Bilge keels have a disadvantage as they produce an increment of the ship's hydrodynamic resistance. Their advantage is the ease of installation (welded onto the hull), and they do not occupy internal spaces, like the stabilizing fins do when they are retracted. These fins have a hydrodynamic profile, similar to an airplane's wings, whose angle of attack is adjustable through a hydraulic mechanism. Therefore, by changing this angle, the optimum stabilizing moment is obtained. This stabilizer type is effective only at ship speeds higher than 6 knots, much lower than usual cruiser velocity. The fins are extracted out of the hull only in case of necessity. When they are not in use, they do not affect the hull's hydrodynamic properties.

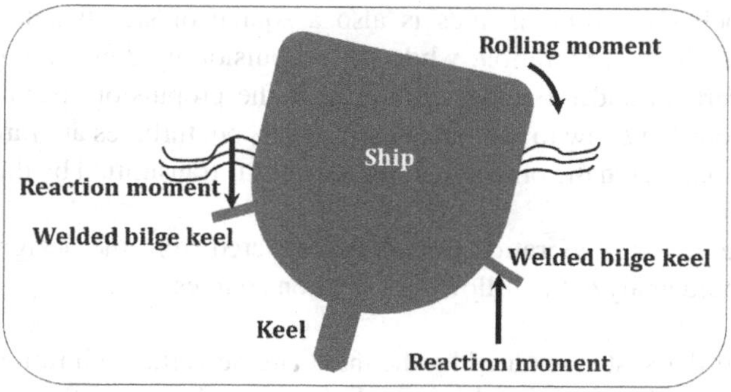

FIGURE 7.6 Bilge keel.

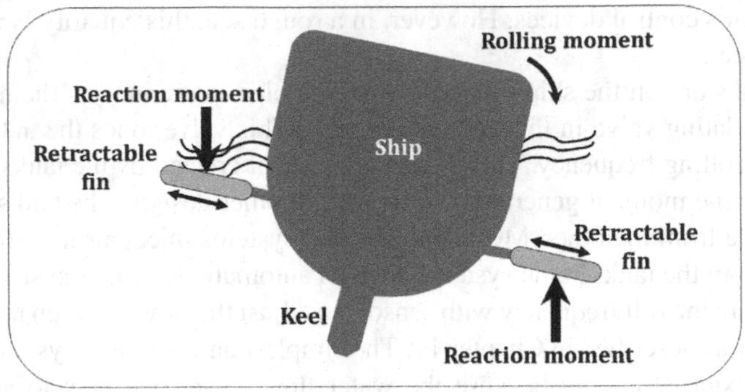

FIGURE 7.7 Retractable fin and bilge keel.

The stabilizer systems have different efficiencies. The best of them is the active fin system because it reduces the roll angle by 90%. The passive anti-roll tanks reduce the angle amplitude by 65%. The bilge keel performance is low. They reduce only 35% of the roll angle and require significant maintenance because they are always under salty seawater. The other systems require moderate mechanical maintenance. Of course, active systems are costly; the bilge keel is the less expensive to install[6].

7.4 SHIP'S MECHANICAL VIBRATION

Ships are elastic bodies subject to harmonic excitation forces from several sources, such as diesel motors or turbines coupled to gearboxes, shafts and propellers, pumps, electrical generators, and other moving devices. In summary, onboard, any rotating machine or any part moving at a constant or quasi constant frequency produces vibration. As with other types of machines or structures, this vibration intensity depends on a) the magnitude of the exciting forces, b) the ship's structure rigidity and inertia, and c) the proximity of the excitation frequency to any of the ship's resonance frequencies.

The propeller's hydrodynamics is also a source of significant vibration. It produces longitudinal vibration, while the propulsion by diesel motor produces transverse vertical and torsional vibrations. If the propulsion machine is a turbine, vibrations have low to moderate amplitudes, but turbines are rarely used for marine propulsion. On the other hand, the vibrations transmitted by diesel engines are significant.

Therefore, the ship vibration must be considered from the design stage. The designer should analyze the following vibration sources:

- Hull girder system excited by the main engine vertical vibration.
- Stern's hydrodynamic shape and distances between the hull and the propeller.

- Shafting longitudinal vibration produced by the propeller and the non-uniform flow.
- Vibration due to the propeller cavitation and alternating thrust.
- Induced vibration in the superstructure.
- Induced vibration by the cannons' firing in warships.
- Strong impact at airplane take-off or landing in aircraft carriers.
- Icebreaker ships do not ram the ice pack, but they rather mount on top of it to break it. When the ice breaks, the ship "falls," undergoing a strong impact that produces vibrations in its structure.

Some types of ships cannot tolerate vibration. Warships and submarines must be silenced to prevent being detected by the enemy's sonar. Tourism vessels must provide passengers and crew with a comfortable non-vibrating floor and a lack of noise. Fishing ships cannot emit sounds because any vibration propagated inside the seawater drives away the fish. Thus, the identification of vibrating sources and methods to mitigate them starts with the design stage.

A more detailed list of onboard vibration sources, briefly described above, are:

- An anti-roll fin or rudder could vibrate due to unsteadiness of the flow (von Karman street vortices. See Section 9.4)
- Propellers are submerged in a non-uniform viscous flow of water around the hull plates, which induces vibration in the hull. Some propellers have adjustable blade angles to optimize the ship's velocity.
- Cavitation is a serious concern for the propeller's life because of the vibra tion induced by an outgoing violent and unsteady flow.
- Diesel engines have strong inertial forces due to accelerated masses (pistons, connecting rods and crankshaft) produced by combustion inside the cylinders. On occasions, vibration is produced by inadequate combustion in their cylinders. However, modern engines are becoming more efficient due to the use of sophisticated monitoring systems.
- Other auxiliary machines are vibration sources, for example, the electrical power plant, pumps, HVAC, electric motors, compressors, etc.

These vibration sources produce three types of vibration.

- Axial or longitudinal vibration.
- Lateral vibration.
- Torsional vibration.

The hull is not a perfectly rigid structure; therefore, it undergoes several vibration modes. Induced vibrations by the main engine and the propeller are transmitted to and propagated through the ship's structure, producing discomfort to the crew and passengers, and mechanical fatigue to the equipment and structures. These vibrations must be properly mitigated during the design stage or the ship's construction and commissioning.

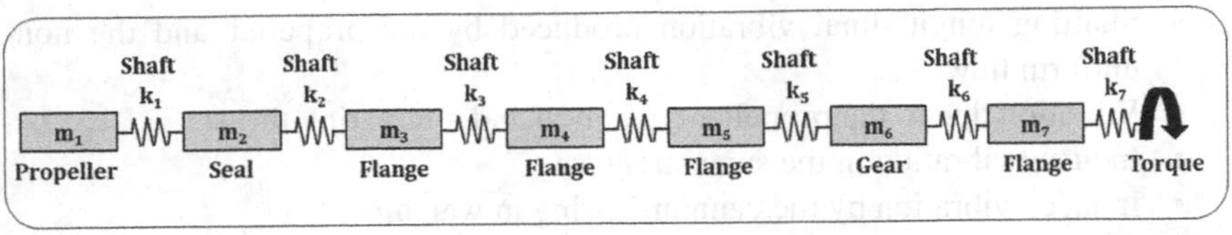

FIGURE 7.8 System's model for the study of longitudinal vibration in a propulsion plant.

Figure 7.8 shows the equivalent mass-spring system of the propulsion plant of Figure 7.1. It has seven rotating masses linked by springs; then, it is a seven-degrees of freedom system. Though rigidity and masses are distributed properties, they are considered concentrated properties to ease the propulsion plant's model. Therefore, the best first approach to the plant's vibration problem model is the multi-degree of freedom system explained in Chapter 2 using the matrixial method. This model may be later improved by the finite element method (FEM) if more abundant and exact results are needed.

In marine propulsion plants, damping is negligible ($\zeta<0.1$); so, only springs and masses are considered in vibration models. Springs stand for the shafts' rigidity, and flanges, gearbox and propellers stand for masses or mass moments of inertia. The bearings' masses are static devices for the shaft support and alignment purposes, so they are not considered a dynamic part.

The propulsion system's shafting is significantly long between the engine and the propeller. As infinite rigidity does not exist, this length obliges to install more than one bearing to reduce the shaft deflection as much as possible. Given that a perfectly balanced shaft does not exist, the system is subject to whirling because the shaft gravity center is not exactly located on the ideal rotation axis. Therefore, the bearings' correct location is of utmost importance because it minimizes the shaft deflection to avoid the lateral vibration aftermaths discussed in Chapter 5.

Another matter of importance is the shaft torsional vibration. Intermediate shafts mechanically linked by flanges form the shafting system. These shafts, excited by diesel motors of the propulsion plant, are prone to experiment torsional vibration. Their vibration amplitude may be significant and approach the stress to the steel yield strength because of their length.

7.4.1 Longitudinal Vibration Excited by the Propeller

The hydrodynamic conditions and the propeller's geometry gyrating in a viscous media as water does not produce a uniform inflow and outflow in the propeller. This non-uniform flow makes each blade generate a pulsating thrust. See Figure 7.9.

Hence, there are as many oscillating thrusts per revolution as the number of propeller's blades. The number of blades times the propeller's RPM is the thrust frequency in cycles per minute. This alternating thrust may be as high as 10% of the steady thrust. The propeller then produces a longitudinal vibration that obliges to take some measure to prevent the alternating thrusts propagation along with the

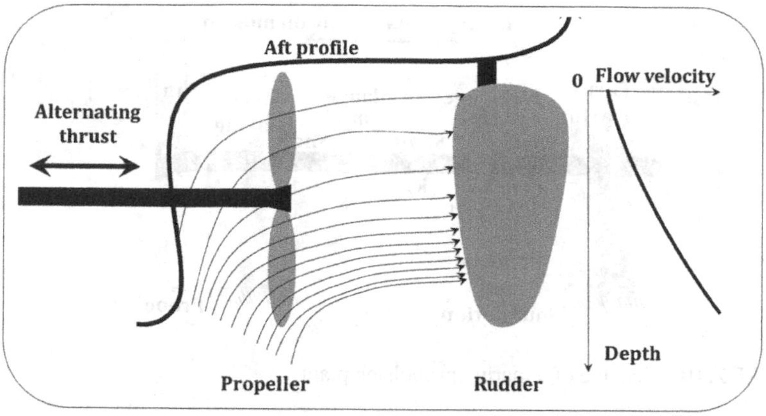

FIGURE 7.9 Non-uniform flow in the propeller produces an alternating thrust.

ship. As the propulsion system has several masses and springs, it has more than one natural frequency. None of these natural frequencies must be close to the propeller's thrusts frequency (Blade quantity × RPM) to prevent resonance.

7.4.2 Isolation of Longitudinal Vibration

The propeller's longitudinal vibration absorption is calculated by pre-setting the shafting natural frequencies away from the thrust frequency. These natural frequencies are obtained by accordingly adjusting the thrust bearing rigidity. Several technical publications recommend a 5% to 20% minimum gap between the thrusts frequency and any of the system's natural frequencies, though this range must be verified before accepting.

The propulsion plant of Figure 7.10 is a linear system with three degrees of freedom and one restriction, which is the thrust bearing. As the system has three degrees of freedom, a sixth-degree frequency equation must be expected. The system scheme is shown in Figure 7.11. For simplicity reasons, the scheme of Figures 7.5 and 7.6 do not consider the gearbox behind the thrust bearing. Therefore, it is ideally assumed that the main engine and the propeller have the same velocity.

The shaft's linear rigidity coefficients are k_1, k_2 and k_3. They are remarkably high because steel elasticity to axial efforts is too low. These coefficients are calculated with $k_1 = A \cdot E/L$, where A is the shaft diameter, E is Young's modulus, and L is the shaft length. For example, a propeller's shaft, made with AISI 304 steel, of 20 m length and 0.3 m diameter, has a 69,600 kg/mm longitudinal rigidity coefficient. It means that 69.6 tonnes thrust shortens the shaft by 1 mm. Coefficients k_2 and k_3 and masses m_1, m_2 and m_3, are known. The coefficient k_1 is unknown, and its determination is the aim of this section. The system is equivalent to the one restriction system of Figure 3.4.

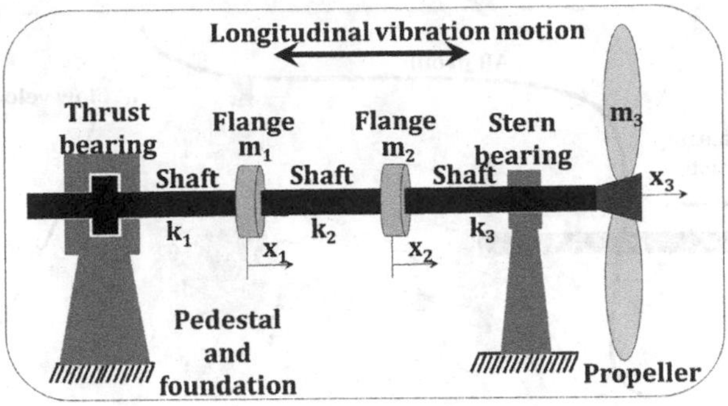

FIGURE 7.10 Scheme of a marine propulsion plant.

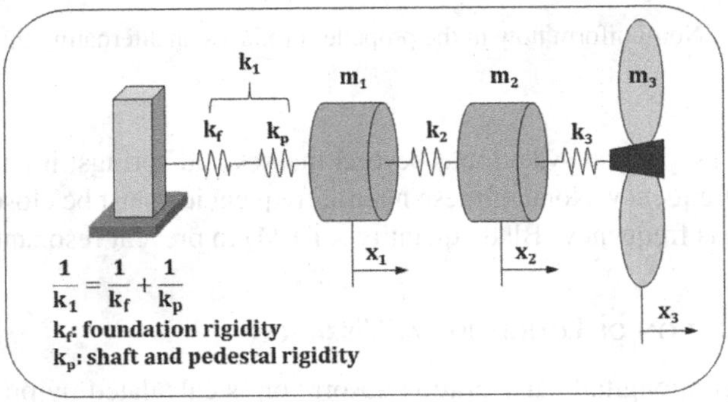

FIGURE 7.11 Model of a propulsion plant for longitudinal vibration study.

In this system, the vibration motion is propagated at a high velocity. As was seen before, the typical propagation velocity of longitudinal waves in steel is 5,000 m/s. Therefore, from the vibration source (the propeller) to the foundation under the thrust bearing, the propagation time is negligible, as in any mechanical system.

The scheme of the propulsion system of Figure 7.10 is illustrated in Figure 7.11. Based on the characteristic determinant ($\mathbf{det}\left[-\mathbf{M}\cdot\omega_n^2+\mathbf{K}\right]=\mathbf{0}$) the frequency equation is derived. With the same procedure used in Section 3.2.2, to derive the equations of motion of a system with two restrictions, the equations of motion for this system with one restriction are derived and solved. The characteristic determinant of the above scheme is shown in Table 7.2.

The expanded form of the characteristic determinant is the frequency Formula 7.21. The a_i coefficients of this equation are obtained from Table 7.2. Therefore, each coefficient a_i is a function of the rigidity coefficients and masses.

Formula 7.21 Frequency equation of three degrees of freedom system

$$a_6\left(k_i,m_i\right)\cdot\omega_n^6+a_4\left(k_i,m_i\right)\cdot\omega_n^4+a_2\left(k_i,m_i\right)\cdot\omega_n^2+a_0\left(k_i,m_i\right)=0$$

TABLE 7.2
Characteristic Determinant of a Three Degrees system

Marine Propulsion System of Figures 7.10 and 7.11
Characteristic Determinant
Longitudinal Vibration

$k_1 + k_2 - m_1 \cdot \omega_n^2$	$-k_2$	0		0
$-k_2$	$k_2 + k_3 - m_2 \cdot \omega_n^2$	$-k_3$	$=$	0
0	$-k_3$	$k_3 - m_3 \cdot \omega_n^2$		0

As the propulsion system is excited by the propeller's intermittent thrust, its natural frequency must be different from this excitation to prevent resonance in the propulsion system. Therefore, the frequency ω_n is pre-set with a different value from the excitation frequency and replaced in the frequency equation. All values of k_i and m_i are known, except k_1, the rigidity of the thrust bearing pedestal and its foundation. Therefore, clearing k_1 from the resulting equation is given the rigidity coefficient of the thrust bearing pedestal and foundation frame. This rigidity makes the natural frequency of the propulsion system equal to the pre-set frequency. In general, this pre-set value is 80% or less of the propeller thrust frequency.

The rigidity k_1 is calculated by introducing the auxiliary variable $\lambda = \omega_n^2$ The result is divided by a_6; so, the frequency equation is converted into the k_1 Formula 7.22, where k_1 is the unknown.

Formula 7.22 Equation to calculate the rigidity coefficient k_1

$$\lambda^3 + b_2(k_1) \cdot \lambda^2 + b_1(k_1) \cdot \lambda + b_0(k_1) = 0$$

Coefficients b_i linearly depend on the bearing pedestal and foundation rigidity k_1. The value of λ is equal to ω_n^2; it is known because ω_n is pre-set as was said before. The first step to solve Formula 7.22 is to calculate the values of the desired natural frequency, obtained from $\omega_n = 0.8 \times \omega_p$, where ω_p is the thrust frequency produced by the propeller. Therefore, the value of $\lambda = \omega_n^2$ is introduced in Formula 7.22 as a constant. Algebraic manipulation of Formula 7.22 demonstrates that the $b_i(k_1)$ coefficients are given the following formulas[7].

Formula 7.23 Factors b_i of the frequency Formula 7.22

$$b_2(k_1) = \frac{k_3}{m_3} + \frac{k_3}{m_2} + \frac{k_2}{m_2} + \frac{k_2}{m_1} + \frac{k_1}{m_1}$$

$$b_1(k_1) = \frac{k_3}{m_3} \cdot \left(\frac{k_2}{m_2} + \frac{k_2}{m_1} \right) + \frac{k_3}{m_2} \cdot \frac{k_2}{m_1} + \left(\frac{k_3}{m_2} + \frac{k_2}{m_2} + \frac{k_3}{m_3} \right) \cdot \frac{k_1}{m_1}$$

$$b_0(k_1) = \frac{k_1}{m_1} + \frac{k_2}{m_2} + \frac{k_3}{m_3}$$

In these formulas, only k_1 is unknown. Hence, introducing Formulas 7.23 in Formula 7.22 returns an equation with only one unknown, the required rigidity coefficient k_1. Solving this equation for k_1, the result is:

Formula 7.24 Thrust bearing and foundation rigidity coefficient

$$k_1 = m_1 \cdot \frac{\lambda \cdot \left(\lambda^2 + c_1 \cdot \lambda - c_2\right)}{\left(-\lambda^2 + c_3 \cdot \lambda - c_4\right)}$$

Coefficients c_i are calculated with the following formulas.

Formula 7.25 Bearings' rigidity coefficients of Formula 7.26

$$c_1 = \frac{k_3}{m_3} + \frac{k_3}{m_2} + \frac{k_2}{m_2} + \frac{k_2}{m_1}$$

$$c_2 = \frac{k_3}{m_3} \cdot \left(\frac{k_2}{m_2} + \frac{k_2}{m_1}\right) + \frac{k_2}{m_3} \cdot \frac{k_2}{m_1}$$

$$c_3 = \frac{k_3}{m_3} + \frac{k_3}{m_2} + \frac{k_2}{m_2}$$

$$c_4 = \frac{k_3}{m_3} \cdot \frac{k_2}{m_2}$$

The bearing pedestal and the foundation's rigidity act as two springs in series. See Figure 7.11. Therefore, the total rigidity k_1, formed by k_p (pedestal) and k_f (foundation), is lower than any of them. The foundation rigidity k_f is given by the hull girders' structure rigidity and cannot be modified. A typical value of the foundation rigidity is 24,000 kg/mm. After the k_1 rigidity coefficient has been determined with Formula 7.24, the pedestal rigidity is calculated with the following expression[8].

Formula 7.26 Pedestal rigidity for a pre-set natural frequency

$$k_p = \frac{k_1 \cdot k_f}{k_f - k_1}$$

This formula shows that the foundation and thrust bearing set rigidity k_1 is lower than its two components, k_p and k_f of pedestal and foundation, respectively.

7.4.3 Isolation of Shaft Torsional Vibration

It was seen in Chapter 2 that long shafts are prone to experience torsional vibration of significant amplitude and high frequency. Ships have a remarkably long shaft, the propeller's shaft, which mechanically links the main engine with the propeller.

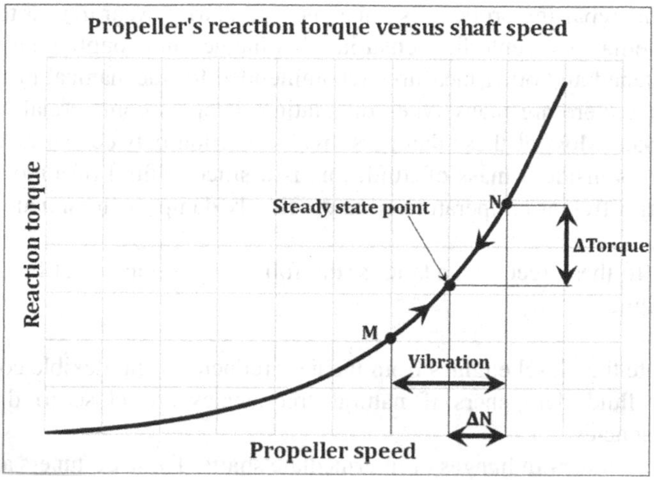

FIGURE 7.12 Propeller's reaction torque versus propeller's speed.

In the past, the main engine was in the middle of the ship, and propulsion shafts were too long. However, modern cargo and oil tankers have the main engine near the stern; therefore, their shafts are shorter than in the past; nonetheless, the propeller's shaft is still significantly long and prone to experience torsional vibration.

Fortunately, the propeller is an effective torsional vibration damper because the steady-state relation between the propeller's reaction torque and shaft velocity is a square or cubic parabola[9] due to the water viscosity. See this parabola in Figure 7.12. This curve is computed during the ship test, measuring the torque at the engine and the corresponding shaft RPM. As torsional vibrations are superimposed to the steady-state rotation, vibration amplitudes imply a velocity change that at once creates a torque variation that opposes the velocity change. This mechanism is a sort of negative feedback that tends to keep a constant shaft velocity, almost free of torsional vibrations.

Figure 7.12 shows how the propeller torque changes due to superimposed vibration to the steady-state rotation. The reaction ΔTorque is opposed to the propeller's speed increment ΔN. Conversely, the speed decrease tries to move the steady-state point to point M. In this case, the lower reaction torque at point M allows re-establishing the speed and torque to the steady-state point. The equilibrium is not permanently reached because all these changes respond to transient regimes, but the net result is that vibration is damped though not cancelled. The energy dissipated in this process is transferred to the seawater.

The model of the three degrees of freedom system of the last section applies to torsional vibration by analogy. Linear system's formulas hold for rotating systems, where linear rigidity coefficients k_i must be substituted by torsional rigidity coefficients k_{ti}, and masses m_i are replaced by polar mass moments of inertia J_i.

There are procedures to isolate or damp torsional oscillations such as: running the engine velocity far away from its natural frequencies, balancing the main engine, repairing propeller's imbalances, repairing wear in gear teeth faces, aligning the flanges' centerline, ensuring that flanges and coupling bolts are correctly tightened and other measures recommended for mechanical maintenance. If necessary, there are many types of rotating dampers commercially available in the market. Most of these dampers involve friction between surfaces. Others are flywheels inside a mass of fluid; this is a silicone fluid of almost constant viscosity at different temperatures that efficiently damp the torsional crankshaft vibration.

Further to these recommendations, the following actions must be considered by the designer.

1. Isolate the diesel engines from the gear reductor using flexible couplings.
2. Add fluid dampeners if natural frequencies are close to disturbance frequencies.
3. Use de-tuners in flanges of intermediate shafts. These de-tuners are formed by springs compressed when the vibration starts, therefore increasing intermediate shafts rigidity.
4. Use propellers with an adjustable pitch to change the system rigidity.

These action courses must be complemented with a good vibration monitoring system, as described in Section 10.3.

7.4.4 Diesel Motors Excitation

Diesel motors convert pressure into pistons' reciprocating motion. They are widely used in naval propulsion because of their high fuel consumption efficiency. However, the pistons' motion produces forces and moments that easily generate vertical vibrations that propagate to the hull and other ship's parts. They act on the motor's crosshead; therefore, their reactions are pulsating transverse forces. These frequency of these forces depends on the number of cylinders and the shaft velocity. Vibratory motions are mitigated by strong braces that link the upper motor part to the hull's frame. The forces' values for diesel motors are high because their moving parts are massive; therefore, their inertial forces are significant. Models of these forces are determined based on the D'Alembert's principle explained previously.

Due to the amplitude of harmonic forces and moments created by diesel reciprocating motors, they are mounted on elastic seats. These elastic seats must isolate the hull and frames from the vibration of at least the first six natural frequencies. The isolation design must include the excitation coming from the propeller. Special attention must be paid to moments produced by the second component of vibration (second-order moments), which are in a vertical plane with a frequency equal to two times the machine speed. As these dynamic efforts are important, the designer must calculate them. However, a better choice is to request their values from the motor's manufacturer. Second-order moments of six or more cylinder motors may produce the hull's girders resonance with the propeller's shaft velocity.

Experience reveals that the hull girders' vertical vibration's natural frequency is approximately twice the shaft velocity. This resonance deforms the hull as if it were a beam with free-free supports. This hull's deformation is mitigated or solved by adding compensator masses to the engine's crank. The indicator that defines whether compensators are needed is the ratio between the second-order vertical moments in N•m and the kW engine power. This ratio is called Power Related Unbalance (PRU). The following table is recommended to use when deciding the mass compensators installation.

The motor manufacturer is also the moment compensators' manufacturer. These devices are reciprocating masses added to the crankshaft that create moments opposing the second-order vertical moments.

In the design stage of a ship propelled by diesel motors, it must be verified that the number of cylinders and propeller's blades should not be multiples of any integer number to prevent resonance between the motor and the propeller's blades. Therefore, a motor with an even number of cylinders is only used with an odd number of propeller's blades.

7.5 BEAM SHIP VIBRATION

In a simplified conception, a ship structure may be calculated as a beam with free-free supports. The hull's girders support the loads, such as weights of propulsion engines, bulkheads, auxiliary machines, power plants, artillery in warships, and many other vertically acting loads. According to Archimedes' principle, these forces are counteracted by the buoyant force. A curve is used to represent the difference between load and buoyant forces from aft to bow. As these forces are vertical and opposed, their differences are the shear efforts endured by the ship. The integral of the shear efforts curve returns the flexural moments' curve. As the ship is not a perfectly rigid body, these efforts and moments deform the hull's girders according to the Mechanics of Materials science laws.

Ship motions discussed in Section 7.3 are studied as rigid body vibrating at mode 1, that is, that roll, yaw, pitch, surge, heave, and sway motions do not produce ship deformation. Instead, hull elastic deformations are produced in two nodes mode and an even higher number of modes. Usually, the mode shapes are studied at a frequency double the propeller velocity.

7.5.1 Beam-Ship Natural Frequencies

Vertical vibrations produced by the propeller or diesel motors may become a serious problem for the hull girders' structure because of resonance at low frequencies. This resonance is typical of motors or propellers at a slow speed. In the propeller operating velocity, the ship's girders' structure is not easily excited. This low-level vibration may excite the ship superstructure at its natural frequency and produce harm due to resonance at the propeller frequency.

There are several methods applied to calculate the beam-ship natural frequencies of vertical vibrations. This section will present two of them: a) the Euler-Bernoulli equation and b) Kumai's[10] and Johannssen-Ossark's (1980) formulas.

7.5.1.1 Natural Frequencies by Euler-Bernoulli Equation

The Euler-Bernoulli beam theory allows calculating the efforts, moments, and beam-ship's natural frequencies[11], as discussed in Section 3.4. Its application requires the ship's discretization to describe the transverse vibration's behavior along the ship's length. The Euler-Bernoulli equation is reproduced below.

$$\frac{\partial^4 \delta(\mathrm{x,t})}{\partial x^4} + \frac{\mu}{\mathrm{E} \cdot \mathrm{I_x}} \cdot \frac{\partial^2 \delta(\mathrm{x,t})}{\partial t^2} = 0$$

This equation applies to the beam ship to calculate its transverse vibration as a function of time and the distance x to the coordinate's origin, usually located in the stern. The Euler-Bernoulli equation has a constant parameter, designated μ, representing the mass distribution along the ship's centerline in terms of mass per unit length (μ = ρ·A). This property is formed by the ship's mass and the associated water mass in the hull's proximities, flowing at the same velocity as the ship. This "mass association" makes the inertial forces significantly higher because the associated mass is 25% to 33% of the ship displacement. However, for a beam-ship, Kumai's formula of the first natural frequency returns a higher water mass added to it. This added mass of water is explained in the following section.

An important derivation of the Euler-Bernoulli equation solution is the natural frequencies Formula 3.45, reproduced below. This formula demonstrates that the ship's natural frequencies are inversely proportional to the square root of mass per unit length. Therefore, the water mass associated with the ship mass decreases the ship's natural frequencies by 12% to 15% with respect to the natural frequencies with no associated water mass.

$$\omega_{\mathrm{nj}} = \frac{\alpha_{\mathrm{j}}}{\mathrm{L}^2} \cdot \sqrt[2]{\frac{\mathrm{E} \cdot \mathrm{I_x}}{\mu}}$$

L is the ship's length, and I_x is the ship's longitudinal moment of inertia with respect to the ship's X-axis (see Figure 7.2). μ is the ship's displacement expressed in kg mass over the ship's length. The coefficient α_j depends on the beam's support type. As the ship is floating and the extreme stern and bow points are free to move vertically, it is analogous to a free-free beam. Values of α_j in free-free beams are in Table 3.2 for vibration modes 1 to 5.

The hull's girders system vibration is mainly produced by non-uniform hydrodynamic forces acting on the stern. This phenomenon induces vibrations in the 0.6

TABLE 7.3
Requirement of Compensators in Ship Engines

PRU N.m/kW	Requirement of Moment Compensators
Below 120	Not likely
120 to 220	Likely
Over 220	Most likely

to 6 Hz range. This vibration's amplitude is higher than the amplitude of longitudinal vibration created by the propeller. Nonetheless, the ship-beam structure is so complex that classical vibration theory cannot accurately explain the hull's girders' vibration. Therefore, ship-beams are calculated with empirical formulas or by the finite element model[12]. The hull's shape is presently designed using hydrodynamic software to calculate the optimum geometric shape for the expected water flow around the hull.

7.5.1.2 Hull Girder's Natural Frequencies

Kumai's formula serves to calculate the hull girder's natural frequencies. This formula is valid for a two nodes vibration mode. Reminder: a node is a beam's section where no vibration occurs at the frequency under study. However, said section is vibrating at other frequencies because the beam-ship is a continuous system with, theoretically, infinite natural frequencies, out of which only the first ones are of practical interest.

Formula 7.27 Kumai's formula for beam-ship of 2-nodes shape mode

$$\omega_{n2} = 17.29 \times 10^6 \times \sqrt[2]{\frac{I_z}{a_w \cdot \Delta \cdot L^3}}$$

In this formula, the natural I_y is the ship's cross-section moment of inertia in m^4. The coefficient a_w is the coefficient that increases the ship's inertia for the associated water. This factor is calculated with the following formula.

Formula 7.28 Added water coefficient

$$a_{aw} = 1.2 + \frac{1}{3} \cdot \frac{B}{H_{avg}}$$

The ship's breadth is B, in meters. H_{avg} is the ship's average draft in meters. In general, the B/H_{avg} ratio is higher than 1. A typical value of this ratio is 2 or more, which, replaced in the above formula, increases the displacement by 86%. Figure 7.13 displays the typical ship-beam natural frequency returned by Kumai's formula as a function of the displacement for several ship types.

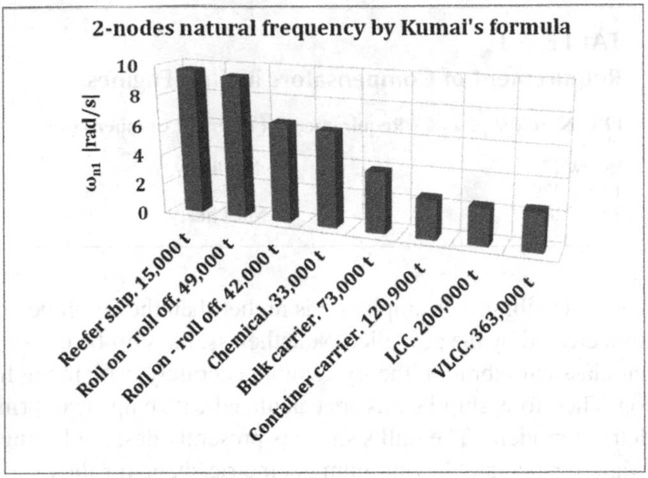

FIGURE 7.13 Ship-beam natural frequencies for vertical vibration[13].

TABLE 7.4
Exponent α of Johannssen and Skaar Formula

Ship Type	α
General cargo	0.845
Bulk carriers	1.000
Tankers	1.020

For higher-order vibration modes, Kumai's formula does not apply; instead, the following formula (from Johannssen and Skaar) gives acceptable results.

Formula 7.29 Natural frequencies of higher-order modes

$$\omega_{n2} \approx \omega_{nj} \cdot (j-1)^{\alpha}$$

Where ω_{n2} is the fundamental vibration form formed by two nodes, the exponent α is specific for each ship type, and j is the number of nodes. The combination of Kumai, and Johannssen and Skaar formulas are valid for vibration modes up to five. For a higher degree of vibration mode, these formulas are not recommended.

Figure 7.14 displays the natural frequencies for vertical vibration as a function of ship displacement and vibration mode. This figure shows that light ships have higher natural frequencies than heavy ships. The second-order vertical vibration of diesel motors has twice the frequency of motor velocity. Therefore, during the design stage, Kumai's formula helps to find potential resonance risks by

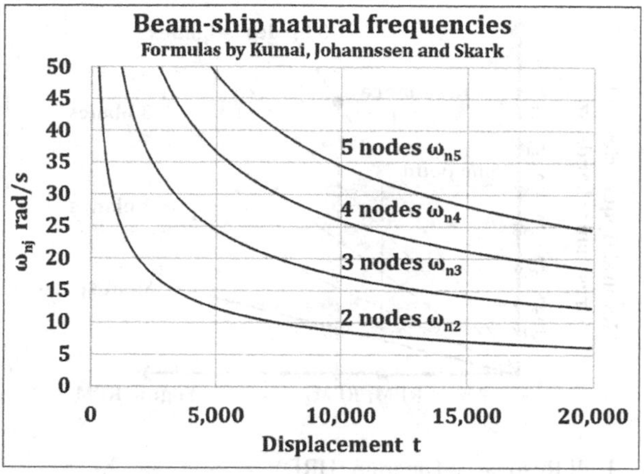

FIGURE 7.14 Example of beam-ship natural frequencies versus displacement.

comparing the calculated natural frequencies with the diesel motor's second-order exciting frequency.

Diesel motors usually have crankshaft counterweights and flywheels that oppose the piston force, reducing the dynamic impact that produces vibration. It is also common in ships to install spring-damper absorbers to isolate the hull girders system from the diesel motor knocking.

For example, a ship with 15,000 tonnes displacement, whose natural frequency curves are shown in Figure 7.14. The propulsion engine is diesel and is operating at a speed of 75 RPM. Therefore, its second-order excitation frequency is 150 RPM or 15.7 rad/s. However, Figure 7.14 shows that the hull's natural frequency of three nodes vibration mode is 14.1 rad/s. This frequency differs by only 11% of the engine excitation frequency, which is unacceptable because the recommended difference is 20%.

7.5.2 The Hull Resonance Diagram

The ship's stern is subject to vertical hydrodynamic forces, which are produced by the propeller. As mentioned previously, this excitation frequency is the same as the propeller's axial forces frequency (RPM×blades).

The hull's girders vibration system is analyzed with a graphical procedure to calculate the number of blades. This procedure requires knowing all the vibration modes produced by transverse forces, vertical forces, and torsional vibration modes. This knowledge allows constructing a chart of natural frequencies versus the main engine velocity, called the Hull Resonance Diagram (HRD). The best choice for the propellers' number of blades is obtained (see Figure 7.15).

The ordinates of Figure 7.15 are the natural frequencies, and the abscissas represent the shaft (or propeller) angular velocity. The horizontal lines are the hull's girders' natural frequencies of vertical, transversal, and torsional vibration modes.

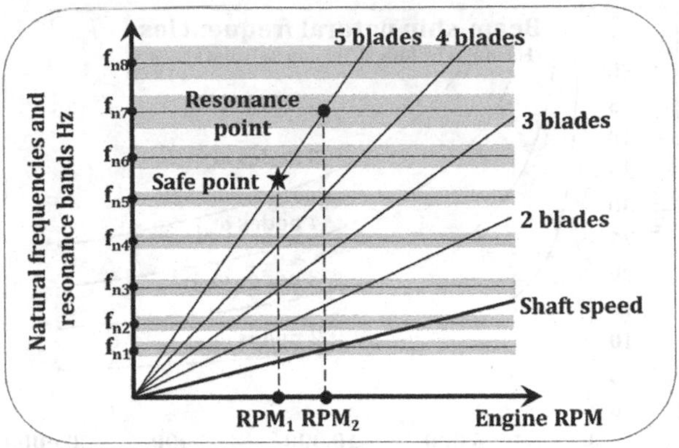

FIGURE 7.15 Hull Resonance Diagram (HRD).

These frequencies must have been previously calculated. The grey bands are the tolerance limits, for example, ± 5% of the corresponding natural frequency. These resonant bands are forbidden areas during the ship's navigation.

As the frequency induced by the propeller is the shaft speed times the propeller's blade quantity (RPM×blades), it is represented by a straight line. The formula to draw the right lines is:

Formula 7.30 Natural frequency of right lines formula

$$f_{ni} = b_i \cdot \frac{RPM_{shaft}}{60}$$

Where b_i is the number of blades, in this formula, the shaft RPM is divided by 60 to convert the RPM to cycles per second. The above formula can be written as a function of the engine's RPM because the shaft RPM is equal to the engine RPM divided by the speed reducer gear ratio n_r.

Formula 7.31 Natural frequency formula versus engine RPM

$$f_{ni} = \left(\frac{b_i}{60 \times n_r} \right) \cdot RPM_{engine}$$

Then, ordinates are measured in cycles/second or Hz, and abscissas represent the engine's RPM. The term between parenthesis is the slope of the straight lines of Figure 7.15.

The HRD chart's objective is to prevent the hull's girders' resonance due to the main engine's excitation. Therefore, the HRD figure is used by designers to choose the most appropriate quantity of propellers' blades for the selected main engine velocity. In the example of Figure 7.15, the selected engine's velocity, RPM_1, has only one option to avoid the resonance bands: a five blades propeller. See the operating point represented by a star. Suppose the engine's velocity is changed from RPM_1 to RPM_2. In that case, the five blades' propellers are no longer acceptable because the operating point, indicated with a black circle, is within a forbidden resonance band. As ships have a changing velocity, this chart is applied for the cruiser velocity, which is supposed to be the velocity at which the ship navigates most of the time.

The HRD figure results are a good approach for designing the propeller's blades to avoid resonances. However, in this procedure, the natural frequencies are calculated with empirical methods based on several assumptions that may have low validity. Therefore, it is recommended to use a ship's finite element method, as required by classification societies.

7.5.3 Finite Element Method. Brief Description

The finite element method (FEM) is a numerical procedure applied to complex structures or sophisticated machinery (like a ship) that is cumbersome to solve with classical methods or poor accuracy. This method calculates deflections, vibration characteristics, reaction forces, and other significant physical properties.

The FEM converts a complex structure or machine into a set of many small elements (finite elements), each with mass, damping and rigidity properties. This discretization allows creating a mathematical model of a ship, where each element has a set of equations of motion. This method requires as many sets of differential equations as elements that integrate the created structure. Solutions to the differential equations describe the structure's behaviour under the applied efforts. This solution is computed by specialized software.

The model provides the ship's dynamic behaviou under free and forced vibration conditions. Natural frequencies, vibration amplitudes, and velocities are calculated for each element. Therefore, the model output for different areas and machinery is compared with tolerances established by classification societies, like the American Bureau of Shipping ("ABS"). Non-compliance cases oblige to redesign the affected element and iterate the procedure until one of the designs complies with the required standards.

7.5.3.1 Ship's Deformation by Torsional Torques

The longitudinal vibration study of the propulsion plant of Figure 7.10 is done with the equivalent system of springs and masses of Figure 7.11. The same scheme can study torsional vibrations by replacing masses m_i by mass moments of inertia J_i and linear stiffness coefficients k_i by torsional stiffness coefficients k_{ti}. As there are so many components, the system is modelled with the matrixial method. As commented before, this method allows calculating the system's natural frequencies, also known as eigenvalues and the deformations undergone by a ship when freely vibrating.

The ship's equation of torsional motion is studied with the Transfer Matrices Method, whose equation of motion is:

Formula 7.32 Ship's matrixial equation of motion

$$\mathbf{M}\cdot\ddot{\vartheta}(\mathbf{t})+\mathbf{D}\cdot\dot{\vartheta}(\mathbf{t})+\mathbf{K}\cdot\vartheta(\mathbf{t})=\mathbf{T}(\mathbf{t})$$

Symbols of the ship's matrixial equation of motion are:

M: diagonal mass matrix
D: symmetrical damping matrix
K: symmetrical stiffness matrix
$\ddot{\vartheta}(\mathbf{t})$: vector matrix of angular acceleration
$\dot{\vartheta}(\mathbf{t})$: vector matrix of angular velocities
$\vartheta(\mathbf{t})$: vector matrix of angular displacements
T(t): vector matrix of harmonic torques

As ships are frictionless systems, matrix **D** is discarded. Therefore, in vector notation, the ship's equation of motion at free vibration is:

Formula 7.33 Matrixial equation of eigenvalues ω_n and ship's mode shape

$$\mathbf{M}\cdot\ddot{\vartheta}(\mathbf{t})+\mathbf{K}\cdot\vartheta(\mathbf{t})=0$$

The ship is deformed during free vibration[14] , and the function that describes the ship's shape during vibration is ϑ_a, called the ship's mode shape. Therefore, the mode shape and the natural frequency are the ship's properties that indicate its structural behaviour under dynamic efforts. It is a qualitative assessment of the ship's shapes under vibration. It is recommended to conduct all calculations with dedicated software because calculating matrices larger than 5×5 is cumbersome and does not provide a good graphical picture of the ship's deformation. Deformation is exaggerated and coloured to create a clear vision of parts more affected by deformation or under risk of rupture.

7.5.4 Vibration Tolerance Standards

Tolerance to vibrations onboard is specified by the ABS Classification Society, ISO 6954 and other standards. These standards' main purpose is to reduce the crew and passengers' vibration effects and preserve machinery and structures from the fatigue effect. The standards refer to vibration frequencies from 0.5 to 80 Hz.

The standards ANSI S2.27 (2002) give the following maximum (RMS) vibration velocities to assess the vibration severity:

- The thrust bearing ≤ 5 mm/s
- Longitudinal vibration of motors, propellers, and shafts ≤ 13 mm/s
- Lateral vibration of shafts couplings ≤ 7 mm/s
- Stern tube and shaft bearing ≤ 7mm/s
- Diesel engine at bearing ≤ 13 mm/s
- Top of slow or medium speed Diesel engines (over 1,000 HP) ≤ 18 mm/s
- Top of high-speed Diesel engines (less than 1,000 HP) ≤ 13 mm/s

Nonetheless, it is highly recommended to consult the standards or contract a specialist if a vibration problem arises.

NOTES

1 Book recommended for Chapter 7. 1. Arquitectura Naval by Antonio Mandeli. Librería y Editorial Alsina. 1985 - 2. Mechanical Vibrations by J.P. Den Hartog. Dover Publications, Inc. New York, 1985, 4th edition – 3. Mechanical Vibration by Haym Benaroya, Mark Nagurka and Seon Han. CRC Press. Taylor and Francis Group, fourth edition 2018 –5. Introduction to Naval Arquitecture by Eric Tuppe. Elsevier Science & Technology Books, 1996.3rd edition.

2 See site: https://magazines.marinelink.com/Magazines/MaritimeReporter/201310/content/rollsroyce-turbine-combat-211593

3 It is recommended reading a description of a marine propulsion system in: https://www.mermaid-consultants.com/ship-propeller-and-shafting-system.html

4 See the paper: TECHNISCHE UNIVERSITEIT Laboratorium vooc Scheepshydromechanlcs Archief Mekelweg R W PEACH, MSE, PE, CEng, .A K. BROOK, BSc, MSc, CEng, British Maritime Technology Ltd. Report on a presentation and discussion 1 6th February 1987.

5 See the papers: a) Anti-Roll Tanks in Pure Car and Truck Carriers, by Björn Windén. KTH Centre for Naval Architecture, Stockholm 2009, Master's Thesis, http://www.diva-portal.org/smash/get/diva2:233558/FULLTEXT01d
b) On the Development of Ship Anti-Roll Tanks, by Reza Moaleji, Alistair R. Greig. Elsevier, Science Direct.
https://www.engr.mun.ca/~bachmayer/engr6055/assignments/reading-assignments/roll_control.pdf

6 See Different Types of Roll Stabilization Systems Used for Ships https://www.marineinsight.com/naval-architecture/roll-stabilization-systems/

7 See book of Ship Vibration. American Bureau of Shipping (ABS), 2018.

8 This formula is derived from the general expression applicable to n springs in series.

$$\frac{1}{k_{total}} = \sum_{i=1}^{n} \frac{1}{k_i}$$

9 Mechanical Vibrations by J.P. Den Hartog. Dover Publications, Inc. New York, 1985, 4th edition, Section 5.8.

10 See: Kumai T (1968) On the estimation of natural frequencies of vertical vibration of ships. Report of Research Institute for Applied Mechanics.

11 Timoshenko's beam theory is also a useful mathematical tool to calculate ships' natural frequencies. Consult: Predicting Method of Natural Frequency for Ship's Overall Vertical Vibration, by Yumei Yin, Hongyu Cui, Deyou Zhao and Ming Hong, and the paper by Hakar Ucan in site:
https://www.researchgate.net/publication/305721123_Ship_Hull_Girder_Vibration
12 See the book: Ship Hull Vibration, by F.H. Todd, 1961.
13 See the book: Guidance Notes on Ship Vibration. American Bureau of Shipping. May 2021. See in Figure 7.16 an interpretation of Table 1 in section 4g.
14 There are many sites in internet showing ships deformation by the finite element method that is recommendable to visit.

Part III

Vibration Control Systems

8 Vibration Isolation

8.1 INTRODUCTION TO TRANSMISSIBILITY OF FOUNDATIONS

Chapter 6 discussed the vertical forces transmitted by lateral vibration and demonstrated that these forces are maximum when the centrifugal force points downward. Figure 8.1 shows a polar plot of centrifugal forces combined with the shaft rotor weight versus different angular positions ϑ of the shaft gravity center. These forces are transmitted to pedestals and foundations.

Three different scenarios form the study of transmitted forces, namely: a) the foundation cannot vibrate because it is perfectly rigid, b) the foundation has one vertical degree of freedom; therefore, it vibrates because it is not stiff, and c) the foundation is an off-land installation such as a ship's hull that easily transmits vibration.

Every piece of equipment in industries such as power plants, overhead lines and similar installations need a foundation to endure its weight and dynamic efforts. In inland installations, these foundations are usually made of concrete. In general, a metal frame is laid between the machine and the concrete partially buried underfloor. This layout is usual in turbines, fans, hydraulic presses, large electric motors, diesel motors, ball mills and much other heavy equipment. Typically, the foundation mass is recommended to be between three and five times heavier than the machine mass. This type of foundation may be considered perfectly rigid, which means that they do not vibrate even if they are excited by a harmonic force. Anyway, it is necessary to isolate the foundation from harmonic efforts transmitted by the machine to prevent damages.

Isolating devices are first-order systems formed by springs and dampers, which support the machine and partially absorb its weight and dynamic efforts transmitted to the foundation. See Figure 8.2. A harmonic force $F_a(t)$ acts on the machine. $F_{t1}(t)$ and $F_{t2}(t)$ are the transmitted forces to the foundation through the spring-damper sets. The damper material is an appropriate rubber, though, in the past, cork was used. Springs are made of alloy steel.

The system of Figure 8.2 has a mass "floating" on springs; so, part of the rotor-shaft weight is not transmitted to the foundation due to the springs' reaction. If no springs and dampers are laid under the pedestals, the harmonic force $F_a(t)$ is 100% transmitted to the foundation. As this is perfectly rigid, it does not vibrate; however, forces produced by the machine vibration are transmitted to the ground. Most concrete foundations of large machines may be considered perfectly rigid. Spring-damper sets installed under the machine pedestals change the natural frequency, so that vibration frequency keeps away from the machine's natural frequency and reduces the vibration amplitude.

DOI: 10.1201/9781003175230-11

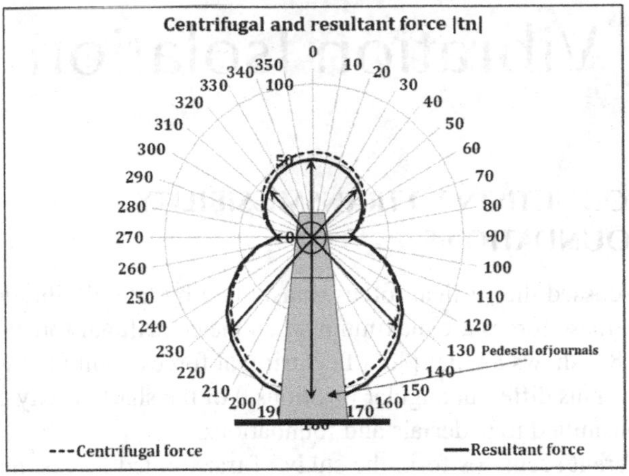

FIGURE 8.1 Centrifugal forces combined with the rotor-shaft weight.

8.2 TRANSMISSIBILITY OF RIGID FOUNDATION

8.2.1 Mechanical Impedance Definition

It is convenient to define the mechanical impedance beforehand to facilitate any foundation type's mathematical model. The mechanical impedance is the physical property opposed to the system motion, formed by mass inertia, spring rigidity and damper friction. The following expression is the formula of the mechanical impedance of a linear system:

Formula 8.1 Impedance of a discrete mechanical system

$$Z(\omega) = \frac{F}{a_v} = k - m \cdot \omega 2 + j \cdot \omega \cdot f$$

F is the amplitude of the harmonic applied force, and a_v is the induced vibration amplitude. Two impedances form this physical property: inertial impedance $Z_1 = m \bullet \omega^2$ and spring-damper set impedance $Z_t = k + j \bullet \omega \bullet f$. Subscript t stands for transmitter impedance. The inertial reaction force is equal to Z_1 times the machine's vibration amplitude a_1. Similarly, the spring-damper set reaction force is equal to Z_t times the spring's net stretching. The net stretching is the difference between the machine and foundation vibration amplitudes ($a_1 - a_2$). In rigid foundations, a_2 is zero.

In summary:

Reaction force = Mechanical Impedance × Vibration Amplitude

On the base of Formula 8.1, the impedance ratio z is obtained by dividing Formula 8.1 by k. Therefore, introducing the damping ratio ζ and the frequency ratio u, the impedance ratio z(u) formula is:

Formula 8.2 Impedance ratio of a discrete mechanical system

$$z(u) = \frac{Z(\omega)}{k} = 1 - u^2 + j \cdot 2 \cdot \zeta \cdot u$$

The above formula indicates that at zero frequency (u = 0), the impedance is only the spring stiffness k; then, the impedance ratio is equal to 1. As excitation frequency increases, the impedance's inertial part also increases against the spring stiffness and damper friction. In Chapter 1, it was demonstrated that friction force is 90° ahead of the stiffness reaction and 90° behind the inertial force. This configuration indicates that rigidity and inertial reactions share the same direction, but they are opposing forces[2]. Therefore, the absolute value of the total impedance and impedance ratio formulas are:

Formula 8.3 Absolute value (modulus) of the impedance and impedance ratio

$$|Z| = \sqrt[2]{\left(k - m \cdot \omega^2\right)^2 + 4 \cdot \omega^2 \cdot f}$$

$$|z| = \sqrt[2]{\left(1 - u^2\right)^2 + 4 \cdot \zeta^2 \cdot u^2}$$

Note: in successive uses of Z(ω) or z(u), they are noted as Z and z for simplicity. The concept and formulas of impedance allow a convenient analogy between mechanical and electrical systems to analyze more complex mechanical configurations than that represented in Figure 8.2.

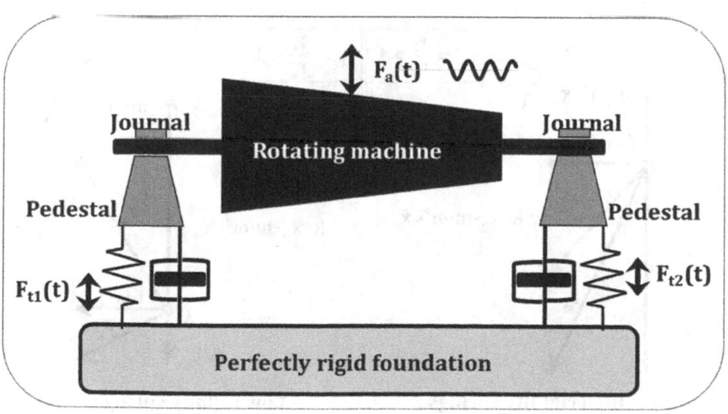

FIGURE 8.2 Scheme of transmissibility of a disturbance force to the foundation.

8.2.2 Transmissibility Ratio

In the mechanical system of Figure 8.2, the applied and the transmitted forces are calculated with:

Formula 8.4 Complex expressions of the applied and transmitted forces

$$F_{applied}(\omega) = \left(k - m \cdot \omega^2 + j \cdot f \cdot \omega\right) \cdot a_1 = Z_1 \cdot a_1$$

$$F_{transmitted}(\omega) = \left(k + j \cdot f \cdot \omega\right) \cdot a_1 = Z_t \cdot a_1$$

For the usual case of having n identical springs or dampers, the equivalent coefficients are calculated with $k = k_i \cdot n$ and $f = f_i \cdot n$ for springs and dampers, respectively, where k_i and f_i are the coefficients of the spring and damper i.

The vector interpretation of Formula 8.4 is displayed in Figure 8.3.

The application of this vector interpretation is valid for applied and transmitted forces of Formulas 8.4. Figure 8.3 shows two different cases: the left diagram corresponds to a system with a transmissibility ratio equal to 1.8. It means that the transmitted force is 80% higher than the applied force, which is undesirable. The transmitted force is higher than the applied force; this happens because of the excitation frequency's proximity to the system's resonant frequency. The right diagram corresponds to a system with a transmissibility ratio equal to 0.5. Therefore, the transmitted force is half the applied force, which could be acceptable.

The vector modulus of applied and transmitted forces of Figure 8.3 are:

Formula 8.5 Modulus of applied and transmitted forces

$$F_a(\omega) = \sqrt[2]{\left(k - m \cdot \omega^2\right)^2 + \left(f \cdot \omega\right)^2} \cdot x_a$$

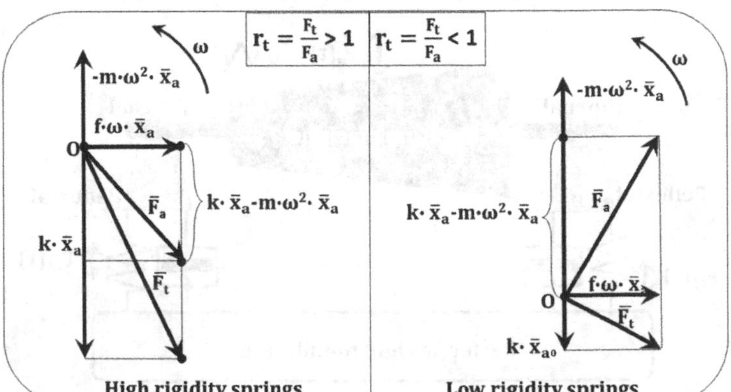

FIGURE 8.3 Vector interpretation of applied and transmitted forces for high and low springs rigidity.

$$F_t(\omega) = \sqrt[2]{k^2 + (f \cdot \omega)^2} \cdot x_a$$

The upper expression of Formulas 8.5 describes the applied and the three resisting forces given by springs, dampers, and machine mass. The second formula represents the springs' and dampers' reactions transmitted to the foundation. Therefore, clearing x_a, from the $F_a(\omega)$ formula, the vibration amplitude is returned as:

Formula 8.6 Vibration amplitude after isolators installation

$$x_a = \frac{F_a(\omega)}{\sqrt[2]{(k - m \cdot \omega^2)^2 + (f \cdot \omega)^2}}$$

This formula demonstrates an intuitive concept: by increasing the rigidity or damping coefficients, the vibration amplitude is reduced, but this formula does not say how much force has been transmitted to the foundation. The answer to this question is the transmissibility ratio $r_t = F_t(\omega)/F_a(\omega)$. This ratio is the proportion of the applied force transmitted to the foundation. It is also defined as the % of the applied force transmitted to the foundation. This ratio depends on the frequency and may be higher or lower than 1, but it can never be zero.

Formula 8.7 Transmissibility ratio of two degrees of freedom system

$$r_t = \sqrt[2]{\frac{k^2 + (f \cdot \omega)^2}{(k - m \cdot \omega^2)^2 + (f \cdot \omega)^2}}$$

By replacing ω with ($u \cdot \omega_n$) and introducing the damping ratio formula, the transmissibility ratio is returned with the following easy-to-use, non-dimensional formula:

Formula 8.8 Transmissibility ratio of a perfectly rigid foundation

$$r_t(u) = \frac{F_t(u)}{F_a(u)} = \sqrt[2]{\frac{1 + 4 \cdot \zeta^2 \cdot u^2}{(1 - u^2)^2 + 4 \cdot \zeta^2 \cdot u^2}}$$

As the foundation is rigid, it does not vibrate; therefore, its physical parameters do not participate in the transmissibility ratio formulas. Figures 8.4 and 8.5 show an example of the transmissibility ratio performance as a function of the springs' rigidity coefficient and the dampers' absorption coefficient.

The springs and dampers design's objectives are to make this transmissibility ratio lower than 1 as much as possible. If no dampers are used, the transmissibility formula is reduced to $r_t = k/(k - m \cdot \omega^2)$ or $r_t = 1/(1 - u^2)$, demonstrating the existence of an undesired resonance at u = 1. With Formula 8.7, charts of Figures 8.4 and 8.5 are generated, showing the transmissibility ratio versus the rigidity coefficient or the damping coefficient. They help to visualize the transmissibility ratio

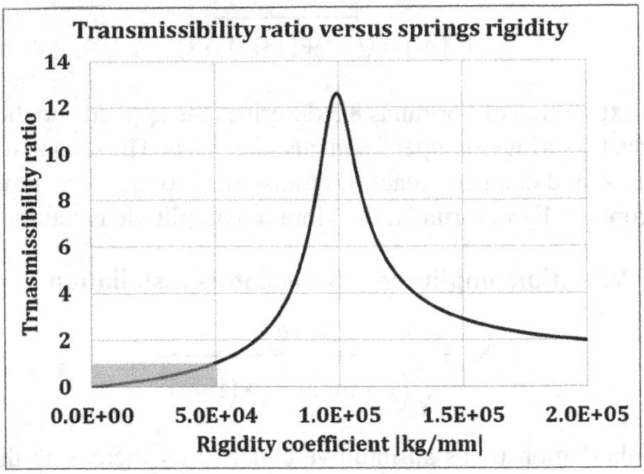

FIGURE 8.4 Transmissibility ratio versus k.

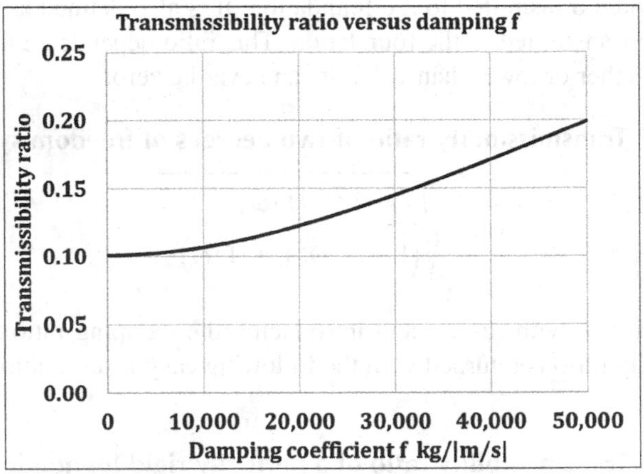

FIGURE 8.5 Transmissibility ratio versus f.

behavior. The frequency used to compute Figures 8.4 and 8.5 with Formula 8.7 is the disturbance frequency ω to obtain a design according to the system's real conditions.

A combination of springs and masses' properties define the system's dynamics; therefore, they can produce an acceptable low transmissibility value and avoid an undesired resonance. Instead, dampers form a passive element that does not produce resonance but is useful for reducing the system's natural frequency, absorbing the vibration energy, and mitigating the vibration amplitude by increasing the damping coefficient. See Figure 8.5. At a high damping coefficient, dampers tend to act as solid bars between the engine and the foundation, then applied forces on the engine could reach 100% transmission.

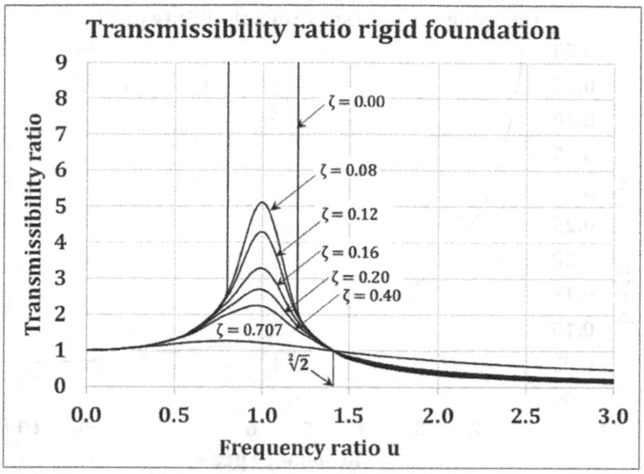

FIGURE 8.6 Transmissibility ratio of a perfectly rigid foundation.

The curve $r_t(k)$ of chart 8.6 has a critical point at which the transmissibility ratio equals 1. As the design's goal is to place the operating point below 1, the shaded area shown in Figure 8.4 must be the operating point's location. At transmissibility values higher than 1, the curve shows a resonance peak. It does not make sense to study this part of the curve to design isolating springs because, in that range, there is no force attenuation.

Figure 8.5 shows, for this case, the curve of the transmissibility ratio versus the damping factor f. It shows that the higher the damping factor, the higher the transmissibility ratio, theoretically infinite at infinite damping factor. In this case, the machine and the foundation are locked. As no more isolation exists, both machine and foundation vibrate with the same amplitude and frequency.

In practice, the transmissibility ratio is sometimes expressed as a percentage of the applied force. The inverse of r_t is known as the attenuation factor, a less used designation than the transmissibility ratio. Curves of r_t versus frequency ratio u for different damping ratios are in Figure 8.6.

The frequency at which the resonance peak occurs is obtained by the classical Calculus derivative of $r_t(u)$ with respect to u and equating the result to zero. This procedure demonstrates that the peak is produced at u =1. Therefore, operating at u = 1 is the worst operating condition for the machine and foundation mechanical integrity. The formula to calculate this peak is readily obtained by substituting u by 1 in Formula 8.8. Figure 8.7 shows the formula and curve of damping ratio versus resonance peak.

Formula 8.9 Resonance peak of the transmissibility ratio

$$r_{peak} = \sqrt[2]{1 + \frac{1}{4 \cdot \zeta^2}}$$

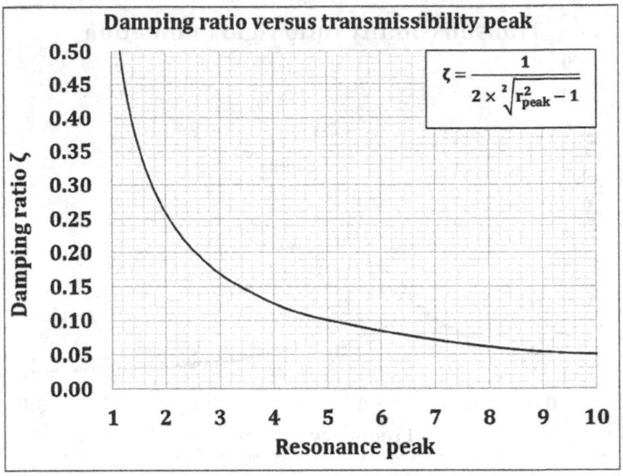

FIGURE 8.7 Transmissibility ratio peak versus damping ratio.

In practice, r_{peak} is specified for the spring-damper set design; in this case, the damping ratio is calculated for such specified peak value. Thus, to estimate the desired damping ratio ζ, it is cleared from Formula 8.9, and the result is Formula 8.10. Figure 8.7 shows the damping ratio ζ versus the resonance peak.

Formula 8.10 Damping ratio versus transmissibility peak

$$\zeta = \frac{1}{2 \times \sqrt[2]{r_{peak}^2 - 1}}$$

Figure 8.6 shows that for frequency ratios higher than $\sqrt[2]{2}$ transmissibility ratios are lower than 1. This critical value of frequency ratio is obtained from the solution to the equation $r_t(u) = 1$.

The attenuation region is magnified in Figure 8.8. This figure helps to estimate the desired operating point. The spring-damper designer must note that the transmissibility ratio is lower in the attenuation region, the lower the damping coefficient ζ. It does not happen the same for frequency ratios lower than $\sqrt[2]{2}$. Therefore, the attenuation region of applied forces in Figure 8.6 is at frequency ratios higher than $\sqrt[2]{2}$. This frequency range defines the acceptable region for the operation of any machine.

In summary: an acceptable approach to protecting the foundation is to make the machine operate in the attenuation region of Figure 8.6 in the high-frequency ratio range (recommended $u > 3$). As the frequency ratio is inversely proportional to the natural frequency, the machine-spring-damper set should have a low natural frequency to assure the frequency ratio has a value larger than 3. The operation in this region is achieved with soft springs. Dampers are added to hasten the transient vibration decay. They are especially useful in the vicinity of the resonance zone.

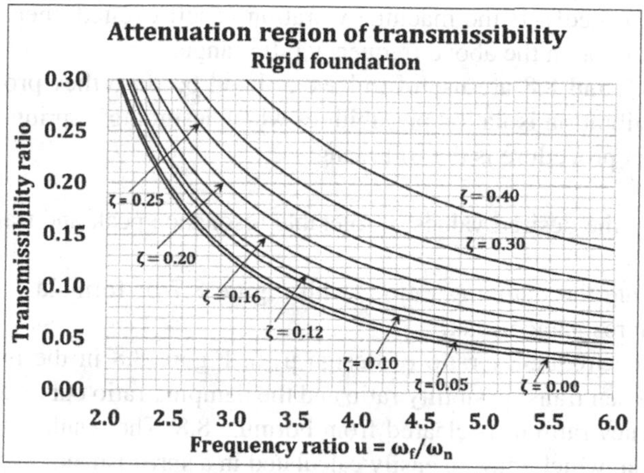

FIGURE 8.8 Transmissibility ratio of perfectly rigid foundations. Attenuation region.

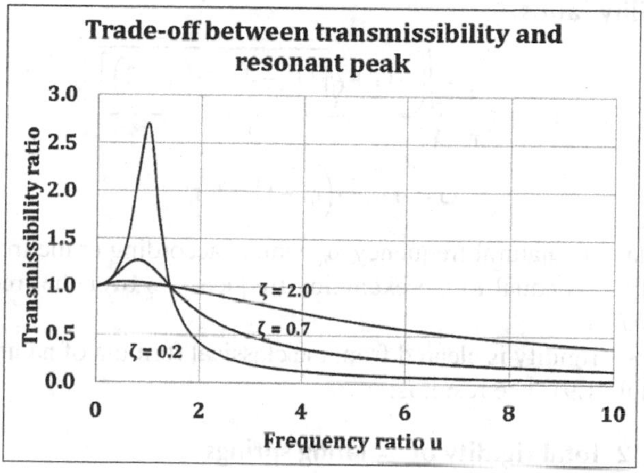

FIGURE 8.9 Low transmissibility is obtained with high resonance peak.

The rotor operation in the attenuation region has a caveat because it passes through the startup's resonance zone; therefore, machines operating in this zone have flexible rotors. See Section 4.1. Nonetheless, it is possible to mitigate this peak transmission by reducing the damping ratio. However, decreasing the damping ratio makes the resonance peak increase; therefore, there is a trade-off situation between transmissibility and resonance peak during startup. The designer has to decide between the two. This trade-off is evident by analyzing Figure 8.9.

8.2.3 Spring-Damper Set Design

It is recommended that the spring-damper design sets the operating frequency ratio between 3 and 5 to assure that the engine will work in the attenuation region (and far from resonance!). If the machine vibration is self-excited, then the machine velocity should be in the above frequency ratio range.

Figures 8.7 and 8.8 are useful to have at hand because they provide a quick insight into vibration isolation possibilities when designing a spring-damper set.

The design procedure is the following:

- Choose the desired values of tolerable resonance peak and transmissibility ratio.
- Calculate the damping ratio ζ with Figure 8.7 or formula 8.10 using the chosen resonance peak.
- Find the frequency ratio of design u_d in Figure 8.8 in the intersection of the chosen transmissibility ratio and the damping ratio curve. Alternatively, frequency ratio u_d is cleared from Formula 8.8. The result is the following formula, which may be easily calculated in a spreadsheet.

Formula 8.11 Operating frequency ratio versus damping and transmissibility ratios

$$u_d = +\frac{1}{r_t}\cdot\sqrt[2]{\frac{\left(-G+\sqrt[2]{\left[G^2-4\cdot r_t^2\cdot\left(r_t^2-1\right)\right]}\right)}{2}}$$

$$G = 4\cdot\zeta^2\cdot\left(r_t^2-1\right)-2\cdot r_t^2$$

- Calculate the natural frequency ω_n, which, according to the frequency ratio definition, is equal to the excitation frequency ω over the frequency ratio ($\omega_n = \omega/u_d$).
- Springs's rigidity is cleared from the classical formula of natural frequency (Formula 1.9). The result is:

Formula 8.12 Total rigidity of isolating springs

$$k = m\cdot\omega_n^2$$

This formula returns the rigidity of all springs. As more than one spring is necessary to support a rotating machine, the rigidity per spring is the total rigidity divided by the spring quantity. Usually, the machine manufacturer decides the support points' location on the machine where each spring-damper set must be installed.

- As ζ, k and m are known, the critical friction coefficient is calculated with Formula 1.19. Then the friction coefficient to be requested to dampers is:

Formula 8.13 Critical total friction coefficient

$$f = \zeta\cdot 2\cdot\sqrt[2]{k\cdot m}$$

The following example illustrates a practical calculation of calculating springs and dampers coefficients according to specified transmissibility and tolerable peak. The procedure is the same as described above.

8.2.4 Practical Assessment of Transmissibility Attenuation. Perfectly Rigid Foundation

Calculate the springs and dampers coefficients and vibration amplitude for a turbine with the following technical specifications.

Technical specifications:

1. Desired transmissibility ratio: $r_t = 0.15$
2. The tolerable peak of the transmissibility ratio: $r_{t\ peak} = 2.75$
3. Disturbance frequency: $\omega = 200$ rad/s
4. Machine weight: W = 12,500 kg
5. Applied force: $F_a = 2,000$ kg
6. Six spring-damper sets are required.

Calculation of springs' rigidity and dampers' friction coefficients:

1. Damping ratio: $\zeta = 1/[2 \times (2.75^2 - 1)^2 = 0.195$
2. Frequency ratio at the operating point: $u_d = 3.49$. Calculate with Formula 8.11.
3. Natural frequency of the turbine-springs-damper set: $\omega_n = 200/3.49 = 57.3$ rad/s
4. Machine mass: m = 12,500/9.81 = 1,274 kg mass
5. Springs total rigidity: $k = 1,274 \times 57.3^2 = 4,182,745$ kg/m
6. Rigidity per spring: $k_i = 4,182,987/6 = 697,124$ kg/m
7. Critical friction factor: $f_c = 2 \times 1, 274 \times 57.3 = 146,014$ kg/(m/s)
8. Dampers total friction coefficient: $f = 146, 014 \times 0.195 = 28,498$ kg/(m/s)
9. Friction factor per damper = 28,498/6 = 4,750 kg/(m/s)
10. Transmitted force: $F_t = 0.15 \times 2,000 = 300$ kg
11. Static deflection due to the machine's weight: $x_0 = W/k = 12, 500/4,182,745 \times 1,000 = 2.99$ mm
12. Vibration amplitude calculation:

Mechanical impedance after isolators installation:

$$Z = \sqrt[2]{\left(k - m\cdot\omega^2\right)^2 + \left(f\cdot\omega\right)^2}$$

$$= \sqrt[2]{\left(4,182,987 - 1,274\times 200^2\right)^2 + \left(28,499\times 200\right)^2}$$

$$= Z = 47,131,552 \text{ kg/m}$$

Vibration amplitude:

$$x_a = F_a / Z = 2,000 / 47,131,552 \times 1,000 = 0.042 \text{ mm}$$

The transmitted force is 15% of the applied force plus the machine weight. The practical procedure to reduce the transmissibility ratio is to install looser springs.

Let us assume that the transmissibility ratio must be equal to 0.10. Recalculation of springs rigidity returns a value of 392,240 kg/m per spring. Therefore, by reducing the transmissibility ratio to 33% (which is desirable), the spring rigidity descends 44%.

8.3 TRANSMISSIBILITY OF A NON-RIGID FOUNDATION OF KNOWN MASS

An acceptable model of a non-rigid foundation is a system with one degree of vertical freedom. Lateral motions rarely exist or are much smaller; thus, the foundation model does not consider them. With one vertical degree of freedom, harmonically excited non-rigid foundations vibrate with an amplitude inversely proportional to its mass. As these foundations vibrate, the spring-damper set reaction is in equilibrium with the foundation inertial reaction, equal to $m_2 \cdot \omega^2 \cdot x_{a2}$, where m_2 is the foundation mass, and x_{xa2} is its vibration amplitude. The machine, spring-damper sets and foundation assembly is two degrees of freedom system.

The mechanical system of Figure 8.10 is an example of a non-rigid foundation on spring-damper sets. Both the rotor and the machine frame are vibrating due to the applied force $F_a(t)$. In this non-rigid case, the frame is free to move with minimal vertical displacements. This freedom means that the frame vibrates when it is excited by a harmonic force.

In general, non-rigid foundations are in places where it is not possible to place a massive foundation. Horizontal and vertical harmonic forces, explained in Chapter 6, are applied to the ground. The ground reacts as a set of springs and dampers distributed around the foundation. The vibration energy is radiated along the soil under and at the foundation's sides, encompassing a cone-shaped distribution of efforts whose vertex is in the foundation. Horizontal forces produce shear effort waves, and vertical forces create compression and stretching effort waves.

Before the isolators' installation, a non-rigid foundation vibrates with the same amplitude and frequency as the machine. There are neither springs nor dampers between the machine and the foundation. The vibration amplitude for a frictionless

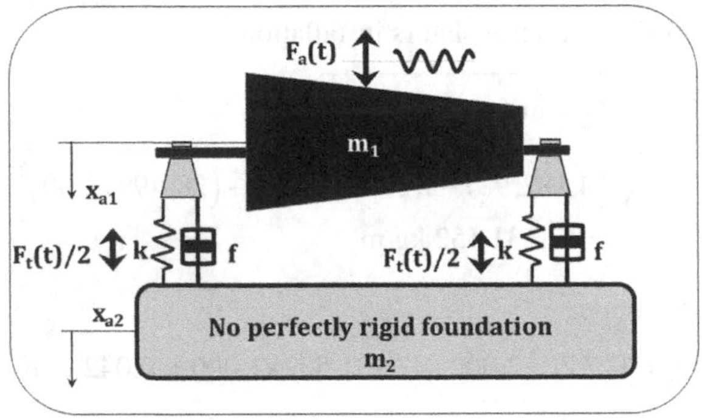

FIGURE 8.10 Frictionless transmission on a non-rigid foundation.

system of the non-rigid foundation before installing an isolator is calculated with the following formula, where F_a is the applied harmonic force, and ω is the disturbance frequency:

Formula 8.14 Vibration amplitude with no isolators between machine and foundation

$$x_a = \frac{F_a}{Z_1 + Z_2} = \frac{F_a}{(m_1 + m_2)\cdot\omega^2}$$

As there is no isolator between the machine and the foundation, they are solidly linked; therefore, they vibrate with the same amplitude at the disturbance frequency.

A non-rigid foundation's motion model, with an isolator between them, describes the vibration of both the machine mass (m_1) and the foundation mass (m_2). Thus, there are two motion equations: one for the machine mass and another for the foundation. See Formulas 8.15. The first equation is the equality between the harmonic applied force and the machine inertia plus the spring-damper set reaction. The second equation represents the equilibrium between the spring-damper set force on the foundation (transmitted force) and the foundation inertia[3].

Formula 8.15 Equations of motion of a non-rigid foundation of known mass

$$-m_1\cdot\omega^2\cdot x_{a1} + (k + j\cdot\omega\cdot f)\cdot(x_{a1} - x_{xa2}) = F_a$$

$$(k + j\cdot\omega\cdot f)\cdot(x_{xa2} - x_{a1}) - m_2\cdot\omega^2\cdot x_{xa2} = F_t$$

Where ω is the disturbance frequency of the harmonic applied force $F_a(t) = F_a\cdot\sin\omega\cdot t$.

The model discussed below assumes that the foundation mass is known. The applied harmonic force is transmitted to the foundation through the spring-damper set net stretching. As spring-damper sets are in mechanical parallel, for simplicity reasons, equations of motion consider they form only one set, with rigidity and damping equivalent to all of them. The spring-damper set's net stretching is $(x_{a2} - x_{a1})$ on the foundation side and $(x_{a1} - x_{a2})$ on the machine side, where x_{a1} and x_{a2} are the machine and foundation's vibration amplitude, respectively.

In Formulas 8.15, F_a is the amplitude of the harmonic applied force and amplitudes x_{a1} and x_{a2} are the vibration amplitudes of masses m_1 and m_2, respectively. In impedance notation, the equations of motion are as follows.

Formula 8.16 Equations of motion in impedance notation

$$Z_1\cdot x_{a1} + Z_t\cdot(x_{a1} - x_{a2}) = F_a$$

$$Z_2\cdot x_{a2} + Z_t\cdot(x_{a2} - x_{a1}) = 0$$

All impedances are a function of the disturbance frequency ω.

Formula 8.17 Mechanical impedances of a masses and spring-damper set

$$Z_1 = -m_1 \cdot \omega^2$$

$$Z_2 = -m_2 \cdot \omega^2$$

$$Z_t = k + j \cdot \omega \cdot f$$

Z_1 is the machine inertial impedance, Z_2 is the foundation inertial impedance, and Z_t is the spring-damper set impedance or transmitter impedance. Grouping coefficients of x_{a1} and x_{a2} of Formulas 8.16 they are expressed as:

Formula 8.18 Equation of motion of non-rigid foundation

$$\left(Z_1 + Z_t\right) \cdot x_{a1} - Z_t \cdot x_{a2} = F_a$$

$$-Z_t \cdot x_{a1} + \left(Z_2 + Z_t\right) \cdot x_{a2} = 0$$

After solving equations of Formula 8.18, the machine and foundation's vibration amplitude formulas are returned.

Formula 8.19 Vibration amplitudes

$$x_{a1}(\omega) = \frac{\left(Z_2 + Z_t\right) \cdot F_a}{Z_1 \cdot Z_2 + Z_t \cdot \left(Z_1 + Z_2\right)} \quad x_{a2}(\omega) = \frac{Z_t \cdot F_a}{Z_1 \cdot Z_2 + Z_t \cdot \left(Z_1 + Z_2\right)}$$

The foundation inertia opposes the force transmitted to it. Thus, this force is equal to the foundation impedance times the vibration amplitude x_{a2}.

Formula 8.20 Transmitted force to a non-rigid foundation

$$F_t = Z_2 \cdot x_{a2}(\omega)$$

Therefore, the transmissibility ratio F_t/F_a is obtained by replacing Formula 8.19 in Formula 8.20:

Formula 8.21 General formula of transmissibility ratio for rigid and non-rigid foundations

$$r_t(\omega) = \frac{F_t}{F_a} = \frac{Z_2 \cdot Z_t}{Z_1 \cdot Z_2 + Z_t \cdot \left(Z_1 + Z_2\right)}$$

This a general formula to calculate rigid and non-rigid foundations. In rigid frictionless foundations, discussed in Section 8.1, this formula is $r_t = Z_t/(Z_t+Z_1)$. Like in rigid foundations, a non-rigid foundation's transmissibility ratio may be higher or lower than 1, but it cannot be zero.

8.3.1 The Undamped Non-Rigid Foundation of Known Mass

Assume that the system of Figure 8.10 has no dampers and that the foundation mass m_2 is known. The system disturbance is ω. Therefore, the mechanical impedance of the system and the mass ratio are:

Formula 8.22 Mechanical impedances of a frictionless non-rigid foundation

$$Z_1 = -m_1 \cdot \omega^2$$

$$Z_2 = -m_2 \cdot \omega^2$$

$$Z_t = k$$

The transmissibility ratio arises by replacing these three formulas in the general Formula 8.21:

Formula 8.23 Transmissibility ratio of a frictionless non-rigid foundation

$$r_t(\omega) = \frac{F_t}{F_a} = \frac{-m_2 \cdot k}{m_1 \cdot m_2 \cdot \omega^2 - k \cdot (m_1 + m_2)}$$

Dividing the last formula by $k \cdot (m_1 + m_2)$, it returns:

Formula 8.24 Transmissibility ratio showing the system's natural frequency

$$r_t(\omega) = \frac{m_2}{m_1 + m_2} \cdot \frac{1}{1 - \dfrac{\omega^2}{\omega_n^2}}$$

Where the natural frequency ω_n^2 is equal to the rigidity k over the equivalent mass $m_{eq} = m_1 \cdot m_2/(m_1 + m_2)$, then:

Formula 8.25 Equivalent mass and natural frequency of a non-rigid foundation

$$m_{eq} = \frac{m_1 . m_2}{m_1 + m_2} \quad \omega_n = \sqrt[2]{\frac{k}{m_{eq}}}$$

The transmissibility ratio of Formula 8.24 may be expressed in dimensionless variables. This form allows constructing figures, valid for any non-rigid foundation system, that eases the springs' design. This procedure introduces the mass or weight ratio $b = m_2/m_1 = W_2/W_1$ and frequency ratio $u = \omega/\omega_n$ in Formula 8.24 to obtain the following non-dimensional formula of the transmissibility ratio:

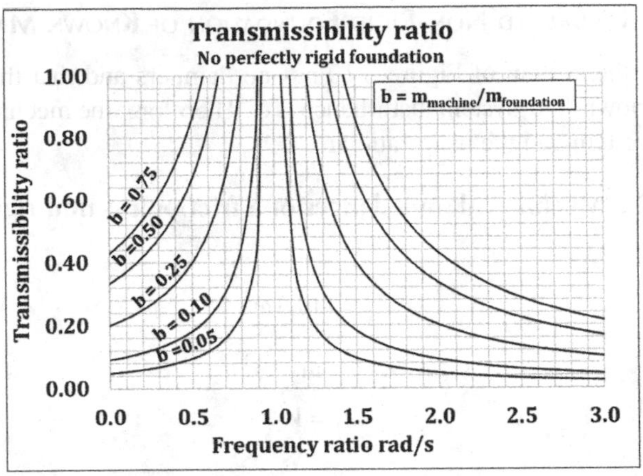

FIGURE 8.11 Transmissibility ratio of non-rigid foundations.

Formula 8.26 Transmissibility ratio versus non-dimensional variables b and u

$$r_t(u) = \frac{F_t}{F_a} = \frac{b}{b+1} \cdot \frac{1}{1-u^2}$$

In practice, the mass ratio b may be as low as 0.1. This ratio is used as the parameter of transmissibility curves versus frequency ratio in Figures 8.11 and 8.12. This figure is only valid for no damper isolators; only springs are considered.

Figure 8.11 displays the transmissibility ratio curves constructed with Formula 8.26, where b is each curve's parameter. It shows a resonance peak at u = 1. The machine must not work in this zone due to the excessively high transmitted forces to the foundation. However, it is impossible to skip this region during the startup of a self-excited turbomachine.

The lowest transmissibility ratios are in the right part of Figure 8.11; then, this is the attenuation region where the system's operating point must be placed. Figure 8.12 shows the amplified diagram of the attenuation region. In the attenuation region, the lower the mass ratio b, the lower the transmissibility ratio. It is recommended to adopt a frequency ratio equal to 2 or 3 times the disturbing frequency or even more. Clearing u from Formula 8.26, the design frequency ratio formula is:

Formula 8.27 Design frequency ratio versus b and r_t

$$u_d = \sqrt[2]{1 + \frac{b}{(b+1) \cdot r_t}}$$

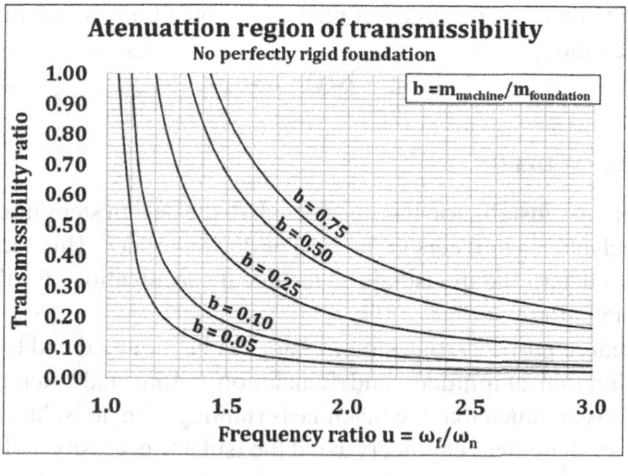

FIGURE 8.12 Transmissibility ratio in the attenuation region of non-rigid foundations.

In summary: the operating frequency ratio is only determined by two constants usually known: the masses ratio b and the wanted transmissibility ratio.

8.3.1.1 Vibration Amplitude Ratios

After isolator springs are installed, the machine and foundation have different vibration amplitudes, calculated with Formulas 8.19. From these formulas, the following amplitudes ratio r_a is derived:

Formula 8.28 Ratio of the foundation to the machine amplitude

$$r_a = \frac{x_{a2}}{x_{a1}} = \frac{Z_t}{Z_t + Z_2} = \frac{k}{k - m_2 \cdot \omega^2} = \frac{1}{1 - u^2}$$

Another indicator of the isolator performance is the ratio of the foundation amplitudes before and after the isolator installation.

Formula 8.29 Foundation amplitudes ratio after/before isolator installation

$$r_{ab} = \frac{x_{a2}}{x_a} = \frac{Z_t \cdot (Z_1 + Z_2)}{Z_1 \cdot Z_2 + Z_t \cdot (Z_1 + Z_2)} = \frac{1}{1 - u_b^2}$$

x_a is the vibration amplitude before the isolator installation when machine and foundation are solidly linked. It is calculated with Formula 8.14. The frequency ratio is not the same as in Formulas 8.26 and 8.28 because the natural frequency is different. In this case, is $\omega_{nb} = \sqrt[2]{k / m_2}$. This natural frequency is different from the ω_n of Formula 8.25.

Finally, note that the formula that relates the force transmissibility ratio r_t and the amplitude ratio r_a:

$$r_t = b/(1+b) \cdot r_a$$

8.3.2 Isolator Design

After the isolators' installation has been decided, the design steps must be planned by expert mechanical engineers in the field or design offices. This section presents a procedure to determine the isolators' technical specifications for their procurement and installation.

This procedure purports to design isolators to be mounted and to estimate the machine's vibration amplitudes and foundation before and after the isolators' installation. It is assumed that the machine is running with no isolators. Therefore, this is the case where field engineers avoid the isolator necessity and calculate the isolator for the procurement department

- **Initial information to be collected in the field**

1. The foundation and machine weights.
2. Measure the machine disturbance frequency and, if possible, the vibration amplitude.
3. Choose an adequate transmissibility ratio to protect the supporting structure of the foundation. It is recommended to consult a civil engineer expert in foundation design.
4. Estimate the force F_a and frequency ω applied by the vibrating machine.

- **Calculation procedure**

1. Mass ratio $b = m_2/m_1 = W_2/W_1$.
2. Design frequency ratio. Formula 8.27.

$$u_d = \sqrt[2]{1 + \frac{b}{(b+1) \cdot r_t}}$$

   This calculation can be done with Figure 8.12. Find the intersection of the desired r_t with the curve of the mass ratio b and read the frequency ratio u_d in the abscissa of this intersection point.
3. Natural frequency. It is equal to the disturbance frequency ω over the desired frequency ratio u_d.

$$\omega_n = \frac{\omega}{u_d}$$

4. Equivalent mass. Formula 8.25.

$$m_{eq} = \frac{m_1 \cdot m_2}{m_1 + m_2}$$

5. Springs'stiffness coefficient. It is cleared from the natural frequency Formula 8.25.

$$k = m_{eq} \cdot \omega_n{}^2$$

Each spring rigidity is the above value of k over the springs' quantity.

6. Mechanical impedances. Use Formulas 8.22 to calculate the numerical values of Z_1, Z_2 and Z_t.

$$Z_1 = -m_1 \cdot \omega^2 \qquad Z_2 = -m_2 \cdot \omega^2 \qquad Z_t = k$$

7. Vibration amplitudes. Formulas 8.19.
   Machine amplitude:

$$x_{a1}(\omega) = \frac{(Z_2 + Z_t) \cdot F_a}{Z_1 \cdot Z_2 + Z_t \cdot (Z_1 + Z_2)}$$

   Foundation amplitude:

$$x_{a2}(\omega) = \frac{Z_t \cdot F_a}{Z_1 \cdot Z_2 + Z_t \cdot (Z_1 + Z_2)}$$

8. With Formulas 8.28 and 8.29, calculate:
   Foundation to the machine amplitude ratio:

$$r_a = \frac{x_{a2}}{x_{a1}} = \frac{k}{k - m_2 \cdot \omega^2}$$

   Foundation amplitudes ratio before and after isolators installation:

$$r_b = \frac{x_{a2}}{x_a} = \frac{k}{k + m_{eq} \cdot \omega^2}$$

In the next section, this procedure is applied to a practical case.

8.3.2.1 Practical Assessment of Spring Rigidity for a Non-Rigid Foundation

Consider a rotating machine installed on a non-rigid foundation, transmitting a vibratory force to the foundation. Therefore, it is requested to isolate the machine with $n_s = 6$ springs and reduce the transmitted forces to 5% of its present value. The estimated applied force is F_a = 1,000 kg and the disturbance frequency is

$\omega = 120$ rad/s. The machine's weight is $W_1 = 19{,}000$ kg, and the foundation weight is $W_2 = 9{,}000$ kg.

It is requested to calculate:

1. Rigidity coefficient of springs to be procured and installed.
2. Machine and foundation's vibration amplitudes, before and after springs installation.

Calculations:

1. Machine and foundation masses:

$$m_1 = W_1 / g = 1{,}937 \text{ kg mass}$$
$$m_2 = W_2 / g = 917 \text{ kg mass}$$

2. Desired transmitted force:

$$F_t = 0.05 \times 1{,}000 = 50 \text{ kg}$$

3. Mass ratio:

$$b = m_2 / m_1 = 0.47$$

4. Design frequency ratio:

$$u_d = \sqrt[2]{1 + \frac{b}{(b+1) \cdot r_t}} = 2.73$$

5. Natural frequency of machine and foundation:

$$\omega_n = \frac{\omega}{u_d} = 44.0 \, \text{rad/s}$$

6. Equivalent mass:

$$m_{eq} = \frac{m_1 \cdot m_2}{m_1 + m_2} = 623 \, \text{kg mass}$$

7. Rigidity coefficient of 6 springs to be installed:

$$k_{total} = m_{eq} \cdot \omega_n^2 = 1{,}206{,}775 \, \text{kg/m}$$

8. Rigidity coefficient per spring:

$$k = k_{total} / n_s = 201{,}129 \text{ kg/m}$$

9. Mechanical impedances modulus:

$$Z_1 = m_1 \cdot \omega^2 = 27{,}889{,}908 \text{ kg/m}$$

$$Z_2 = m_2 \cdot \omega^2 = 13{,}211{,}009 \text{ kg/m}$$

$$Z_t = k_{total} = 1{,}206{,}775 \text{ kg/m}$$

10. Vibration amplitude of machine and foundation before springs installation:

$$x_a = \frac{F_a}{Z_1 + Z_2} \times 1{,}000 = 0.024\,\text{mm}$$

11. Vibration amplitudes of machine and foundation after springs installation:

$$x_{a1} = \frac{(Z_2 + Z_t) \cdot F_a}{Z_1 \cdot Z_2 + Z_t \cdot (Z_1 + Z_2)} \times 1{,}000 = 0.070\,\text{mm}$$

$$x_{a2} = \frac{Z_t \cdot F_a}{Z_1 \cdot Z_2 + Z_t \cdot (Z_1 + Z_2)} \times 1{,}000 = 0.003\,\text{mm}$$

12. Amplitude ratios:

$$r_a = \frac{x_{a2}}{x_{a1}} = 0.043$$

$$r_b = \frac{x_{a2}}{x_a} = 0.125$$

Conclusion: due to the springs' installation, the transmitted force has been reduced 20 times, from 1,000 kg to 50 kg. Before installing the springs, the foundation vibration amplitude was equal to the machine's vibration amplitude (0.024 mm). However, after installing the springs, the machine amplitude is 0.07 mm, and the foundation amplitude is only 0.003 mm, or 12.5% of the amplitude before the isolator's installation. Ratio r_a indicates that the foundation amplitude is 4.3% of the machine vibration amplitude.

8.4 TRANSMISSIBILITY OF OFF-LAND Z FOUNDATION

Off-land foundations are those mounted in ships, oil and gas offshore platforms and other similar installations. In ships, these foundations are a steel frame supported by ribs. The hull and ribs mathematical model is almost impossible to calculate by looking at CAD drawings or evaluating the hull rigidity and friction by the classical methods explained in this text. Therefore, it is necessary to model this foundation based on in situ tests that return the site's frequency response where the machine is or will be installed. This section describes such a test, convenient as a first approach to the vibration study because it returns the foundation impedance. If more accuracy and details are needed, further studies may be done with a finite element method software. In this case, hiring a specialized consulting company or technician is a good option.

Typical off-land installations are airplanes, ships, offshore hydrocarbon platforms, or any other aerial or marine installation type. In these plants, rotating machines transmit vibrations not only to their frames but also to nearby structures. In naval propulsion plants, the rotating machines are installed on racks attached to the hull's ribs, which react with inertial, rigid, and damping forces. The ship's ribs, frame, hull and other mechanically linked nearby parts supporting the machine are considered only one foundation, identified in this text as Z foundation.

In the naval industry, it is essential to prevent vibration, not only because of materials fatigue but also because vibration waves propagate in the water, with undesirable results. The aftermath of this propagation for fishing vessels is that fish flee from the ship because of vibratory waves. For the same reason, surface warships must not emit vibrations to be undetected by enemy submarines. At the same time, submarines should not be detected by surface ships and antisubmarine airplanes. Submarines and warships must not emit vibratory waves of any kind, a significant technical challenge.

The frequency test is simple because it uses a vibration generator to excite the Z foundation and a vibrometer to measure the structure frequency response at different nearby places. Therefore, this test returns the Z foundation frequency response curve. With the results of this test, it is possible to model the foundation impedance formula. The method to obtain Z(ω) is explained below for three typical frequency responses. See in the scheme of Figure 8.13 the mechanical impedance Z(ω) replacing the foundation of Figure 8.10. The typical Z foundation reaction is a combination of inertial, rigidity and damping forces. Then it is expressed as:

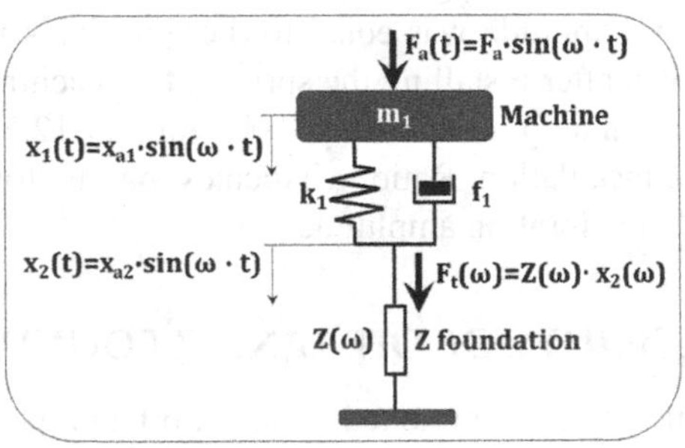

FIGURE 8.13 Scheme of an off-land installation.

Formula 8.30 Mechanical impedance formed by three reaction components

$$Z_z(\omega) = \frac{F(\omega)}{x_z(\omega)} = k_z - m_z \cdot \omega^2 + j \cdot f_z \cdot \omega$$

The calculation of k_z, m_z and f_z is made with the test's frequency response curve. In the usual case of a machine supported by a spring-damper set, the equations of motion are Formulas 8.16, where the impedance Z_z of the foundation replaces Z_2. In systems with Z foundation, the general Formula 8.21 is also valid for calculating their transmissibility ratio.

8.4.1 Frequency Response Test of a Z Foundation

A Z foundation's frequency response test returns the vibration amplitude curve versus the vibrator frequency used to excite the Z foundation. The test results are plotted in the Bode plot, named after its creator. The magnification factor is in decibels (1 dB = 20 · log Q) versus frequency in a logarithmic scale in this figure. The chart in Figure 8.14 represents the Bode plot for several types of Z foundations. At low frequencies, the curve shape is dominated by the spring stiffness, and as frequency increases, the foundation response is more dependent on the foundation inertia. Finally, the mass reaction declines to zero because the system inertia cannot follow a high-frequency excitation. It means that inertial forces dominate the system response at high frequencies by severely reducing the vibration amplitude.

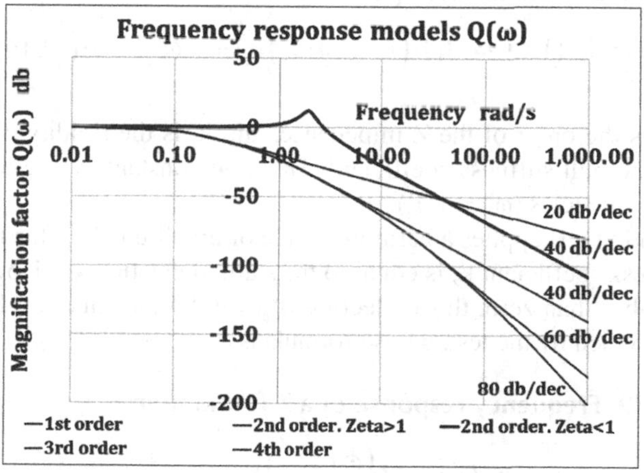

FIGURE 8.14 Frequency response of several Z foundations.

The thick curve of Figure 8.14 presents a resonance peak Q_r of 12 dB at ω_r = 2.1 rad/s, which corresponds to systems with a damping ratio lower than 0.707. Curves of thin lines have no resonance peaks because their damping ratio is higher than 0.707. These curves correspond to systems formed by several first-order systems. The curve decline rate allows the determination of the Z foundation system order. A first-order system has a slope of –20 dB/decade at high frequencies, and a second-order system has a slope of –40 dB/decade (regardless of its damping ratio); a third-order system has a slope of –60 dB/decade and so on. In general, several first-order systems form the Z foundation model; consequently, the curve continuously declines with increasing slopes as frequency rises.

The high-frequency slope is approximated by right lines asymptotes of the curve[4]. This approximation simplifies the analysis of the test results. These asymptotes allow calculation of the foundation impedance formula's parameters, as demonstrated in the next section. Frequencies at which the asymptote slope changes are called breakpoints or corner frequencies ω_b. For example, if the highest slope is –80 dB/decade, the Z foundation is formed by four first-order systems, which means that the curve has four breaking frequencies. These frequencies are identified in the Bode plot, as shown in Figure 8.14 for a second-order system. This identification is made graphically on the base of the curve returned by the test. There are as many corner frequencies as the system order of the Z foundation.

8.4.1.1 Frequency Response With No Resonance Peak

The above discussion concludes that if no resonance peak arises in the test curve, the product of several first-order impedances forms the Z foundation impedance.

Therefore, the general formula of a Z impedance of a non-resonant system is:

Formula 8.31 Z foundation impedance for $\zeta > 0.707$

$$Z_z(\omega) = k_z \cdot (1 + j \cdot \omega \cdot T_1) \cdot (1 + j \cdot \omega \cdot T_2) \cdot (1 + j \cdot \omega \cdot T_3) \cdots (1 + j \cdot \omega \cdot T_n)$$

where n is the order of the Z impedance, and k_z is the rigidity coefficient. In ships, k_z is the hull stiffness coefficient. The time constants T_i are the inverse of the corner frequencies ($\omega_{bi} = 1/T_i$).

The vibration test applies a harmonic force of amplitude F_v. This force divided by the stiffness coefficient k_z is equal to the static deflection x_{z0}. However, at frequencies higher than zero, this deflection is given by the curve of the frequency response provided by the test, whose formula is:

Formula 8.32 Frequency response of a Z foundation

$$Q_z(\omega) = \frac{x_z(\omega)}{x_{z0}} = \frac{F_v}{Z_z(\omega)} = \frac{1}{z_z(\omega)}$$

In this formula, the impedance ratio is $z_z(\omega) = Z_z(\omega)/k_z$.

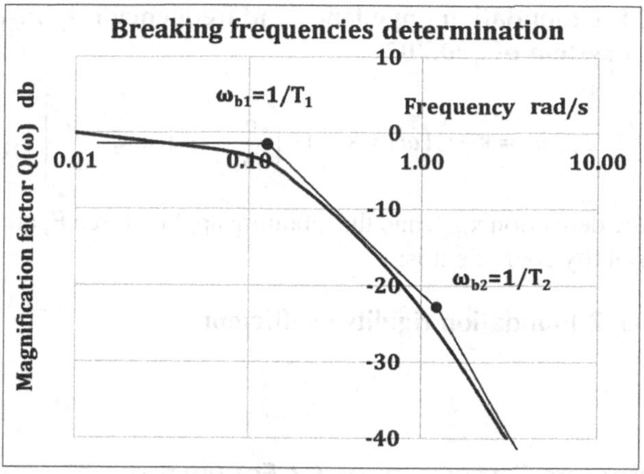

FIGURE 8.15 Graphical determination of break frequencies ω_b in a second-order Z foundation.

As an example of a second-order Z foundation for $\zeta > 0.707$, see Figure 8.15. This figure shows two asymptotes and two breaking frequencies. Therefore, this Z foundation has two first-order impedances, whose time constants are T_1 and T_2.

Two lines of −20 and −40 dB/decade are drawn tangent to the curve at different frequencies (see the thin straight lines). The black points show the break frequencies. In this example, they are at 0.1 and 1.0 rad/s. x_{z0} is read at the lowest possible frequency because logarithmic scales do not have zero. In this case, x_{z0} in dB is read at 0.01 rad/s. Its magnitude is 0 dB, which means $x_{z0} = 1$.

8.4.1.2 Frequency Response With Resonance Peak

Chapter 1 states that an underdamped second-order system has two parameters: natural frequency ω_n and damping ratio $\zeta < 1$. Both are a function of the resonance peak Q_r and the peak frequency ω_r. Therefore, knowing these two parameters makes it possible to calculate the Z foundation's damping ratio and the natural frequency. Q_r and ω_r are read in the Bode plot produced by the frequency test, and the damping ratio and natural frequency are calculated with Formulas 8.33. These formulas are derived from Formulas 1.73 and 1.74.

Formula 8.33 Damping ratio and natural frequency from the frequency test

$$\zeta = \sqrt[2]{\frac{1-\sqrt[2]{1-\dfrac{1}{Q_r}}}{2}} \rightarrow \omega_n = \frac{\omega_r}{\sqrt[2]{1-2\cdot\zeta^2}}$$

With the ω_n and ζ values, the Z foundation impedance formula is:

Formula 8.34 Z foundation impedance and frequency response as a second-order system of $\zeta<0.707$

$$Z_z(\omega)=k_z \cdot z_z(\omega)=k_z \cdot \left[1-\left(\frac{\omega}{\omega_n}\right)^2+j\cdot 2\cdot \zeta \cdot \frac{\omega}{\omega_n}\right]$$

As the static deflection x_{z0test} and the vibrating applied force F_v are known from the test, the rigidity coefficient is:

Formula 8.35 Z foundation rigidity coefficient

$$k_z=\frac{F_v}{x_{z0test}}$$

8.4.2 Impedances Calculation of a Z Foundation

Mechanical impedance properties (mass, stiffness, and damping) of a Z foundation are obtained from the comparison of the $x_{ztest}(\omega)$ curve given by the vibration test with the theoretical expression of $x_z(\omega)$. This comparison allows calculating the foundation properties m_z, k_z and f_z.

8.4.2.1 Z Model of First-Order

The Z foundation of the first-order has a neglectable mass, but its rigidity and damping properties cannot be neglected. Therefore, the Z foundation is equivalent to a spring-damper set whose properties f_z and k_z are unknown. The amplitude (F_v) of the force applied by the vibrator is known. Time constant T_{test} and x_{z0test} are also known because they are read in the Bode plot of the frequency response test. Next, the test result is compared with a first-order system's theoretical model to obtain f_z and k_z. This comparison is made by looking at formulas of $x_{ztest}(\omega)$ and $x_{zmodel}(\omega)$.

$$x_{ztest}(\omega)=\frac{x_{z0test}}{(1+j\cdot\omega\cdot T_{test})} \quad x_{zmodel}(\omega)=\frac{F_v}{k_z+j\cdot f_z\cdot\omega}=\frac{x_{z0model}}{1+j\cdot\omega\cdot\frac{f_z}{k_z}}$$

Because $x(\omega)z_{test}=x_{zmodel}(\omega)$, the following formulas arise:

$$x_{z0test}=x_{z0model} \quad T_{test}=\frac{f_z}{k_z}$$

Therefore, k_z and f_z are:

Formula 8.36 Rigidity and friction of a first-order Z foundation as a function of test results

$$k_z=\frac{F_v}{x_{z0test}} \quad f_z=T_{test}\cdot k_z$$

And the Z foundation impedance formulas are:

Formula 8.37 Mechanical impedance of a first-order Z foundation

$$Z_z(\omega) = \frac{F_v(\omega)}{x_z(\omega)} = k_z + j \cdot f_z \cdot \omega \quad |Z_z(\omega)| = \sqrt[2]{k_z^{\ 2} + (f_z \cdot \omega)^2}$$

8.4.2.2 Z model of Second-Order

The test reveals a second-order system because the high-frequency slope is −40dB/dec. Therefore, the foundation model is formed by two first-order systems in series (ζ>0.707) or by one undamped second-order system (ζ<0.707)

8.4.2.3 Frequency Response Curve with No Peak (ζ>0.707)

Parameters returned by the test results are T_1, T_2, x_{z0test} and F_v. Where T_1 and T_2 are the inverses of corner frequencies, read in the Bode plot, both the test frequency response and the model frequency response must be equal to obtain the Z model parameters based on the frequency test results:

$$x_{ztest}(\omega) = \frac{x_{z0test}}{(1 + j \cdot \omega \cdot T_1) \cdot (1 + j \cdot \omega \cdot T_2)} = \frac{x_{z0test}}{1 - T_1 \cdot T_2 \cdot \omega^2 + j \cdot (T_1 + T_2) \cdot \omega}$$

$$x_{zmodel}(\omega) = \frac{F_v}{k_z - m_z \cdot \omega^2 + j \cdot f_z \cdot \omega} = \frac{x_{z0model}}{1 - \frac{\omega^2}{\omega_{nz}^2} + j \cdot \frac{f_z}{k_z} \cdot \omega}$$

By equating the last two formulas, the following expressions show the calculation sequence:

Formula 8.38 Z impedance properties as a second-order system of ζ>1

$$x_{z0test} = x_{z0model} \quad \omega_{nz}^2 = \frac{1}{T_1 \cdot T_2} \quad k_z = \frac{F_v}{x_{z0\,test}} \quad m_z = \frac{k_z}{\omega_{nz\,z}^2} \quad f_z = (T_1 + T_2) \cdot k_z$$

In a ship, these are the physical properties of a discrete system equivalent to the hull distributed system. Note that the inclusion of $\omega_{nz}^{\ 2}$ in the above formulas does not mean that the system oscillates. It is just the inverse of the time constants product ($T_1 \cdot T_2$). With the above formulas of k_z, f_z and m_z, the mechanical impedance Zz(ω) formula is written.

Formula 8.39 Z foundation formulas as a second-order system of ζ>1

$$Z_z(\omega) = \frac{F_v}{x_z(\omega)} = k_z - m_z \cdot \omega^2 + j \cdot f_z \cdot \omega \quad |Z_z(\omega)| = \sqrt[2]{(k_z - m_z \cdot \omega^2)^2 + (f_z \cdot \omega)^2}$$

8.4.2.4 Frequency Response Test with Peak (ζ<0.707)

The frequency response given by the vibration test exhibits a resonance peak and a −40 dB/decade slope at high frequencies. Therefore, the Z foundation reacts as a second-order system of $\zeta<0.707$. The known parameters returned by the test are F_v, x_{z0}, M_p and ω_p. With the same comparison procedure used in the last case, formulas of ζ_z, ω_{nz}, k_z, m_z and f_z are shown below in the calculation sequence:

Formula 8.40 Z foundation properties as a second-order system of $\zeta<0.707$

$$x_{z0test} = x_{z0model} \quad k_z = \frac{F_v}{x_{z0test}} \quad \zeta_z = \sqrt[2]{\frac{1-\sqrt[2]{1-\frac{1}{Q_r}}}{2}}$$

$$\omega_{nz} = \frac{\omega_p}{\sqrt[2]{1-2\cdot\zeta^2}} \quad m_z = \frac{k_z}{\omega_{nz}^2} \quad f_z = \frac{2\cdot\zeta}{\omega_{nz}}$$

These formulas determine the same expression for $Z(\omega)$ and $|Z(\omega)|$ than the case of $\zeta>0.707$. See Formulas 8.39.

The study of vibration transmission in ships, oil platforms, airplanes and other similar installations off-land is challenging. However, specialized software can create acceptable models to simulate vibrations and their propagation through the hull, piping, machines, and structures.

8.4.3 Example of Spring Calculation to Isolate a Z Foundation

A ship's hull was subject to a frequency response test to assess a turbine installation place that weighs 25,000 kg and runs at a velocity of 600 RPM. The frequency response curve shows two corner frequencies at 0.2 and 1.0 rad/s. After the lowest corner frequency (0.2 rad/s) the slope of the curve is −20 dB/dec and after the second one (1.0 rad/s) the slope is −40 dB/dec. The test's applied force was 900 kg, which produced a hull's static deflection of 1.13 mm.

It is expected that the machine will be self-excited; therefore, it is required to isolate the machine from the hull by installing springs between the turbine and the hull.

The identification of the Z foundation order and calculation of the hull's properties, it is needed the following data:

1. Rigidity k_z
2. Natural frequency ω_{nz}
3. Damping coefficient f_z
4. Mass m_z
5. Critical friction coefficient f_{cz}
6. Damping ratio ζ_z
7. To calculate the spring rigidity for a transmissibility ratio equal to 0.10 at the rated turbine's speed.
8. To plot the transmissibility ratio curve versus the turbine velocity to assess potential risks at velocity other than the rated speed.

Order of the Z foundation and disturbance frequency

As the curve slope at high-frequency is -40dB/decade, the foundation is second-order. The curve does not present a peak; therefore, the foundation is formed by two first-order systems. The time constants are:

$T_1 = 1/\omega_1 = 1/0.2 = 5$ s
$T_2 = 1/\omega_2 = 1/1.0 = 1$ s

Calculation of hull's properties:

1. Disturbance frequency $\omega = \pi \cdot N/30 = 62.8$ rad/s
2. Rigidity: $k_z = F_v/x_{z0} = 900/1.13 \times 1{,}000 = 796{,}460$ kg/m
3. Friction factor: $f_z = k_z \cdot (T_1 + T_2) = 796,460 \times (5 + 1) = 4{,}778{,}761$ kg/(m/s)
4. Natural frequency: $\omega_{nz} = \sqrt[2]{\omega_1 \cdot \omega_2} = \sqrt[2]{0.2 \times 1.0} = 0.45\,\text{rad/s}$
5. Mass: $m_z = k_z/\omega_{nz}^2 = 796,400/0.45^2 = 3{,}982{,}301$ kg mass
6. Critical friction coefficient:
   $fcz = 2 \times \sqrt[2]{k_z \cdot m_z} = \sqrt[2]{796{,}460 \times 3{,}982{,}878} = 3{,}561{,}878\,\text{kg}/(\text{m/s})$
7. Damping ratio: $\zeta_z = f_z/f_{cz} = 4{,}778{,}761/3{,}561{,}878 = 1.34$
8. Calculation of mechanical impedances at the disturbance frequency
   a. Turbine mass: $m_1 = W/g = 25{,}000/9.8 = 2{,}548$ kg mass
   b. Turbine: $Z_1 = m_1 \cdot \omega^2 = 2{,}551 \times 62.8^2 = 10{,}060{,}759$ kg/m
   c. Z foundation:

$$Z_2 = k_z \cdot \sqrt[2]{\left[\left(1+\omega^2 \cdot T_1^2\right)\cdot\left(1+\omega^2 \cdot T_2^2\right)\right]}$$
$$= 796{,}460 \times \sqrt[2]{\left[\left(1+62.8^2 \times 5^2\right)\cdot\left(1+62.8^2 \times 1^2\right)\right]}$$

$$Z_2 = 15{,}723{,}564{,}417 \text{ kg/m}$$

   d. Springs mechanical impedance Z_t for $r_t = 0.10$ is cleared from Formula 8.21. This operation returns the following expression.

$$Z_t = k_z = \frac{r_t \cdot Z_1 \cdot Z_2}{(1-r_t)\cdot Z_2 - r_t \cdot Z_1}$$

Formula 1.41 Transfer impedance as a function of Z_1, Z_2 and r_t

$$Z_t = k_z = \frac{0.10 \times 10{,}072{,}025 \times 15{,}723{,}564{,}417}{(1-0.10)\times 15{,}723{,}564{,}417 - 0.10 \times 10{,}071{,}025} = 1{,}117{,}042\,\text{kg/m}$$

9. Transmissibility ratio curve versus the machine velocity:

Formula 8.21 is used with figures calculated above to see the transmissibility ratio curve: r_t(RPM). The result is Figure 8.16.

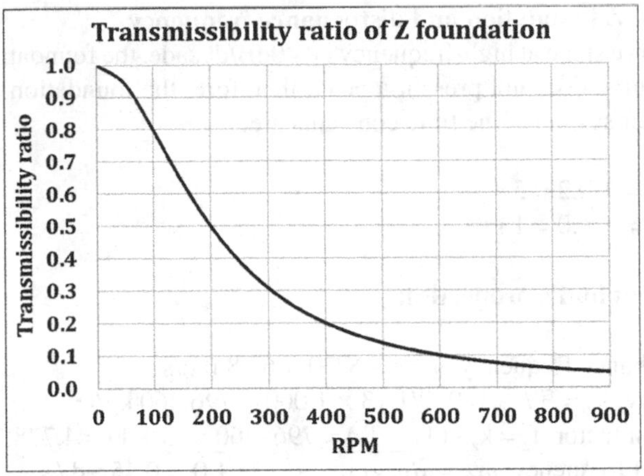

FIGURE 8.16 Transmissibility ratio of a system with second-order Z foundation.

This figure reveals that the system has no resonance; however, transmitted forces are high at low velocities. Therefore, if the machine runs at a slower velocity than the rated speed, the transmitted force will be a problem that will oblige the rotor's balancing.

In summary: there is a good transmissibility ratio at rated speed. However, if the turbine velocity is reduced to 50% of the rated speed, the transmitted forces will be three times higher than forces at the rated speed ($r_t = 0.30$). During startup, the transmitted forces are high, but they are reduced as velocity increases.

NOTES

1 Books recommended for Chapter 8. 1. Mechanical Vibrations by S. Graham Kelly. Schaum's Outline Series. McGraw Hill, 1996 – 2. Fundamentals of Vibration Engineering by Isidor Bykhovsky. MIR Publishers, Moscow. 1st published 1972 – 3. Mechanical Vibrations by J.P. Den Hartog. Dover Publications, Inc. New York, 1985, 4th edition – 4. Mechanical Vibration by Haym Benaroya, Mark Nagurka and Seon Han. CRC Press. Taylor and Francis Group, fourth edition 2018. 5. Marks' Standard Handbook for Mechanical Engineers by Eugene A. Avallone and Theodore Baumeister, Mc Graw Hill, 10th edition

2 It is advisable to revisit Section 1.5.1 about the vector interpretation of the equation of motion of a mass-damper-spring system.

3 See Mechanical Vibrations by J.P. Den Hartog. Dover Publications, Inc. New York, 1985, 4th edition Section 3.6, and Fundamentals of Vibration Engineering by Isidor Bykhovsky. MIR Publishers, Moscow. 1st published 1972 Section 14.

4 See Modern Control Engineering, by Katsuhiko Ogata, Pearson Editorial, 5th edition, 2010. Figure 7-6.

9 Vibration Absorption

9.1 INTRODUCTION TO VIBRATION ABSORPTION

This chapter explains the two most common types of absorbers: the Frahm and the Stockbridge. However, other more sophisticated absorbers exist, including some with automatic feedback control described in Chapter 11.

The German engineer Hermann Frahm (1867–1939) was the pioneer of vibration absorbers, not only for mechanical rotating machines but also for ships. He invented the anti-roll tank (schlingertank) to stabilize ships sailing in a rough sea. Today, tall buildings installed in windy zones are built with an inner pendulum of significant dimensions based on the same principle. The pendulum oscillation is opposed to the building oscillation induced by the wind. Frahm's invention of an absorber composed of a mass and spring tuned with the machines or structure's natural frequency has proven successful in industrial installations.

The other important absorber discussed in this chapter is the Stockbridge, named after its inventor. This absorber is for preventing vibrations produced by Karman vortices in the overhead lines leeward flow.

If the machine is a second-order SDOF system, the machine-absorber assembly is a fourth-order system. This fourth-order system's equation of motion demonstrates that if the absorber is correctly tuned, the host vibration amplitude is zero. It is possible to absorb vibration with only one additional mass attached to the machine. However, this absorption method is poorer than the spring-mass set absorption. In this text, the absorber formed by adding mass to the machine is not considered.

The Frahm's absorber is perhaps the most common vibration absorber used today in machines and structures. The physical principle (see Figure 9.2) of this absorber is used in large industrial fans, diesel motors, hammer mills, high voltage overhead lines, structures, piping, buildings, hand devices such as baseball bats (smart-bat), hairdryers, smart-skis, and many other applications.

According to the sign convention of action and reaction forces, the applied force F_d and the masses inertial reactions are positive and point downward. The spring reaction forces are negative; then, they are pointing upward. The equilibrium of these forces is represented by the equations of motion discussed in Section 9.3.1.

Absorbers have the following uses and acronyms: a) Absorbers, used for self-excited vibration: TMD for Tuned Mass Damper or DVA for Dynamic Vibration Absorber, b) Neutralizers, used for externally-excited vibration: TVN for Tuned Vibration Neutralizer[2]. Any of these devices have a natural frequency equal to the exciting frequency. If this condition is met, the machine vibration amplitude is zero, for reasons explained ahead in this chapter.

Despite this designation difference between absorber and neutralizer, it is common to use the term absorber for neutralizers because they respond to the same

DOI: 10.1201/9781003175230-12

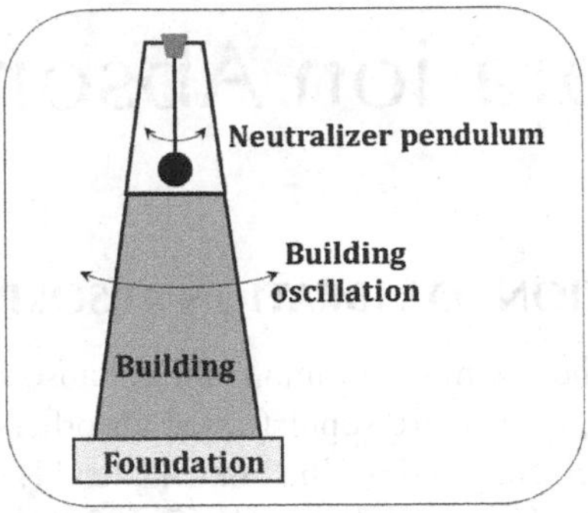

FIGURE 9.1 Building with an oscillation absorber pendulum.

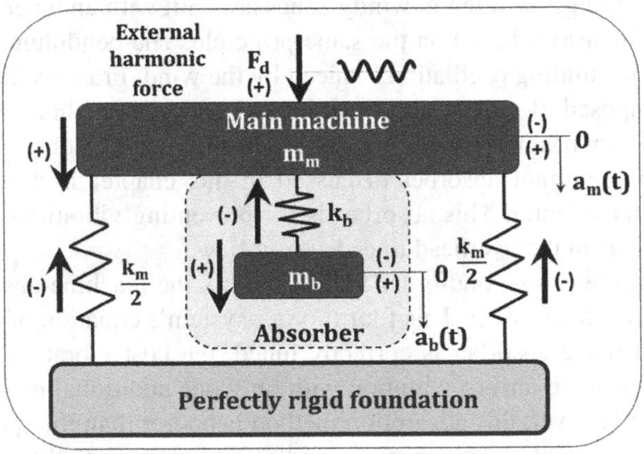

FIGURE 9.2 Scheme of a Frahm's absorber installation.

mechanical configuration. In many technical publications, the term machine is replaced by the term host, more general because it involves piping, structures and other vibrating parts that are not rotating.

9.2 VIBRATION ABSORBERS FOR ROTATING MACHINES

An absorber aims to shape the system frequency response so that the machine vibration has zero amplitude at the disturbance frequency. Figure 9.3 shows the case of a self-excited system with a natural frequency of 175 rad/s. This frequency coincides with the shaft velocity (1,671 RPM), producing an intolerable resonance condition. Therefore, an absorber is installed tuned at 175 rad/s. Figure 9.3 shows two curves of the machine vibration amplitude: before the absorber is installed (fine curve) and after the absorber is installed (thick curve). After the absorber installation, this amplitude falls to zero at 175 rad/s. The chart shows the best operating point (BOP) at that frequency. Note the symmetry of the

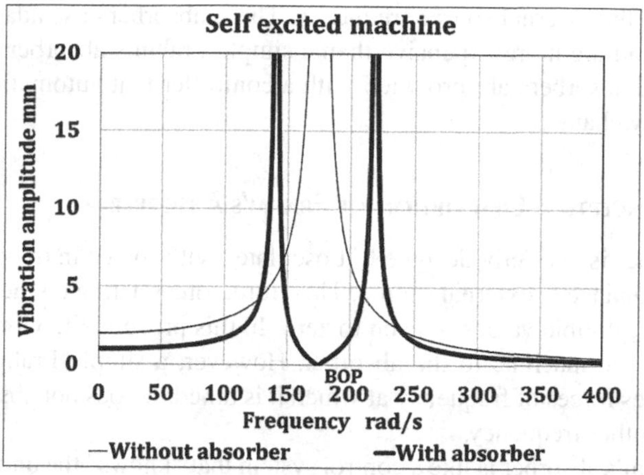

FIGURE 9.3 Absorption produced by an absorber of a self-excited vibration.

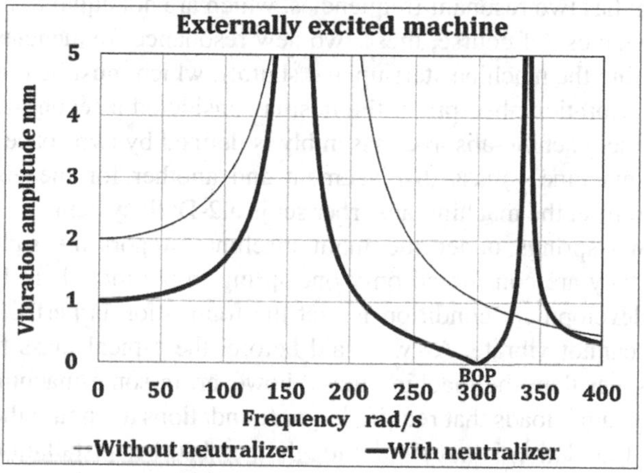

FIGURE 9.4 Absorption produced by a TVN absorber to an external source vibration.

machine-absorber resonance peaks with respect to the natural frequency typical of a self-excited case.

Figure 9.4 shows the same machine excited by an external source of 300 rad/s frequency. Vibration amplitude before the absorber installation (see fine curve) is 1.03 mm at that frequency. After the absorber is installed (see thick curve), the vibration amplitude descends to zero. The machine does no longer vibrates at 300 rad/s. It is noted that the machine resonance peaks are not symmetrical as in the case of self-excited vibration,

Disturbance frequency deviations degrade the absorber performance; the most sensitive to this problem are the neutralizers. Instead, absorbers in machines of constant velocity are less sensitive because their excitation frequency is their velocity. Nonetheless, more sophisticated absorbers exist that adapt their natural frequency to the external source frequency. These absorbers use adaptive control techniques and are more expensive than a simple Frahm's absorber. These more sophisticated absorbers are provided with a controller that automatically adjusts the frequency changes.

9.2.1 Conceptual Description of Frahm's Absorber

If the absorber is correctly designed, it oscillates with forces in opposition to the machine resonance or external source. Therefore, forces acting on the machine are reduced to negligible values or even to zero. In this process, the vibration energy moves from the machine to the absorber. However, a simple Frahm's absorber only mitigates forces of frequency at which it is tuned. It does not absorb transient waves of another frequency.

The Frahm's absorber is like a control system that "knows" the undesired vibration frequency and removes it from the machine, device or structure. The absorber inevitably adds its resonant frequency to the installation. As a result, the integrated system has two resonant frequencies, which are not equal to the machine's natural frequencies. Of course, these two new resonance frequencies may create problems during the machine start-up and shutoff, which must be mitigated.

To model vibration absorption, the host is considered a second-order system. The state of the machine-absorber assembly is defined by two space coordinates, one for the machine's mass displacement and another for the absorber mass motion. Therefore, the machine-absorber set is a 2-DOF system.

As the two springs under the main machine support the same load (see Figure 9.2), they are considered only one spring in the models to be discussed. The model development condition is that the foundation is perfectly rigid, and therefore, it cannot vibrate. As was said before, the typical mass foundation is five times bigger than the machine mass; however, in some machines, there are significant dynamic loads that require larger foundations to endure the transmitted effort. If the foundation is not rigid enough, the absorber installation may fail to remove the main machine's vibration.

The absorber size is smaller than the machine size, but not significantly. Typically, the absorber mass is between 5% and 35% of the machine mass. These percentages reveal that an absorber may be a significantly large device.

9.3 FRAHM'S ABSORBER MODEL

9.3.1 Equations of Motion

The vibration amplitude formula of an SDOF frictionless system with no absorber is (see Chapter 1):

Formula 9.1 Machine vibration amplitude before an absorber installation

$$a_{m0} = \frac{x_{0m}}{1 - u_d^2}$$

The term a_{mo} is the machine vibration amplitude, and x_{0m} is the machine static deflection F/k_m, where F is the amplitude of the acting harmonic force $F(t) = F{\bullet}\sin \omega{\bullet}t$. The zero of a_{m0} stands for a system with no absorber. The frequency ratio u_d is the disturbance frequency ω over the machine natural frequency ω_{mn}. After the absorber is installed, the machine assembly is converted into a frictionless 2-DOF system. Therefore, according to the vector interpretation of Section 1.5.1, two equations of motion are written:

Formula 9.2 Equations of motion of an undamped Frahm's machine-absorber assembly

$$-m_m \cdot \omega^2 \cdot a_m + k_m \cdot a_m + k_b \cdot (a_m - a_b) = F_d$$

$$-m_b \cdot \omega^2 a_b + k_b \cdot (a_b - a_m) = 0$$

Subscript m stands for the machine and b for the absorber. Subscript d refers to the forced disturbance. Terms $a_{m,}$ and a_b are the machine and absorber masses vibration amplitude, respectively. The frequency ω is the disturbance frequency of the applied force.

The first equation represents the equilibrium between the applied force F_d and the system's reaction (mass inertia, springs rigidity and the absorber's spring reaction). The third reaction is equal to the spring coefficient k_b times the net stretching $(a_m - a_b)$. These two equations are coupled by the reaction due to the net stretching. The second equation represents the equality between the absorber mass reaction $(-m_b{\cdot}\omega^2{\cdot}a_b)$ and the absorber spring force, proportional to the net stretching $(a_b - a_m)$.

9.3.1.1 Machine and Absorber Vibration Amplitude

By grouping the factors of a_m and a_b in each equation and using the impedance notation for convenience.

Formula 9.3 Equations of motion of a Frahm's absorber in impedance notation

$$Z_m \cdot a_m - k_b \cdot a_b = F_d$$

$$-k_b \cdot a_m + Z_b \cdot a_b = 0$$

Z_m is the machine mechanical impedance, and Z_b is the absorber mechanical impedance, whose formulas are:

Formula 9.4 Mechanical impedances of the machine and undamped Frahm's absorber

$$Z_m(\omega) = -m_m \cdot \omega^2 + k_m + k_b$$

$$Z_b(\omega) = -m_b \cdot \omega^2 + k_b$$

Equations 9.3 have two unknowns: vibration amplitudes a_m and a_b. The solution to this set of equations is the following formulas, where amplitudes are a function of frequency ω.

Formula 9.5 Vibration amplitude of the machine and absorber masses

$$a_m(\omega) = \frac{Z_b(\omega)\cdot F_d}{Z_m(\omega)\cdot Z_b(\omega) - k_b^2} \quad a_b(\omega) = \frac{k_b \cdot F_d}{Z_m(\omega)\cdot Z_b(\omega) - k_b^2}$$

These formulas are similar to the transmitted force's amplitude seen in Chapter 8.

9.3.1.2 Vibration Absorption Condition

In the typical system, as shown in Figure 9.2, the machine is considered free of vibration if amplitude a_m is zero. However, some limited tolerance exists that may be accepted with extreme care. Formula 9.5 demonstrates that the machine vibration amplitude is zero if the absorber impedance Z_b is zero. Therefore the vibration amplitude formulas must comply with the following absorption condition:

Formula 9.6 Absorption condition of machine vibration

$$Z_b = -m_b \cdot \omega^2 + k_b = 0$$

Therefore, as ω is the disturbance frequency, it is concluded that the absorber natural frequency ω_{nb} must be equal to the disturbance frequency ω.

Formula 9.7 Absorber natural frequency = disturbance frequency

$$\omega_{nb} = \sqrt[2]{\frac{k_b}{m_b}} = \omega$$

If the vibration is self-excited, the disturbance frequency is the rotating velocity; then the absorber's natural frequency must be equal to the velocity. The machine zero amplitude ($a_m = 0$) reduces the equation of motion to Formulas 9.8, revealing the forces' equilibrium with a perfectly tuned absorber.

Formula 9.8 Forces in a perfectly tuned machine-absorber assembly

$$k_b \cdot a_b = F_d$$

$$m_b \cdot \omega^2 \cdot a_b = k_b \cdot a_b$$

Therefore, if the absorber is perfectly tuned, the applied force is equilibrated by the absorber spring reaction, and the absorber inertial force is equal but opposite to the spring reaction. As will be seen in the following examples, the acceleration $\omega^2 \cdot a_b$ can be significant. It means that the mechanical linkage machine-absorber may endure a higher force than the absorber mass weight.

9.3.2 Example of Forces in a Machine-Absorber Assembly

Consider a machine-absorber assembly excited by an external harmonic force of frequency ω. It is requested to calculate forces acting on the machine before and after the absorber installation. As the disturbance frequency may change, the calculation must include the machine and absorber forces at a specified different frequency from the initial tuning.

9.3.2.1 Forces With a Tuned Absorber

- Data:
  Applied force: F = 20,000 kg
  Disturbance frequency: ω = 300 rad/s
  Machine vibration amplitude: $a_m = 0$ mm
  Absorber vibration amplitude: $a_b = 4.44$ mm
  Absorber mass: $m_b = 50$ kg mass
  Static deflection: $x_0 = 1.25$ mm
  Magnification factor without absorber: $Q_{m0} = 1.21$
  Magnification factor with absorber: $Q_m = 0.0$
- Calculation:
  Absorber weight: $W_b = 9.81 \times 50 = 490.5$ kg
  Force on machine without absorber: $F_m = Q_{m0} \cdot F_d = 1.21 \times 20{,}000 = 24{,}200$ kg
  Absorber vibration velocity and acceleration:
  $v_{max} = 4.44/1{,}000 \times 300 = 1.33$ m/s
  $a_{max} = -4.44/1{,}000 \times 300^2 = -400$ m/s²
  Absorber inertial force: $F_{b\,max} = 50 \times 400 = 20{,}000$ kg
  This reaction force is equal to the applied force F_d and equivalent to 40.8 times the absorber mass weight.
  Force on the machine: as $Q_m = 0$, this force is also 0.

9.3.2.2 Forces With an Untuned Absorber

- Data:
  Disturbance frequency: ω = 200 rad/s
  Machine vibration amplitude: $a_m = 2.07$ mm
  Absorber vibration amplitude: $a_b = 1.66$ mm
  Machine mass: $m_m = 325$ kg mass
  Machine magnification factor: $Q_m = 1.66$
- Calculation:

1. Machine vibration velocity and acceleration:
   $v_{max} = 2.07/1{,}000 \times 300 = 0.62$ m/s
   $a_{max} = -2.07/1{,}000 \times 300^2 = -186.61$ m/s²
   Force on machine: $F_{m\,max} = Q_m \cdot F_d = 1.66 \times 20{,}000 = 33{,}200$ kg
   In the perfectly tuned case, this force is zero.

2. Absorber vibration velocity and acceleration:
   v_{max} = 1.66/1,000 × 300 = 0.50 m/s
   a_{max} = -1.66/1,000 × 300² = -149.28 m/s² equivalent to 87.4 times the gravity acceleration.
   Force on absorber: $F_{b\,max}$ = 50 × 149.28 = 7,464.2 kg

This force is acting on the absorber mass through the spring. It is equivalent to 12.2 times the absorber mass weight. Nonetheless, this force is much lower than the same force in the perfectly tuned case. The reason is that if the absorber is not perfectly tuned, it does not absorb as much energy as a perfectly tuned absorber. The machine takes a part of the energy, and for this reason, it vibrates.

9.3.3 Frequency Response of Machine-Absorbers

9.3.3.1 Definition of Non-Dimensional Variables and Parameters

For convenience reasons, the magnification factors Q_m and Q_b are expressed with non-dimensional variables and constant parameters that make it easy to interpret and manipulate formulas and charts independent of the used units. Said variables and parameters are defined below.

- Frequency ratios of the machine and absorber:
  As the machine and absorber have different natural frequencies, their natural frequency ratios are also different.

Formula 9.9 Frequency ratios of the machine and absorber

$$u_m = \frac{\omega_d}{\omega_{nm}} \quad u_b = \frac{\omega_d}{\omega_{nb}}$$

u_m: machine frequency ratio
u_b: absorber frequency ratio
ω_{nb}: absorber natural frequency = $\sqrt[2]{k_b / m_b}$
ω_{nm}: machine natural frequency = $\sqrt[2]{k_m / m_m}$

- Stiffness coefficient and mass ratios:

Formula 9.10 Stiffness and mass ratios

$$\mu_k = \frac{k_b}{k_m} \quad \mu_m = \frac{m_b}{m_m}$$

- Natural frequencies ratio:

Formula 9.11 Frequencies ratio

$$r_f = \frac{\omega_{nb}}{\omega_{nm}} = \sqrt[2]{\frac{k_b}{m_b} \cdot \frac{m_m}{k_m}} = \sqrt[2]{\frac{\mu_k}{\mu_m}}$$

- Machine frequency ratio u_m is proportional to the absorber frequency ratio u_b:

Formula 9.12 Machine frequency ratio against the absorber frequency ratio

$$\frac{u_m}{u_b} = \frac{\omega_d / \omega_{nm}}{\omega_d / \omega_{nb}} = r_f \rightarrow u_m = r_f \cdot u_b$$

- Formula of $k_m + k_b$ as a function of μ_k:

Formula 9.13 Sum of the machine and absorber stiffness coefficients versus μ_k

$$k_m + k_b = k_m \cdot (1 + \mu_k)$$

- Machine impedance ratio z_m:

Formula 9.14 Machine impedance ratio

$$Z_m = k_m - m_m \cdot \omega^2 + k_b = k_m \cdot \left(1 - u_m^2 + \mu_k\right)$$

$$z_m = \frac{Z_m}{k_m} = 1 + \mu_k - u_m^2$$

- Absorber impedance ratio z_b:

Formula 9.15 Absorber impedance ratio

$$Z_b = k_b - m_b \cdot \omega^2$$

$$z_b = \frac{Z_b}{k_b} = 1 - u_b^2$$

9.3.3.2 Vibration Amplitudes and Frequency Response

The conversion of amplitude vibration formulas to impedance ratio notation is done with the variables and parameters described in the previous section. Therefore, by introducing them in Formula 9.5 and dividing numerators and denominators by k_m, the result is the following expressions for the machine and absorber vibration amplitude:

Formula 9.16 Vibration amplitude formulas

$$a_m = \frac{k_b \cdot z_b \cdot F_d}{k_m \cdot z_m \cdot k_b \cdot z_b - k_b^2} = \frac{z_b \cdot (F_d / k_m)}{z_m \cdot z_b - k_b} = \frac{z_b \cdot x_{0m}}{z_m \cdot z_b - \mu_k}$$

$$a_b = \frac{k_b \cdot F_d}{k_m \cdot z_m \cdot k_b \cdot z_b - k_b^2} = \frac{(F_d / k_m)}{z_m \cdot z_b - k_b} = \frac{x_{0m}}{z_m \cdot z_b - \mu_k}$$

In these formulas, z_b is a function of u_b, and z_m is a function of u_m. Therefore, the above vibration amplitude formulas are functions of two independent variables, u_b and u_m. Still, only one independent variable is needed to plot the frequency response in a Q versus u figure. Therefore, in Formula 9.14 of z_m, the frequency u_m is replaced by the $r_f \bullet u_b$ product.

Formula 9.17 Machine impedance ratio as a function of u_b

$$z_m = 1 + \mu_k - r_f^2 \cdot u_b^2$$

If z_b and z_m of Formulas 9.14 and 9.15 are introduced in Formulas 9.16, the vibration amplitude formulas as a function of u_b with non-dimensional variables are returned.

Formula 9.18 Machine and absorber vibration amplitude as a function of u_b only

$$a_m(u_b) = \frac{\left(1 - u_b^2\right) \cdot x_{0m}}{\left(1 - u_b^2\right) \cdot \left(1 + \mu_k - r_f^2 \cdot u_b^2\right) - \mu_k}$$

$$a_b(u_b) = \frac{x_{0m}}{\left(1 - u_b^2\right) \cdot \left(1 + \mu_k - r_f^2 \cdot u_b^2\right) - \mu_k}$$

The frequency response or magnification factors are readily obtained by dividing $a_m(u_b)$ and $a_b(u_b)$ by the static deflection x_{0m}.

Formula 9.19 Magnification factors of a Frahm's machine-absorber assembly

$$Q_m(u_b) = \frac{a_m(u_b)}{x_{0m}} = \frac{\left(1 - u_b^2\right)}{\left(1 - u_b^2\right) \cdot \left(1 + \mu_k - r_f^2 \cdot u_b^2\right) - \mu_k}$$

$$Q_b(u_b) = \frac{a_b(u_b)}{x_{0m}} = \frac{1}{\left(1 - u_b^2\right) \cdot \left(1 + \mu_k - r_f^2 \cdot u_b^2\right) - \mu_k}$$

These formulas prove that at $u_b = 1$, the machine amplitude is zero. The absorber design must then be tuned to run at a frequency ratio equal to 1 to make Q_m equal to zero. The machine frequency response with non-dimensional formulas is seen in Figure 9.5.

Figure 9.6 shows the frequency response of the absorber only.

The machine response curve (thick line) has two resonance peaks, which must be avoided during the operation, and one zero at $u_b = 1$. The machine does not vibrate at this frequency ratio; therefore, this is the BOP. The absorber magnification factor $Q_b(u_b)$ of Formula 9.19 indicates that at $u_b = 1$, the absorber amplitude has the value given by Formula 9.20 (right side). See Figure 9.6 and the following formulas.

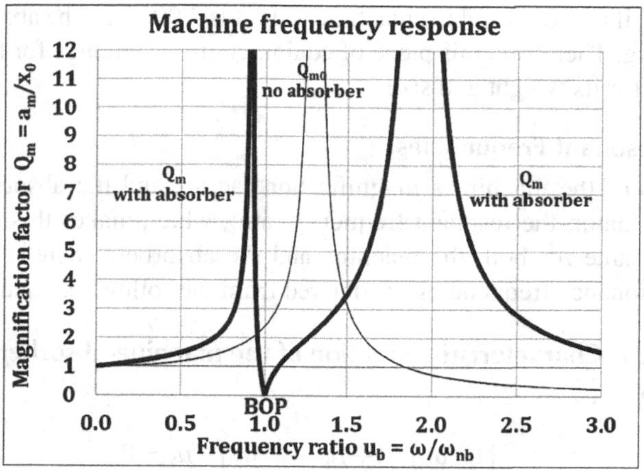

FIGURE 9.5 Machine frequency response with non-dimensional units.

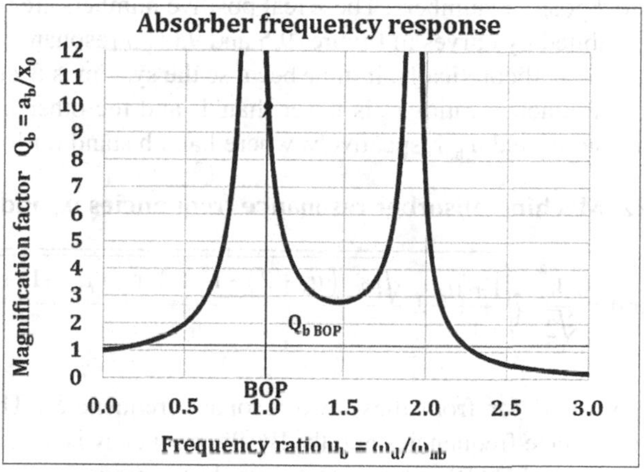

FIGURE 9.6 Absorber frequency response.

Formula 9.20 Absorber magnification factor and vibration amplitude at BOP

$$Q_{b\,BOP}(u_b) = \frac{1}{\mu_k} \rightarrow a_{b\,BOP} = \frac{x_{0m}}{\mu_k}$$

Therefore, at the BOP, the absorber vibrates, and the machine does not. As the absorber takes vibration energy from the machine, it cannot be a small device. The absorber must endure significant mechanical energy transferred from the host without affecting its mechanical integrity. It is possible to have a rough idea of the absorber size with the mass ratio value, which typically is between 0.05 and 0.25. The mass ratio is prefixed according to conditions that are discussed ahead.

For example, if μ_m =0.10 and the machine weight is 2,000 kgs, the absorber weight will be 200 kg. Therefore, this piece of equipment is a challenge for the mechanical designer for its weight and size.

9.3.3.3 Resonant Frequencies

As formulas of the machine's magnification factors and the absorber have the same denominator, the resonant frequency ratio, which makes this denominator zero, is the same for both the machine and the absorber. Therefore, the corresponding resonance frequencies are derived from the following equation:

Formula 9.21 Characteristic equation of the machine-absorber resonance frequencies

$$\left(1-u_r^2\right)\cdot\left(1+\mu_k-r_f^2\cdot u_r^2\right)-\mu_k=0$$

As the above formula is a fourth-degree equation, it has four roots, although only two are real positive numbers. These real positive numbers are the resonance frequencies exhibited by curves in Figures 9.5 and 9.6. At resonance frequencies, Q_m and Q_b peaks are, theoretically, infinite because the system is frictionless. One of the resonant frequency ratios u_r is lower than 1, and the other is higher than 1. They are named u_{rl} and u_{rh}, respectively, where l and h stand for low and high.

Formula 9.22 Machine-absorber resonance frequencies u_{rl} and u_{rh}

$$u_r\left(\mu_k,r_f\right)=\frac{1}{\sqrt[2]{2}}\cdot\sqrt[2]{\frac{1+\mu_k}{r_f^2}\pm\frac{\sqrt[2]{\mu_k\cdot\left(\mu_k+2\right)+r_f^4+2\cdot r_f^2\cdot\left(\mu_k-1\right)+1}}{r_f^2}+1}$$

The BOP must be far from these two resonant frequencies. The difference between the resonance frequencies and the BOP's frequency is the frequency difference to resonance (FDTR), a range that may absorb moderate disturbance frequency deviations. This difference must be considered a safety margin to prevent resonance.

The chart in Figure 9.7 is made with Formula 9.22. It represents the resonant frequency curves as a function of the stiffness ratio μ_k, and natural frequency ratio r_f for any machine with an installed Frahm's absorber. This figure has universal validity. The low resonant frequency ($u_{rl}<1$) readings are on the left ordinate axis, and the high resonant frequency ($u_{rh}>1$) readings are on the right ordinate axis.

This figure proves that as the frequency ratio r_f increases, the high resonant frequency moves away from the BOP and vice versa if the frequency ratio decreases. Conversely, if the frequency ratio increases, the low resonant frequency approaches the BOP and moves away from the BOP if the frequency ratio decreases. Therefore, by changing the frequency ratio, one resonant frequency approaches the BOP, and the other one moves away. The low resonant frequency is moved away from resonance by increasing the rigidity coefficient.

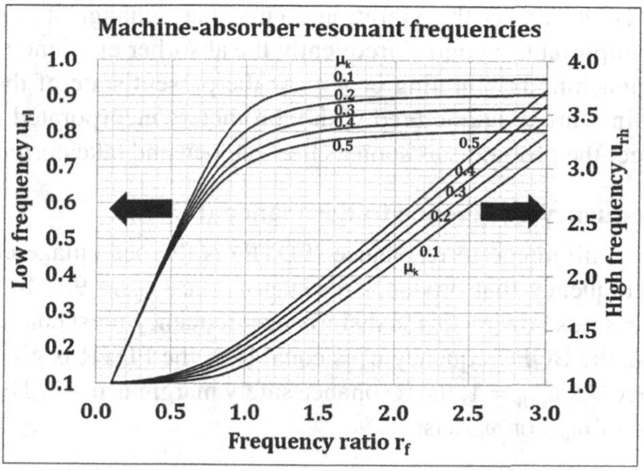

FIGURE 9.7 Resonant frequencies of a machine-absorber assembly.

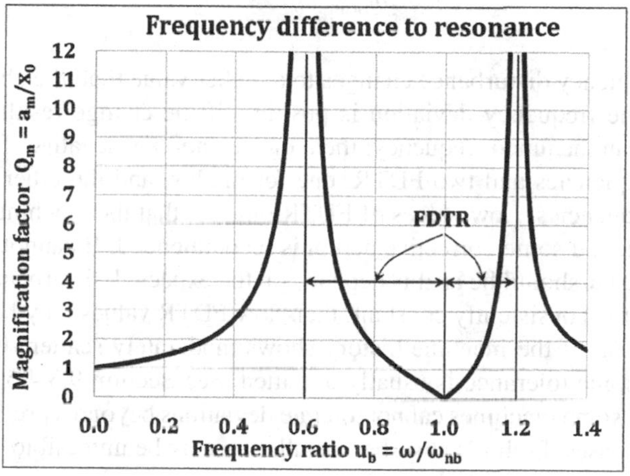

FIGURE 9.8 Graphical definition of FDTR.

However, increasing the rigidity ratio, the high resonant frequency approaches the BOP.

9.3.4 Frahm's Absorber Design and Performance

The absorber design's purpose is to determine its spring stiffness coefficient and mass. It appears a simple task; however, the absorber's mechanical design is a challenge due to space restrictions, environmental conditions, acting forces on the absorber, and other technical issues. Design engineers usually overcome these issues in the design stage; therefore, the absorbers' performance is always satisfactory. However, as the vibration source may change its operating frequency, it is important to control frequently the absorber efficiency by checking whether the machine is vibrating or not. At the present state of the art (2021), critical machines are manufactured with vibrometers incorporated into sensible points to detect the problems as soon as they happen and take corrective actions.

9.3.4.1 Frequency Difference to Resonance (FDTR)

The frequency difference to resonance (FDTR) is the percentage change of the disturbance frequency that produces resonance. See Figure 9.8. In other words: the FDTR measures how much the disturbance frequency must change to produce resonance. As the BOP frequency u_b is equal to 1, the FDTR is given by the u_r - u_{BOP} difference. As $u_{BOP} = 1$, the resonance safety margin is u_r - 1. Then the FDTR in percentage of u_{BOP} or ω_{BOP} is:

Formula 9.23 Machine frequency difference to resonance (FDTR)

$$\text{FDTR}_{\%} = \left(\frac{\omega_r - \omega_{BOP}}{\omega_{BOP}} \right) \times 100 = \left(u_r - 1 \right) \times 100$$

If the frequency disturbance changes to a higher value than the absorber tuned frequency, the frequency deviation is positive. If the change results in a lower frequency than the tuned frequency; then, the deviation is negative. There are two resonant frequencies and two FDTR, one for the low and the other for the high resonant frequencies. Low values of FDTR indicate that the machine is too close to resonance, and some corrective action is recommended. It cannot be said how much the FDTR should be as this depends on the expected disturbance frequency changes. If it is consistently constant; then, low FDTR values may be acceptable, let us say ±5%. If the machine history shows moderately scattered and random deviations, some tolerance is usually admitted. See Section 9.3.4.5. However, it happens that some machines cannot tolerate deviations beyond a predefined value, and in those cases, Frahm's absorber installation may be unfeasible.

9.3.4.2 Absorber Design Procedure

The absorber design starts with its mass and rigidity coefficient calculation according to the expected disturbance frequency. Then, the drawing of the machine-absorber assembly is a mechanical engineering task that may have some challenges because the absorber is not a small piece easy to install on the machine. The initial data of the absorber calculation is the machine weight and the disturbance frequency ω knowledge. With these data, the absorber mass and spring stiffness coefficient are calculated with the following procedure under the absorption condition of $\omega_{nb} = \omega$.

1. Adopt the absorber to machine mass ratio μ_m according to considerations made in the next section. Recommended value: 0.05 to 0.25. However, in some applications, this ratio is in the 0.30 to 0.40 range.

$$\mu_m = \frac{m_b}{m_m}$$

2. Calculate the absorber spring rigidity coefficient for the most probable disturbance frequency by clearing k_b from $\omega_{nb} = (k_b/m_b)^{1/2}$.

$$k_b = m_b \cdot \omega^2$$

3. Alternatively, if the machine is self-excited due to eccentricity, bent shaft or unbalanced rotor, the k_b coefficient may be calculated with the following formula[3]:

$$k_b = k_m \cdot \mu_m$$

Therefore, as $\mu_k = k_b/k_m$, it is $\mu_k = \mu_m$ in the self-excited vibration mode.

4. After the stiffness coefficient is calculated, the designer must verify the dynamic efforts to assure that the mechanical elements are strong enough to support them and decide the space needed for the absorber installation.

9.3.4.3 Mass Ratio Determination

The mass ratio is a sensitive parameter. Though a low mass ratio is desirable because it reduces the need for space and has lower manufacturing and installation cost, an issue arises because the lower the mass ratio, the closer the low resonant frequency to the BOP. See in the chart in Figure 9.9 two frequency responses having different mass ratios. The absorber with a lower mass is represented by the fine curve, whose resonance right peak is closer to the BOP than the frequency response with the higher mass ratio, represented by the thick curve. Note that the fine curve

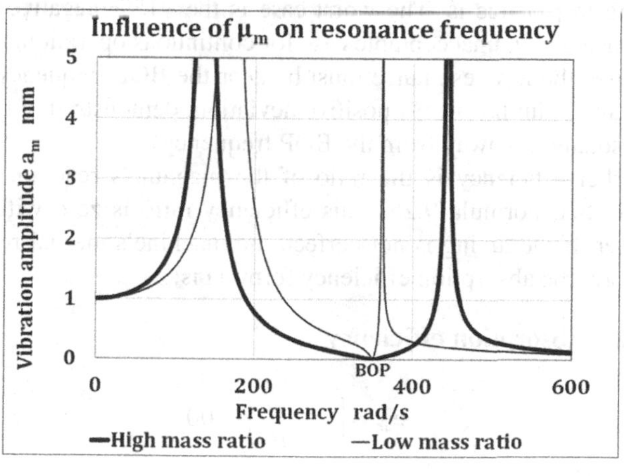

FIGURE 9.9 Machine magnification factor for two different mass ratios.

has a smaller FDTR than the response with the higher mass ratio. Therefore, the designer must keep in mind that the lower the mass ratio, the lower the FDTR.

The chart in Figure 9.9 shows that both resonance frequencies approach the BOP when the ratio mass change is reduced. The mass ratio's recommended value is a trade-off between the absorber size, manufacturing, installation costs, and the FDTR, among others.

The risk severity of a resonant frequency deviation depends on the FDTR. It is recommended to perform several tests under different conditions at different rotational velocities and produce statistics of vibration created by external sources before designing the absorber. In summary: reliable and abundant disturbance frequency statistics are highly recommended to estimate the optimal mass ratio value. Suppose statistics prove that frequency deviations are too scattered and far from the BOP; in this case, it is likely that Frahm's absorber is not applicable. Another solution to mitigate vibration must be worked out.

9.3.4.4 Tuning Error

If the absorber's natural frequency is not exactly equal to the disturbance frequency, a perfect vibration removal does not happen. The tuning error is the percentage difference of the disturbance frequency minus the absorber's natural frequency. See Formula 9.24. The aftermath of this error is that the machine vibrates.

Formula 9.24 Absorber tuning error

$$\text{Tuning error} = \left(\frac{\omega}{\omega_{nb}} - 1\right) \times 100$$

Vibration histograms in Figures 9.10–and 9.12 show three different cases with the same ordinates scale. They provide a graphical visualization of how important correct tuning in practice is. The worst case is the −15% negative deviation of Figure 9.11, usually an unacceptable case for continuous operation. As the deviation is negative, the low resonance must be near the BOP frequency. It does not happen the same with the +15% positive deviation, demonstrating that the high-frequency resonance is away from the BOP frequency.

The absorber efficiency is the ratio of the machine's reaction force to the applied force. See Formula 9.25. This efficiency ratio is zero with a perfectly tuned absorber. If the tuning is not perfect, the machine's inertial reaction is not zero. Therefore, the absorption efficiency formula is:

Formula 9.25 Absorption efficiency

$$e_{ab} = \left(1 - \frac{F_m}{F}\right) \times 100$$

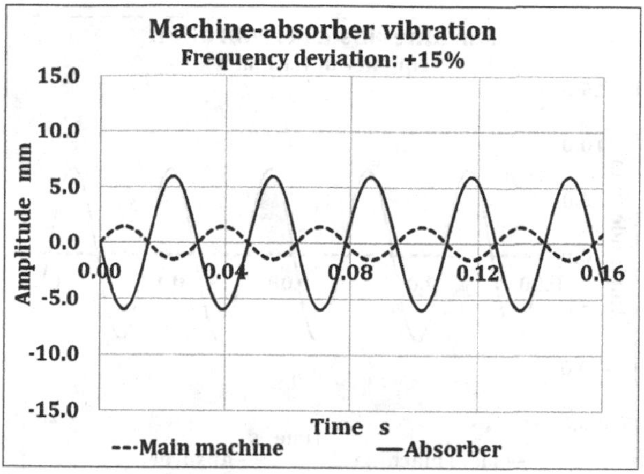

FIGURE 9.10 Untuned case. Deviation = +15%.

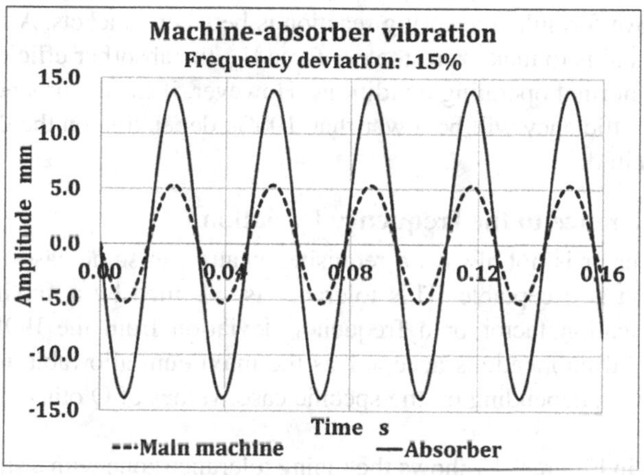

FIGURE 9.11 Untuned case. Deviation = −15%.

F is the applied disturbance force, and F_m is the machine inertial reaction force, equal to the difference between the applied force and the springs' reaction. This net force is given by the following expression derived from the equations of motion:

Formula 9.26 Machine force reaction = disturbance force – springs reaction

$$F_m = -m_m \cdot a_m \cdot \omega_f^2 = F - \left[\left(k_m + k_b \right) \cdot a_m - k_b \cdot a_b \right]$$

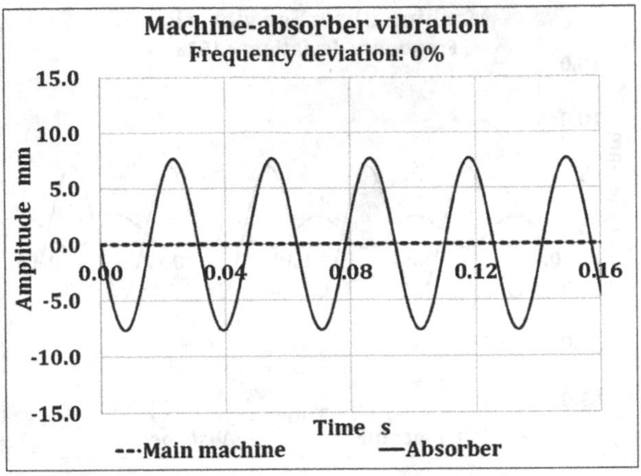

FIGURE 9.12 Correctly tuned case. Deviation = 0%.

In the above formula, the spring reaction is between brackets. As the absorber installation goal is to make zero the net force F_m, the absorber efficiency must be 100% under normal operating conditions. However, if the disturbance frequency deviates, the efficiency will be lower than 100%, depending on the deviation frequency magnitude.

9.3.4.5 Tolerance to the Frequency Deviation

Zero tuning error is not always a requisite because, in some cases, tolerance to mild vibration is acceptable. This tolerance is specified by a maximum admissible magnification factor or a frequency deviation from the BOP frequency. Many technical publications accept 1 as the maximum allowable magnification factor. However, depending on the specific case, values of Q other than 1 may be acceptable.

The chart in Figure 9.13 shows the tuning tolerance zone with a shaded area. In this case, the maximum allowable magnification factor is 1, and the largest allowable deviations from the BOP frequency are $(1\text{-}u_{t1})$ and $(u_{t2}\text{-}1)$.

As the machine vibration amplitude a_m is equal to the $Q_m{\bullet}x_0$ product, accepting $Q_m = 1$ as the largest allowable magnification factor means that the maximum allowable amplitude a_m is equal to the static deflection x_0. Some technical publications support this criterium. The tolerance zone is also called "useful operating bandwidth." However, the name "tolerance zone" may be more representative. Before accepting a tolerance bandwidth, it is recommended to verify the machine's resistance to fatigue stress produced by vibration. Although the conclusion is that the mechanical stress is under the fatigue limit, it is always a clever idea to consult the manufacturer or recognized international technical standards to learn the equipment tolerance zone, if any.

Frequency ratios u_{t1} and u_{t2} of Figure 9.13 are the limits of the tolerance zone. In this section, they are called "extreme frequencies." Therefore, the disturbance

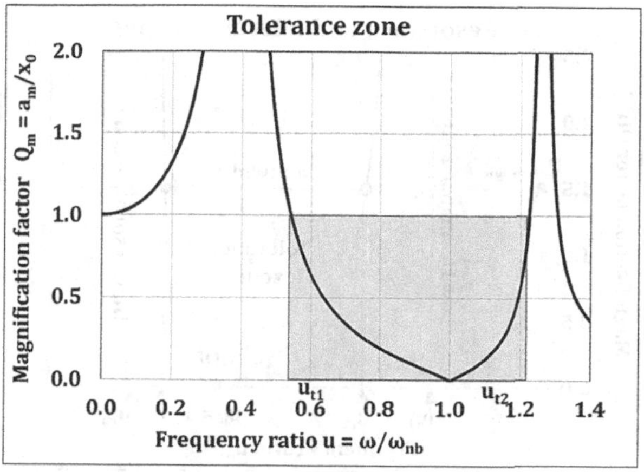

FIGURE 9.13 Tolerance zone in the $Q_m(u_b)$ figure.

frequency must be between these extreme values; otherwise, the machine works beyond the tolerance zone, which is unacceptable.

The equation to calculate the extreme frequencies is derived from Formula 9.19 of $Q_m(u_b)$. Assuming that the allowable magnification factor for the tolerance zone is $Q_m = 1$, the extreme frequencies are derived from the following equation:

Formula 9.27 Equation of extreme frequencies u_{t1} and u_{t2}

$$1 = \frac{\left(1 - u_b^2\right)}{\left(1 - u_t^2\right) \cdot \left(1 + \mu_k - r_f^2 \cdot u_t^2\right) - \mu_k}$$

The solution to this equation returns four roots, out of which only two of them are within the resonant frequency ratios u_{rl}-u_{rh} range. See chart in Figure 9.14. The other two roots are out of this range; therefore, they cannot be within the tolerance zone.

Formulas 9.28 Extreme frequencies u_{t1} and u_{t2} for $Q_{allowable} = 1$

$$u_{t1} = \frac{\sqrt[2]{2}}{2 \cdot r_f} \cdot \sqrt[2]{2 + \mu_k + r_f^2 - \sqrt[2]{r_f^4 + 2 \cdot (\mu_k - 2) \cdot r_f^2 + \mu_k \cdot (\mu_k + 4) + 4}}$$

$$u_{t2} = \sqrt[2]{1 + \mu_m}$$

Formulas to calculate the extreme frequencies for the general case of $Q_{allowable} \neq 1$ are long expressions and cumbersome to manage. Hence, if the adopted $Q_{allowable}$ is different from 1, the author recommends using a graphical solution for a quick calculation of extreme frequencies. The chart in Figure 9.14 illustrates the case of $Q_{allowable} = 1.5$.

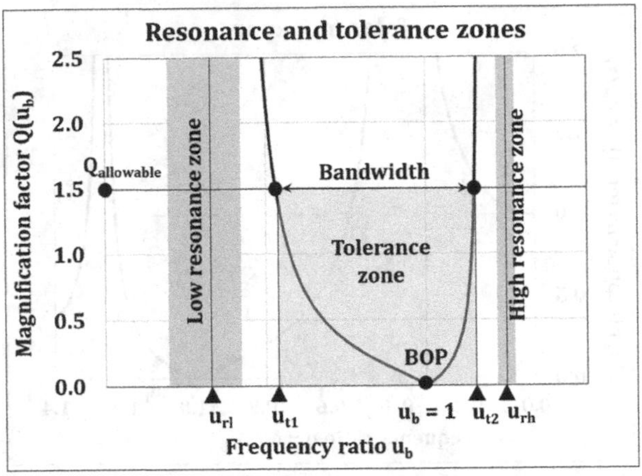

FIGURE 9.14 Resonance and tolerance zones for $Q_{allowable} = 1.5$.

It must be noted in Figure 9.14 that the maximum allowable frequency ratio deviations are $|u_{t1} - u_{BOP}|$ or $|u_{t2} - u_{BOP}|$. Beyond these values, there is resonance risk or at least risk of experiencing a too high vibration amplitude. If statistics prove that potential deviations are larger than the maximum allowable deviation, the absorber designer must re-adjust the absorber mass or change the spring stiffness. If no solution is feasible, a more sophisticated device must be considered because the simple Frahm's absorber has a narrow absorption bandwidth. It is for this reason that some absorbers have adjustable stiffness spring or adjustable mass using interchangeable disks. This problem especially arises in TVN neutralizers, whose purpose is to absorb random frequency changes from external sources. Instead, TMD absorbers are used to absorb the machine's natural frequency that is unlikely to deviate. This deviation is unlikely to occur in constant velocity machines. Therefore, as a TMD is not subject to significant frequency deviations, their resonance and tolerance zone have little or no changes.

9.3.5 Damped Absorption

9.3.5.1 Conceptual Description

As was said before, Frahm's absorber has a narrow bandwidth. Its performance easily deteriorates due to frequency deviations from its original tuning or because the machine runs with variable speed. Therefore, said deviations make the absorber vibrate with high amplitudes and even undergo resonance, especially to machines excited by external sources, where a TVN is used. As a solution to this issue, a damper is added in parallel with the absorber spring to improve the system's performance robustness. See Figure 9.15. This figure shows an alternative location for the absorber damper by linking the damper to the foundation and not the machine mass. This installation is used to overcome space problems, sometimes created by large absorbers.

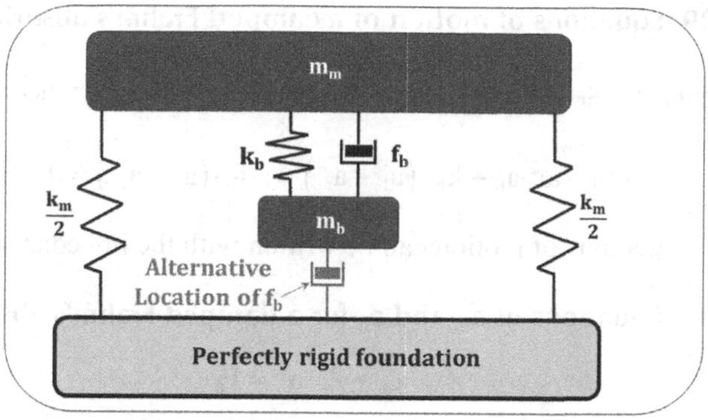

FIGURE 9.15 TVN with damper. The usual and alternative locations.

Therefore, the damper aims to reshape the machine's frequency response curve so that no resonance risk exists and to obtain as much as possible a flat frequency response curve in a preset frequency range. Applying mathematical tools, such as Butterworth polynomials[4], could help design absorbers of low sensitivity to frequency variations. Control engineers use this technique to obtain a flat frequency response curve of automatic control systems because flatness means transient responses with low overshoots.

In the classical design, the frequency response curve is shaped by adjusting two parameters: the natural frequencies ratio ($r_f = \omega_{nb}/\omega_{nm}$) and the damping ratio ζ. Den Hartog elaborated a practical optimization method based on his two fixed-points theory described below[5]. This method is easy to implement in a spreadsheet to visualize and select the best frequency response curve of a damped TVN for a frictionless machine. If the host is damped, Den Hartog's method is not applicable. In this case, the author recommends investigating a more sophisticated procedure like the polynomials mentioned above[6].

If the rotor is flexible, the absorber mass is subject to a considerable amplitude vibration during startup. This large amplitude at startup broadly justifies the use of damped absorbers instead of undamped absorbers. Undamped absorbers reduce the machine amplitude to zero; however, the tuning frequency has little FDTR, which precludes these types of absorbers in variable speed machines. The most common and easy solution to this problem is to add a damper to the absorber. The damped absorber solves the resonance problems due to a low FDTR but does not reduce the machine amplitudes to zero as Frahm's absorber does.

9.3.5.2 Equations of Motion

Equations of motion of the damped absorber are like the undamped absorber equations, but, in the damped model, a term that includes the damper reaction force is included. As was discussed before, this term is imaginary to consider the 90° phase existing between the springs and dampers' reactions. Therefore, according to the vector interpretation of the equations of motion discussed in Chapter 1, the damped absorber equations are vectors represented by the following expressions.

Formula 9.29 Equations of motion of a damped Frahm's absorber

$$-m_m \cdot \omega^2 \cdot a_m + k_m \cdot a_m + k_b \cdot (a_m - a_b) + j \cdot \omega \cdot f_b \cdot (a_m - a_b) = F_d$$

$$-m_b \cdot \omega^2 \cdot a_b + k_b \cdot (a_b - a_m) + j \cdot f_b \cdot (a_b - a_m) = 0$$

The above equations of motion can be written with the impedance's notation.

Formula 9.30 Equations of a_m and a_b for a damped Frahm's absorber

$$Z_m \cdot a_m - Z_t \cdot a_b = F_d$$

$$-Z_t \cdot a_m + Z_b \cdot a_b = 0$$

Where the mechanical impedances Z are:

Formula 9.31 Mechanical impedances of the Frahm's absorber

$$\text{Machine}: Z_t(\omega) = k_b + j \cdot \omega \cdot f_b$$

$$\text{Spring-damper set: } Z_t(\omega) = k_b + j \cdot \omega \cdot f_b$$

$$\text{Absorber: } Z_b(\omega) = -m_b \cdot \omega^2 + k_b + j \cdot \omega \cdot f_b$$

The spring-damper set transfers the vibration energy from the machine to the absorber mass. That is why its impedance has the t subscript.

9.3.5.3 Derivation of Impedance Ratios z

In this section, the mechanical impedances and impedance ratios are derived. The reading of this area is not essential for the understanding of the rest of this chapter. Readers not interested in the mathematical derivation of formulas may skip this derivation.

The ratio $\omega \cdot f/k$ is converted to non-dimensional variables with the following derivation.

$$\omega \cdot \frac{f}{k} = \omega \cdot \frac{f_c \cdot \zeta}{k} = \omega \cdot \frac{\left(2 \cdot \sqrt[2]{k \cdot m}\right) \cdot \zeta}{k} = 2 \cdot \omega \cdot \sqrt[2]{\frac{m}{k}} \cdot \zeta = 2 \cdot \zeta \cdot \frac{\omega}{\omega_n} = 2 \cdot \zeta \cdot u$$

Where f_c is the critical friction coefficient $= 2 \cdot \sqrt[2]{k \cdot m}$

Machine impedance ratio:

$$Z_m = k_m \cdot \left(-\frac{m_m}{k_m} \cdot \omega^2 + 1 + \frac{k_b}{k_m} + j \cdot \omega \cdot \frac{f_b}{k_m} \right) = k_m \cdot \left(-u_m^2 + 1 + \mu_k + j \cdot 2 \cdot \zeta \cdot u_m \right)$$

$$z_m = -u_m^2 + 1 + \mu_k + j \cdot 2 \cdot \zeta \cdot u_m$$

Spring-damper set impedance ratio:

$$Z_t(\omega) = k_b \cdot \left(1 + j \cdot \omega \cdot \frac{f_b}{k_b}\right) = k_b \cdot (1 + j \cdot 2 \cdot \zeta \cdot u_b)$$

$$z_a = (1 + j \cdot 2 \cdot \zeta \cdot u_b)$$

Absorber impedance ratio:

$$Z_b(\omega) = k_b \cdot \left(-\frac{m_b}{k_b} \cdot \omega^2 + 1 + j \cdot \omega \cdot \frac{f_b}{k_b}\right) = k_b \cdot \left(-u_b^2 + 1 + j \cdot 2 \cdot \zeta \cdot u_b\right)$$

$$z_b = -u_b^2 + 1 + j \cdot 2 \cdot \zeta \cdot u_b$$

Non-dimensional variables and parameters of the frequency response discussed in Section 9.3.3 for the undamped absorber are used in the damped absorber model. Therefore, the equations of a_m and a_b are converted into expressions with impedance ratios notation, as explained below.

The solution to the following Formulas 9.32 returns the mass displacements as a function of frequency ω. Therefore, amplitudes a_m, and a_b have complex expressions due to the absorber's damper.

Formula 9.32 Vibration amplitudes a_m and a_b expressed with impedance ratios

$$a_m = \frac{k_b \cdot z_b \cdot F_d}{k_m \cdot z_m \cdot k_b \cdot z_b - k_b^2 \cdot z_a^2} = \frac{z_b \cdot (F_d / k_m)}{z_m \cdot z_b - \frac{k_b}{k_m} \cdot z_a^2} = \frac{z_b \cdot x_{0m}}{z_m \cdot z_b - \mu_k \cdot z_a^2}$$

$$a_b = \frac{k_b \cdot F_d}{k_m \cdot z_m \cdot k_b \cdot z_b - k_b^2 \cdot z_a^2} = \frac{(F_d / k_m)}{z_m \cdot z_b - \frac{k_b}{k_m} \cdot z_a^2} = \frac{x_{0m}}{z_m \cdot z_b - \mu_k \cdot z_a^2}$$

Therefore, the machine and absorber's magnification factors are derived from the above formulas by dividing them by x_{0m}.

Formula 9.33 Frequency responses of the machine and absorber

$$Q_m = \frac{a_m}{x_{0m}} = \frac{z_b}{z_m \cdot z_b - \mu_k \cdot z_a^2}$$

$$Q_b = \frac{a_b}{x_{0m}} = \frac{1}{z_m \cdot z_b - \mu_k \cdot z_a^2}$$

It is important to note that all impedance ratio formulas are complex; hence, Q_m and Q_b of Formulas 9.33 are complex expressions. They have one real part

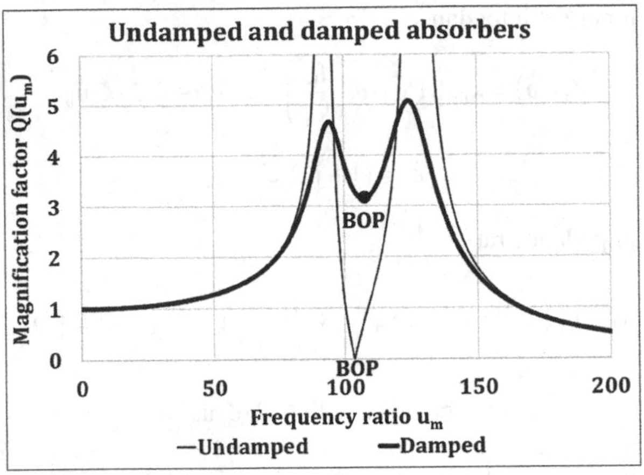

FIGURE 9.16 Comparison of an undamped with a damped neutralizer.

corresponding to the mass and the spring mechanical impedances and one imaginary part representing the damper impedance.

Figure 9.16 compares the machine frequency response of an undamped and damped neutralizer. In this neutralizer, the BOP is the minimum between the two resonance peaks. Instead, in a perfectly tuned neutralizer, the BOP is zero. This chart shows that the damped absorber produces a machine frequency response with two resonance peaks, but it does not create a zero. It means that the machine always vibrates, even at its BOP. In this example, the magnification factor at the BOP is 3.2; therefore, if the static deflection is 1 mm, the machine's vibration amplitude is 3.2 mm, which may be unacceptable due to fatigue consequences.

The damped neutralizer's advantage is that Q peaks are not infinite, like in the undamped. Therefore, deviations from the disturbance frequency are not exposed to extremely high amplitudes. The highest peak in the above chart has a magnification factor of 5, which is 56% higher than the BOP magnification factor. Hence, in the above example, the peak amplitude is 5 mm. If the disturbance frequency is constant, the undamped absorber is a good option. However, if frequency variations are expected, the damped absorber is a better solution because it prevents the risk of extremely high amplitudes. The magnification factor of a machine with a damped absorber is derived by separating the real (R_e) and imaginary (I_m) part of Formulas 9.33 and calculating the module of the resulting expression with the classical formula of complex numbers: $|Q| = \sqrt[2]{R_e^2 + I_m^2}$. This procedure returns the following formula for the machine magnification factor with a damped absorber:

Formula 9.34 Machine magnification factor with a damped absorber

$$Q_m(u_m) = \frac{a_m}{x_{0m}} = \sqrt[2]{\frac{4\cdot\zeta^2\cdot u_m^2 + \left(u_m^2 - r_f^2\right)^2}{4\cdot\zeta^2\cdot u_m^2\cdot\left[\left(\mu_m+1\right)\cdot u_m^2 - 1\right]^2 + \left\{\left[u_m^2\cdot\left(\mu_m - 1\right)+1\right]\cdot\left(u_m^2 - r_f^2\right)\right\}^2}}$$

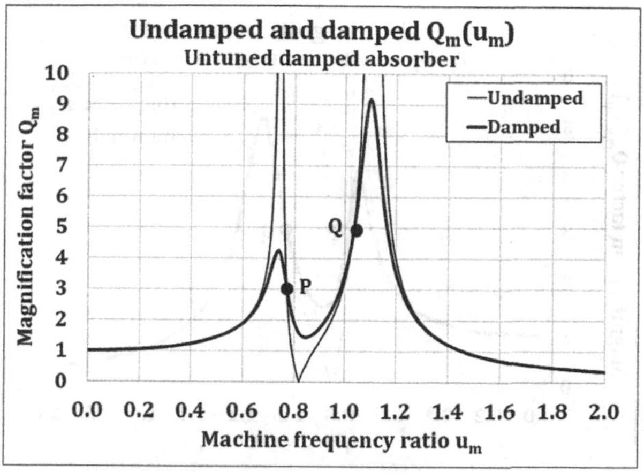

FIGURE 9.17 Frequency response curves of an undamped and untuned damped absorber.

The first significant conclusion of Formula 9.34 is that the magnification factor Q_m depends on one independent variable, the frequency ratio u_m, and three constant ratios: r_f, μ_m and ζ.

Same as undamped absorbers, the first design stage is to select the mass ratio, according to considerations of Section 9.3.4. After that, the mass and damping ratios must be adjusted to obtain the best performance with Den Hartog's method explained in the next section.

9.3.5.4 Den Hartog's Method

This method, proposed by Den Hartog, improves the frequency response curves of a damped absorber by tuning it with the best values of r_f and ζ. Formulas of these best values, credited to Den Hartog, are calculated using the equations of motion 9.29 or 9.30. The frequency response of an undamped and a damped absorber is compared in this section for three different scenarios: a) the damped absorber is untuned, b) the damped absorber is tuned with the best r_f and c) the damped absorber is tuned with the best r_f and ζ. These scenarios are respectively shown in Figures 9.18 and 9.19.

The first scenario is in the chart in Figure 9.17. It shows the curves for $r_f = 0.8$ and $\zeta = 0.15$, which are not the best values proposed by Den Hartog. Points P and Q are at the intersection of the damped and the undamped response curves. These intersections have different magnification factors. Therefore, there is no significant advantage regarding the undamped case because the damped curve is not as flat as desired.

Den Hartog derived and proved that Formula 9.35 returns an r_f value that assures that intersection points P and Q have the same magnification factor. See chart in Figure 9.18.

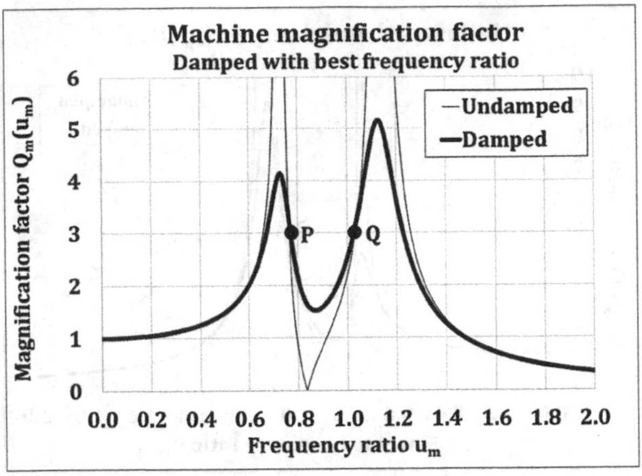

FIGURE 9.18 Damped system with the best frequency ratio.

Formula 9.35 Condition of same Q_m value for P and Q points

$$r_{f\,best} = \frac{1}{1+\mu_m}$$

Furthermore, there are two formulas to calculate the best damping factor to improve the curve flatness of chart 9.18. One is exact, and the other is a simpler one [7]. See the following formulas:

Formula 9.36 Best damping ratio to approach the frequency curve to an almost flat shape

$$\zeta_{best\ exact} = \sqrt[2]{\frac{\mu_m}{8\cdot(1+\mu_m)^3}\cdot\left(3+\sqrt[2]{\frac{\mu_m}{\mu_m+2}}\right)} \quad \zeta_{best\ approximate} = \sqrt[2]{\frac{3}{8}\cdot c\frac{\mu_m}{(1+\mu_m)^3}}$$

The curve changes to a flatter shape after introducing the best damping ratio, as shown in Figure 9.19, though not perfectly flat. The example of the damped absorber design given below demonstrates that the magnification factor difference between the exact and the approximate damping ratio is not significant.

If formulas of r_{fbest} and ζ_{best} are replaced in Formula 9.34, it returns the P and Q magnification factors. The formula given by this replacement is the following:

Formula 9.37 Magnification factor at points P and Q

$$Q_{best} = \sqrt[2]{1+\frac{2}{\mu_m}}$$

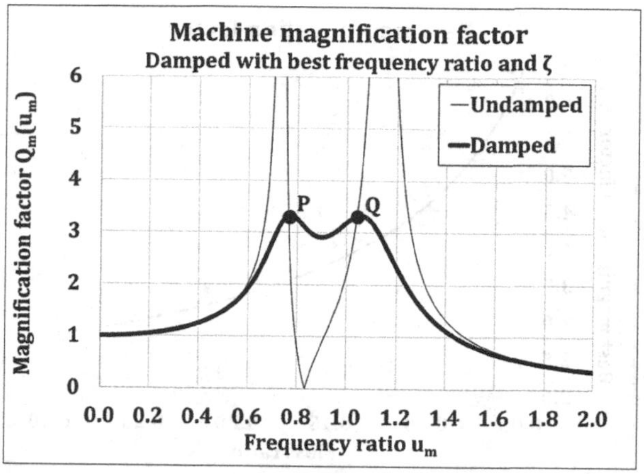

FIGURE 9.19 Damped frequency curve with the best values of frequency and damping ratios.

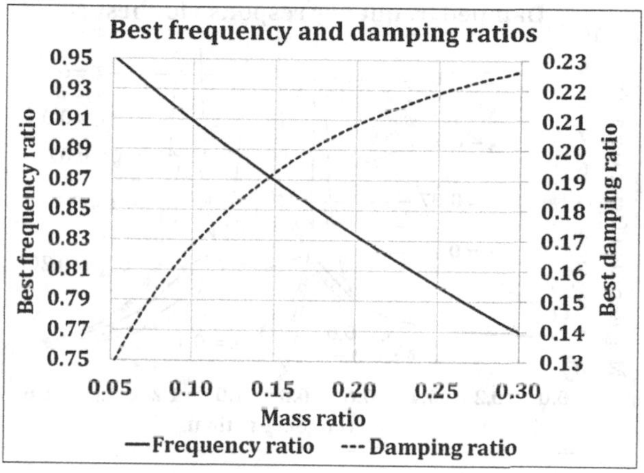

FIGURE 9.20 Best frequency and damping ratios.

Den Hartog's best parameters formulas $r_{f\ best}$, ζ_{best} and Q_{best} are graphically represented in charts of Figures 9.20 and 9.21. The displayed charts allow calculating the best parameters of a damped absorber as a function of the mass ratio. It is noted that the Q_{best} at P and Q is not an input to the calculation procedure but a consequence of the adopted mass ratio. The higher the mass ratio, the lower the Q_{best} values. It means that the mass ratio may be selected according to the Q_m desired value at P and Q. However, limits to the Q_m reduction are the absorber size, cost and space to install a big mass. Theoretically, the mass ratio could be 10 for example, which would result in a flat and low magnification factor. Still, the absorber would be 10 times bigger than the machine, which is an absurd costly design.

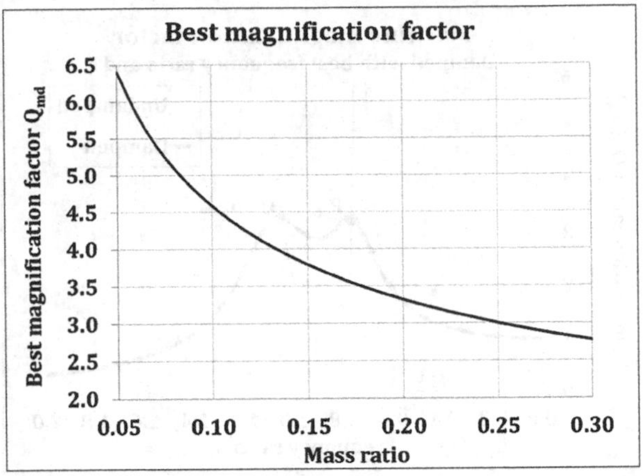

FIGURE 9.21 Best magnification factor.

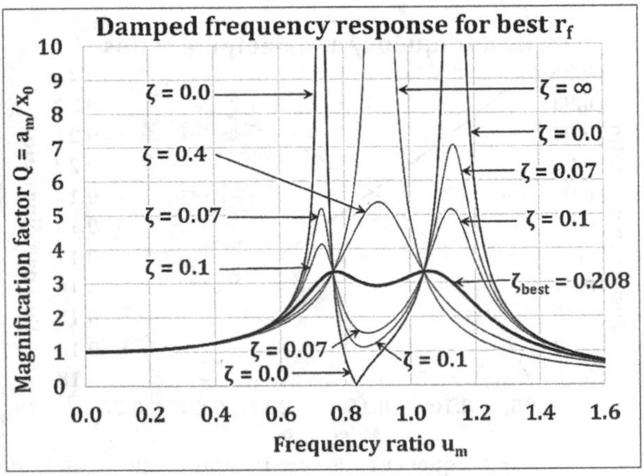

FIGURE 9.22 Family of Q(u) curves for best r_f and $\mu_m = 0.15$.

The chart in Figure 9.22 is an example of the family's curves given by Formula 9.34, where the best frequency ratio ($r_{fbest} = 0.833$) was used. One of the curves was drawn with the best damping ratio ($\zeta_{best} = 0.208$). All curves correspond to a mass ratio μ_m equal to 0.20. It is readily seen that the flatter curve (thick line) is given by ζ_{best}. For values of damping ratio higher than the best, the curves have only one peak. Of course, damping ratios lower or higher than 0.208 are not a good option because of their resonant peaks.

All curves intersect in points P and Q. The curve of the infinite damping ratio is the case of locked masses. As they act as only one body, the system has only one resonance peak like any undamped SDOF system. The curve of $\zeta = 0$ is the undamped absorber discussed previously.

Den Hartog's idea was that between $\zeta = 0$ and ∞, there is always a curve of maximum flatness if the best r_f and ζ formulas are used. Curves with a damping ratio lower than the best have more protruding peaks. See curves of $\zeta = 0.00$, 0.07 and 0.10. As the damping ratio increases over the best, the central peak also increases. See curves of $\zeta = 0.4$ and infinite.

9.3.6 Practical Assessment of a Fan Vibration Neutralization

9.3.6.1 Scenario

In a cement plant, the kiln's induced draft fan of variable speed is abnormally vibrating. Measurements performed at the site with a vibrometer show a vibration amplitude of 6.77 mm at a frequency of $\omega = 200$ rad/s. The expected frequency deviation is $\omega \pm 52$ rad/s. The machine history file shows that the shaft has a static deflection $x_0 = 1.25$ mm. A bump test reveals that the fan's natural frequency is $\omega_{nm} = 221.5$ rad/s. It is requested to design a TVN neutralizer to mitigate the fan's vibration according to the following data.

Fan weight: W = 3,200 kg

Adopted value of mass ratio: $\mu_m = 0.20$.

9.3.6.2 Undamped Absorber Design

1. Q_{m0} at disturbance frequency: $Q_{m0} = 1/[1 - (\omega/\omega_{nm})^2] = 5.41$
2. Vibration amplitude without damper: $a_{m0} = Q_{m0} \cdot x_0 = 6.76$ mm
3. Fan's mass: $m_m = W/g = 326.2$ kg mass
4. Absorber's mass: $m_b = \mu_m \cdot m_m = 65.2$ kg mass
5. Absorber's natural frequency: $\omega_{nb} = \omega = 200$ rad/s
6. Absorber's rigidity coefficient: $k_b = m_b \cdot \omega_{nb}^2/1{,}000 = 2{,}609.6$ kg/mm
7. Frequency ratio: $r_f = \omega_{nb/}\ \omega_{nm} = 0.903$
8. Rigidity coefficients ratio: $\mu_k = r_f^2 \cdot \mu_m = 0.163$
9. Frequency disturbance range
   a. Low frequency: 200 - 52 = 148 rad/s
   b. High frequency: 200 + 52 = 252 rad/s
10. Magnification factor at the extreme frequency's deviation. Formulas 9.19.
    a. Low frequency Q(148) = 2.81
    b. High frequency Q(252) = 6.84
11. Low and high resonant frequencies. Formula 9.22.
    a. Low frequency = 169.4 rad/s
    b. High frequency = 261.4 rad/s
    c. The chart in Figure 9.23 shows four magnification factor curves:
    d. Machine with no absorber. Dotted curve
    e. Machine with the undamped absorber. Fine line curve
    f. Machine with the best-damped absorber. With the approximate formula of ζ_{best}
    g. Machine with the best-damped absorber. With the exact formula of ζ_{best}

After adding the damper, the vibration amplitude decreased from 6.77 mm to 0 mm. The expected frequency deviation range is inside the resonant frequencies

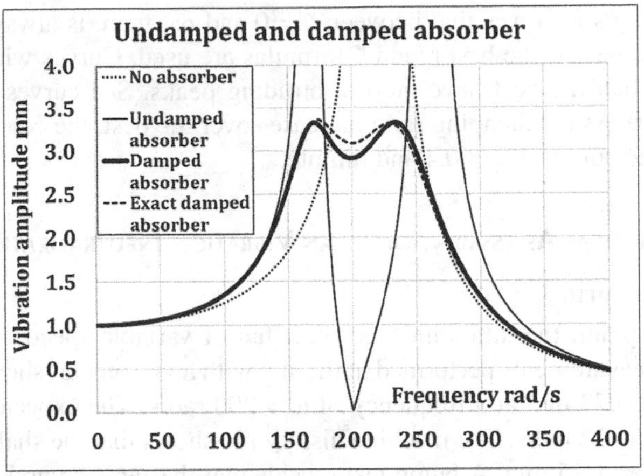

FIGURE 9.23 Fan frequency response with undamped and damped TVN.

range. No risks of resonance exist. However, the magnification factors at extreme frequency deviations (2.81 and 6,84) put the undamped TVN installation project in doubt. Therefore, it is recommended to apply a damped TVN or adopt a more sophisticated absorber like a feedback control system with adjustable absorber mass or rigidity.

9.3.6.3 Damped Absorber Design

Best frequency ratio $r_{f\ best} = 1/(1 + \mu_m) = 0.83$

Absorber's natural frequency $\omega_{nb} = r_{f\ best} \cdot \omega_{nm} = 184.6$ rad/s

Spring rigidity $k_b = m_b \cdot \omega_{nb}^2 = = 2{,}222.2$ kg/mm

Critical friction coefficient $f_c = 2 \cdot (k_b \times 1{,}000 \cdot m_b)^{1/2} = 24{,}084.3$ kg/ | m/s²|

Best damping ratio with approximate formula 9.36. $\zeta = [3/8 \cdot \mu_m/(1 + \mu_m)^3]^{1/2} = 0.208$

Best damping ratio with exact Formula 9.36. $\zeta = 0.222$

Damper friction coefficient $f = \zeta \cdot f_c = 5{,}016.9$ kg/ | m/s²|

Magnification factor at P and Q points: $Q_{md} = (1 + 2/\mu_m)^{1/2} = 3.32$

Vibration amplitude at points P and Q: $a_{md} = Q_{md} \cdot x_0 = 4.15$ mm

The chart in Figure 9.23 shows that the difference between the Q(ω) calculated with the exact and the approximate ζ is not significant. However, the exact formula returns a flatter curve. The damped absorber reduces the initial magnification factor from 5.41 to 2.91 at the BOP, placed between the P and Q points. This reduction is equal to 46% of the initial deflection. Besides, risks of resonance have been significantly mitigated.

- **Alternative with a lower mass ratio:**
  If the absorber's physical size is too big for the available space, it is possible to reduce its size by decreasing the mass ratio, for example, to

0.05; however, the lower value of mass ratio anticipates that the vibration amplitude will increase. See Figure 9.22. This mass ratio produces the same higher P and Q magnification factors. Therefore, with a mass ratio of 0.05 instead of 0.20, the best magnification factor increases from $Q_m = 3.32$ to 6.40, and the vibration amplitude increases from 4.15 mm to 8.00 mm. The lower mass ratio produces a smaller and lighter absorber, but the vibration amplitude is almost duplicated, which is not admissible. In summary: the lighter the absorber, the higher the vibration amplitude.

9.4 ABSORPTION OF OVERHEAD LINES VIBRATION

It is especially important to apply vibration absorbers to electrical overhead lines because the leeward wind excites the conductors. Therefore, this fluid excitation may subject the conductors to severe fatigue stress. The most significant vibration induced by the wind is a) gallop or conductor dancing and b) aeolian motion duc to Karman vortices.

Galloping is due to sleet deposits on the conductor during winter. It is common in northern areas like Canada or Scandinavian countries, or south of Patagonia. Large amplitudes and low-frequency (0.1 to 10 Hz) characterize galloping; therefore, it contributes poorly to the conductor fatigue stress.

Aeolian vibration has a short amplitude and frequencies between 5 and 150 Hz. Therefore, this phenomenon is the main cause of fatigue stress in overhead line conductors. Karman vortices produce it[8] on the conductor leeward side. See Figure 9.24. These leeward vortices are formed around a cylinder for Reynolds numbers (R_e) in the 47 to 100,000 range. Under $R_e = 47$, the flow is laminar though some small vortices are formed for Reynolds numbers 10 to 40. For a higher Reynolds number, the flow is turbulent.

As velocity increases, the flow is no longer laminar, forming small vortices until a von Karman street is fully developed before reaching the turbulence regime. Von

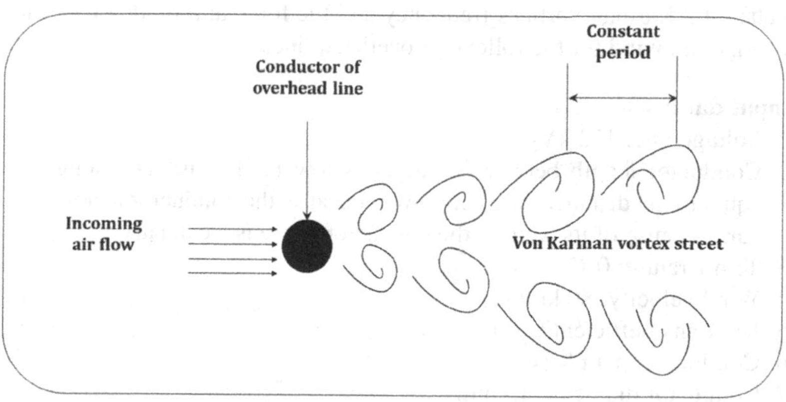

FIGURE 9.24 Von Karman vortex street formation around a cylinder.

Karman vortices are not a laminar flow but, the period between them is constant, suggesting a harmonic excitation on the conductors. This regime induces the conductors' vibration but not due to the incoming flow because it is not harmonic. Still, as the conductor vibrates, it means that the conductor is self-excited at its natural frequency[9] because outcoming Karman vortices are harmonic. Their natural frequency is derived from the Strouhal number defined in Fluid Mechanics.

Formula 9.38 Natural frequency of conductor vibration induced by Karman vortices

$$f = 0.185 \times \frac{V}{D}$$

Where f is the vortices vibration frequency, V is the wind velocity in units of length per second, and D is the conductor diameter in the same units of length. The constant 0.185 is the Strouhal number for typical line conductors, which does not depend on the units used for V and D. The harmonic force on the conductor is calculated with the following formula given in Fluid Mechanics:

Formula 9.39 Force on overhead line conductors produced by Karman vortices

$$F_K = \left[\frac{1}{2} \cdot C_K \cdot \frac{\delta_{air}}{g} \cdot V^2 \cdot A\right] \cdot \sin(2 \cdot \pi \cdot f \cdot t)$$

The term between brackets is the harmonic force amplitude. The Karman coefficient C_K depends on the Reynolds number. For R_e between10^2 and 10^7, a value of $C_k = 1$ is typical. This value corresponds to an air specific-weight of 1.22 kg/m^3 at atmospheric pressure and a temperature of 0°C. The velocity V is given in m/s. A is the conductor sidewise projected area in m^2. The force F_K is in kg. Letter f stands for the frequency calculated with Formula 9.38. As a reasonable approach, it is considered that the force is equally distributed along the conductor.

9.4.1.1 Example of Force Produced by Karman Vortices

Calculate the Karman vortices frequency and the force amplitude on conductors of a horizontal wind for the following overhead lines:

Input data:

1. Voltage line: 132 kV
2. Conductor length between contiguous towers: 100 m. This length is not equal to the distance between towers because the conductor is not a straight bar. Because of its weight, the conductor curve is a catenary.
3. Temperature: 0°C
4. Wind velocity: 80 km/h
5. Karman coefficient C_K: 1
6. Conductors per phase: 2
7. Conductor diameter: 15 mm
8. Conductor natural frequency: 150 Hz

Calculation of frequency and force on the conductor

9. Karman vortices frequency: f = 0.185 × 22.22/0.15 = 27.4 Hz
10. Wind velocity: 80,000/3600 = 22.2 m/s
11. Sidewise projected area: A = 100 × 15/1,000 = 1.5 m^2
12. Force amplitude per conductor: F_K = 1/2 × 1 × 1.22/9.81 × 22.222 × 1.5 = 46 kg
13. Total amplitude force on the tower:
14. Conductors' quantity: 3×2 = 6
15. Force amplitude per tower: 6×46 = 276 kg

This force is not forcefully harmful to the tower's mechanical integrity; however, as it is vibrating at 150 Hz, it may produce resonance due to the tower rigidity and mass and so create a risk. Furthermore, the conductors are also subject to fatigue stress, which may produce a catastrophe. This example proves that conductors of overhead lines need vibration absorption to prevent fatigue stress and resonance problems in the tower due to the conductor's harmonic pulling.

9.4.2 Stockbridge Absorbers

Overhead line absorbers are known as Stockbridge after his inventor, an American engineer who designed and patented the conductor absorbers in 1928. See Figure 9.25.

The absorber has two weights installed on each extreme of a short conductor and a clamp for installing the absorber on the main conductor. Like Frahm's absorber, the Stockbridge is a second-order system with no damping. This absorber has proved to be very efficient. Using it in an overhead line makes the conductors' vibration amplitude to be almost negligible.

The short absorber conductor, parallel to the main conductor, is the rigidity (spring) element. The clamp transmits the main conductor vibration to the absorber, which reacts, reducing the main conductor vibration. There are also sophisticated models, such as asymmetric or unequal weights, that absorb a wider vibration frequency range.

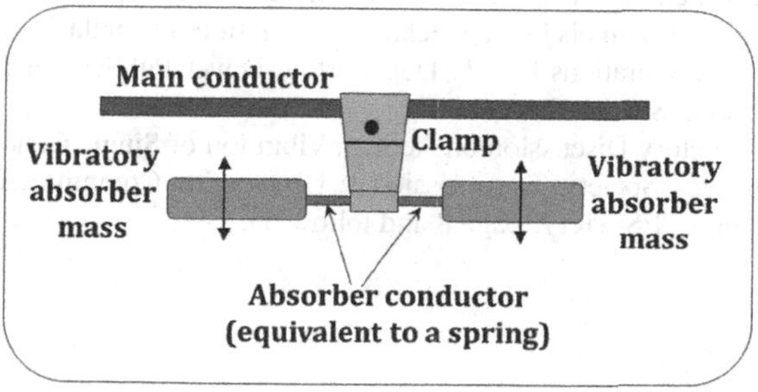

FIGURE 9.25 Stockbridge absorber in electrical overhead lines.

Transmission overhead line owners or constructors usually install absorbers and vibration recorders for as long as needed to identify aeolian vibration characteristics and their severity and determine the wave's antinodes (maximum oscillation amplitude) location. The antinodes are nearby the conductor insulators. This type of absorber has proven successful; hence it is intensively used worldwide in transport overhead lines.

NOTES

1 Books recommended for Chapter 9: 1. Mechanical Vibrations by Kelly –2. Mechanical Vibrations by Den Hartog – 3. Mechanical Vibration by Benaroya, Nagurka and Han – 4. Marks' Standard Handbook for Mechanical Engineering by Avallone, Baumeister. 1. Mechanical Vibrations by S. Graham Kelly. Schaum's Outline Series. McGraw Hill, 1996 – 2. Mechanical Vibrations by J.P. Den Hartog. Dover Publications, Inc. New York, 1985, 4th edition – 3. Mechanical Vibration by Haym Benaroya, Mark Nagurka and Seon Han. CRC Press. Taylor and Francis Group, fourth edition 2018 – 4. Marks' Standard Handbook for Mechanical Engineers by Eugene A. Avallone and Theodore Baumeister, Mc Graw Hill, 10th edition.

2 See a description of applications to tall buildings in book Encyclopedia of Vibration, Chapter Absorbers, Vibration by V. Steffen, Jr and D. Rade, Federal University of Uberlandia, Uberlandia, Brazil

3 This formula derives from the condition: $\omega_{nb} = \omega$. As in the self-excited case it is $\omega = \omega_{nm}$, it results that:
$\omega_{nb} = (k_b/m_b)^{1/2} = (k_m/m_m)^{1/2}$. Clearing k_b from this expression it is obtained:
$k_b = k_m \cdot m_b / m_m = k_m \cdot \mu_m$

4 Eng. Walter Monsberger from University of Cordoba, Argentina authored a paper describing the Chebychev polynomials application used by control engineers of Siemens company in Germany: Practical Method to Optimize Linear Control Systems (in Spanish) 1968.

5 See Mechanical Vibrations by J.P. Den Hartog. Dover Publications, Inc. New York, 1985, 4th edition, Section 3.3.

6 See paper Optimal Design of Damped Dynamic Vibration Absorber for Damped Primary Systems, by Liu and Coppola, Lakehead University, ON, Canada, Transactions of the Canadian Society for Mechanical Engineers, Vol. 34 No. 1. Site: www.tcsme.org/Papers

7 Den Hartog credits the exact formula to John E. Brook's paper: Note on the Damped Vibration Absorber. Transactions ASME 1946, A284. The approximate formula is proposed by Den Hartog in his book Mechanical Vibration, Formula 3.36.

8 See: Mechanical Vibrations by J.P. Den Hartog. Dover Publications, Inc. New York, 1985, 4th editiong, Section 9.6,

9 See: An Introductory Discussion on Aeolian Vibration of Single Conductors, by IEEE Power and Energy Society, Transmission & Distribution Committee Overhead Lines, Technical Report PES-TR17. Page 8 and following.

10 Vibration Control Techniques

10.1 INTRODUCTION TO TECHNIQUES TO REDUCE VIBRATION

When designing a solution to a machine vibration problem[2], the first step identifies the vibration source and the most suitable fundamental form to the observed vibration. One of the four fundamental vibration forms discussed in Section 1.11 must be selected. Identification of the source can be made in the field using a vibrometer or studying the instrumentation displays installed in the machine. For example, suppose vibration frequency coincides with the rotor velocity or a multiple of this speed. In that case, the most fitting vibration form is self-excited (this is the easiest vibration form to identify). A powerful method to identify the vibration source is the frequency spectrum. Many frequency spectrum patterns allow this identification by comparing the actual spectrum with the patterns.

It is almost impossible to design an exact procedure to solve vibration problems unless the most fitted fundamental vibration form is identified. Once the fundamental vibration form is defined, it is possible to do a preliminary solution design, at least as a first approach, and discuss it with specialists or the plant technical team.

A vibration problem may be solved with any of the following basic techniques or a combination of them. These techniques are not listed in order of importance.

1. Technique 1. Increase rigidity and damping in externally-excited machines.
2. Technique 2. Increase rigidity and damping in self-excited machines.
3. Technique 3. Reduce rigidity and damping in base-excited machines.
4. Technique 4. Reduce rigidity and damping in force transmitted machines.
5. Technique 5. Install spring-damper sets to isolate the foundation from harmonic forces coming from the machine.
6. Technique 6. Install a passive vibration absorber or neutralizer or an active control system.
7. Technique 7. Balance the rotor.

Section 10.3 describes the general procedure to calculate springs and dampers to move the current operating point to another pre-selected point in the magnification factor versus frequency plot. This pre-selected point corresponds to two pre-fixed attributes: vibration amplitude and frequency difference to resonance (FDTR). The calculation sequence of the general procedure applies to Techniques 1 through 5. Figure 10.1 conceptually shows the operating point's displacement direction by acting on the system rigidity and the damping.

DOI: 10.1201/9781003175230-13

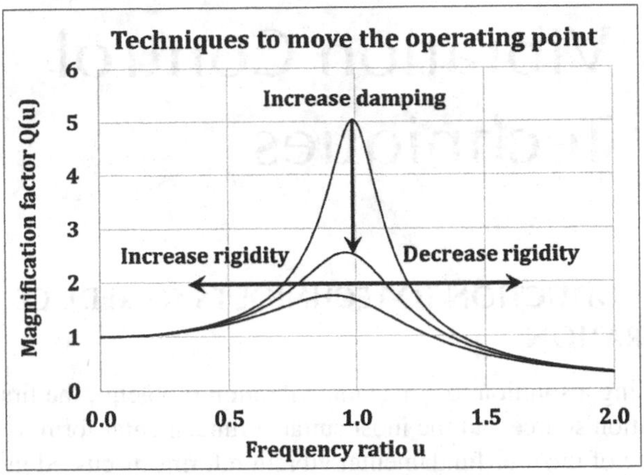

FIGURE 10.1 Effect of rigidity and damping changes in the operating point position.

In general, vibration amplitude depends on the damping ratio and the exciting frequency. These two variables correspond to a unique point on a Q(u) plot. If vibration amplitude at this point is unacceptable, the operating point must be displaced to a position such that vibration amplitude and FDTR have an acceptable value.

The application of these mitigating techniques is not always a crystal-clear decision unless vibration responds to a unique fundamental vibration form like an SDOF second-order system. However, in practice, vibration problems may respond to more than one fundamental vibration form. Therefore, the design solution might consider a combination of these basic mitigating techniques. Note that some of these techniques produce opposite effects, and the result may be confusing.

10.2 CONTROL VIBRATION PHILOSOPHY

The control vibration philosophy is graphically summarized in Figures 10.2–10.5, one for each fundamental vibration form. In a Q(u) plot, these figures show the initial operating point 1 and the expected final point 2. The arrows show the displacement direction of the operating point. The vibration amplitude reduction is achieved by moving the operating point to a lower magnification factor curve of a higher damping ratio. This vertical displacement is obtained by increasing the damping. Simultaneously, the final operating point must be away from the resonant frequency, and that is why the FDTR is a technical requirement. Note that the initial operating point 1 is close to resonance in the four cases; therefore, the expected FDTR will move the operating point horizontally. This horizontal motion is achieved with higher rigidity. However, this is not the case of the base-excited mode, where the final operating point must be in the attenuation zone, obtained by decreasing the natural frequency. As rigidity is proportional to the square of the natural frequency ($k = m \cdot \omega_n^2$), new springs must be looser than springs of point 1.

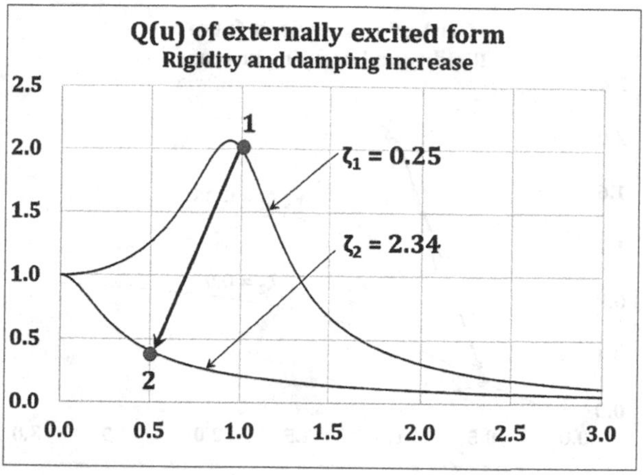

FIGURE 10.2 Mitigation procedure increasing rigidity and damping.

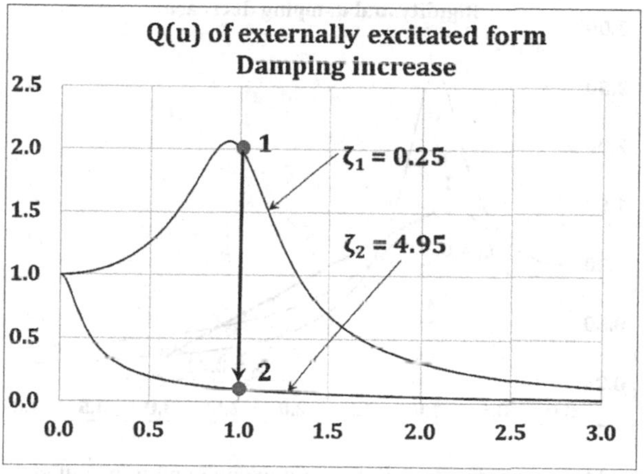

FIGURE 10.3 Mitigation procedure increasing damping only.

The next section's general procedure calculates the magnification factor and damping ratio according to the specified vibration amplitude and FDTR.

10.3 TECHNIQUES GENERAL PROCEDURE

The calculation procedure described below is valid for Techniques 1 to 5. A numeric example based on the same input data (initial scenario) has been prepared and presented below in a spreadsheet form to ease understanding the procedure.

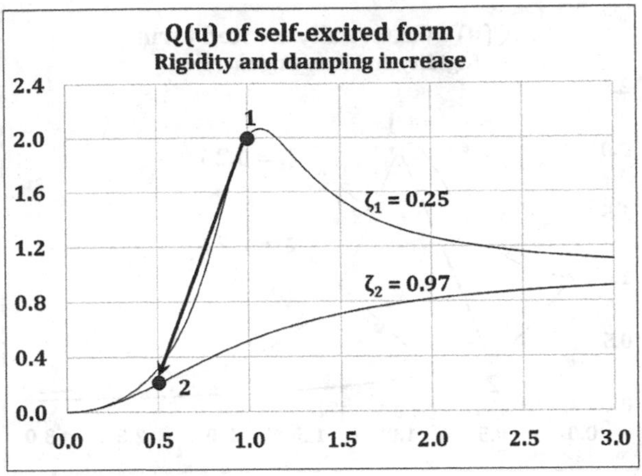

FIGURE 10.4 Mitigation procedure of a self-excited machine increasing rigidity and damping.

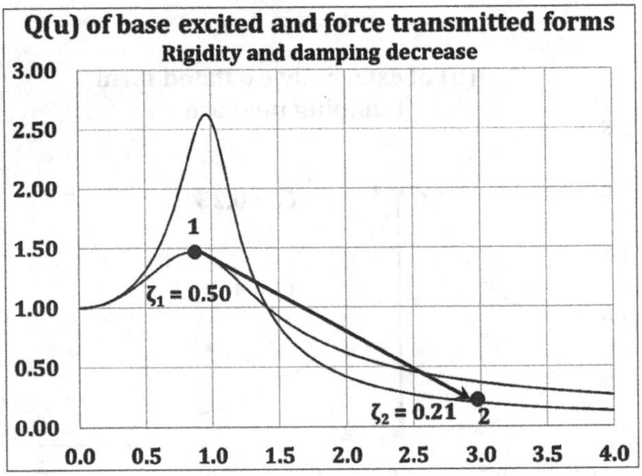

FIGURE 10.5 Mitigation procedure of a base-excited and transmitted force decreasing rigidity and damping.

10.3.1 Scenario Description

This case uses the general procedure based on information commonly available in the field or design office. It is assumed that one turbomachine, installed on a spring-damper set over a perfectly rigid foundation, vibrates with an amplitude higher than the acceptable value. Therefore, it is requested to change the isolating spring-damper set to achieve a vibration amplitude according to ISO standard 7919 and a pre-fixed difference between the disturbance and the resonant frequency (FDTR).

The calculation procedure and formulas correspond to any of the four fundamental forms of Section 1.11. Therefore, there are four scenarios for the same machine, one for each fundamental vibration form. It could be considered that there are four identical turbomachines with four different vibration forms; hence, four different spring-damper sets are calculated.

The information obtained in the field is in the following list.

1. Machine weight W.
2. Disturbance frequency is ω.
3. Static deflection x_0.
4. Vibration amplitude a_1.
5. Machine velocity RPM.
6. The machine is mounted on spring-damper sets whose rigidity and damping coefficients are unknown.
7. The foundation is considered perfectly rigid. If this is not the case, the procedure used below is not valid.
8. Expected vibration amplitude and FDTR with the new spring-damper set.

The admissible vibration amplitude a_l is discussed in Section 10.3.

10.3.2 General Calculation Procedure

The following formula sequence is valid for the scenario described in Section 10.3.1. If the initial scenario has different data, another procedure must be designed following the guidelines given in this section. The calculation sequence described below is implemented in the spreadsheet of Table 10.2 in four columns, one for each fundamental vibration form.

10.3.2.1 Initial Scenario. Point 1 Calculation

1. Machine mass.|kg mass|

$$m = \frac{W}{9.81}$$

2. Magnification factor.

$$Q_1 = \frac{a_1}{x_{01}}$$

3. Rigidity k. |kg/mm|

$$k_1 = \frac{W}{x_{01}}$$

4. Machine natural frequency. |rad/s|

$$\omega_{n1} = \sqrt[2]{\frac{k_1}{m}}$$

5. Frequency ratio at disturbance frequency.

$$u_1 = \frac{\omega}{\omega_{n1}}$$

6. Damping ratios.
   Fundamental forms 3 and 4 have the same formulas.
   Warning: in the fundamental vibration forms 3 and 4, the requested value of Q cannot be lower than the curve of $\zeta = 0$ in the chart in Figure 8.8. Suppose the requested Q falls below this curve. In that case, the above formula of ζ for any of these two vibration forms does not return positive real numbers, meaning that the requested value of Q is impossible to meet at the used frequency ratio. Therefore, a higher frequency ratio must be used. As the design frequency ratio is u = 1+FDTR, a higher FDTR must be used or produce a graphical solution with a chart in Figure 8.8. This issue is produced by an incompatible set of [Q, u] values. For example, see in the chart in Figure 8.8 that at u = 5, it is impossible to obtain a transmissibility ratio lower than 0.04.
7. Friction coefficient |kg/(m/s)|

   Clear f from the damping ratio formula; $\zeta = f / f_c = f / \left(2 \cdot \sqrt[2]{k \cdot m}\right)$, and replace k by $m \bullet \omega_n^2$, to see:

$$f_1 = 2 \cdot \zeta \cdot m \cdot \omega_{n1}$$

   With f_1 and k_1 are the existing spring-damper set properties. They are identified for the four vibrating forms. The point 1 location is given by Q_1 and u_1. Table 10.2 shows Q_1 and u_1 values.

10.3.2.2 Final Scenario. Point 2 Calculation

8. Frequency ratio.
   Forms 1, 2 and 3: If point 2 must be to the left of the resonant frequency, then use:

$$u_2 = 1 - \text{FDTR}$$

   Form 4. If point 2 must be to the right of the resonant frequency, then use:

$$k_2 = m \cdot \omega_{n2}^2$$

9. Natural frequency. |rad/s|

$$\omega_{n2} = \frac{\omega}{u_2}$$

10. Rigidity coefficient. |kg/mm|

$$k_2 = m \cdot \omega_{n2}^2$$

11. Static deflection after the new spring's installation. |mm|

$$x_{02} = \frac{W}{k_2}$$

12. Magnification factor.

$$Q_2 = \frac{a_2}{x_{02}}$$

13. Damping ratio. Use formulas of Table 10.1.
14. Friction coefficient.

Same as was done to obtain ζ_1 in calculation line 7, the damping ratio in point 2 is:

$$f_2 = 2 \cdot \zeta_2 \cdot m \cdot \omega_{n2}$$

Step 14 completes the k_2 and f_2 calculations. k_2 and f_2 must be divided by the total quantity of springs to install, to obtain the rigidity coefficient per spring and the friction coefficient per damper.

10.3.2.3 Design Ratios

The following ratios help to assess the machine performance after installing the previously designed spring-damper set.

a. Frequency ratio.

$$r_u = \frac{u_2}{u_1}$$

This ratio measures how much the operating point was horizontally moved from its initial position (point 1). As the a_2 value is pre-fixed, the above

TABLE 10.1
Magnification and Damping Factors of the Four Fundamental Vibration Forms

Fundamental Vibration Mode	Q(u) Formula	ζ Formula Cleared From Q(u)
1. Externally excited	$Q(u) = \frac{1}{\sqrt[2]{(1-u^2)^2 + 4 \cdot \zeta_1^2 \cdot u^2}}$	$\zeta = \frac{\sqrt[2]{1 - Q^2 \cdot (1-u^2)^2}}{2 \cdot Q \cdot u_1}$
2. Self-excited	$Q(u) = \frac{u^2}{\sqrt[2]{[1-u^2]^2 + 4 \cdot \zeta^2 \cdot u^2}}$	$\zeta = \frac{\sqrt[2]{u^4 - Q^2 \cdot (1-u^2)^2}}{2 \cdot Q \cdot u}$
3 and 4. Base-excited and force transmitted	$Q(u) = \sqrt[2]{\frac{1 + 4 \cdot \zeta^2 \cdot u^2}{(1-u^2)^2 + 4 \cdot \zeta^2 \cdot u^2}}$	$\zeta = \sqrt[2]{\frac{Q^2 \cdot (1-u^2)^2 - 1}{4 \cdot u^2 \cdot (1-Q^2)}}$

procedure assures that vibration amplitude a_2 will be according to specifications. Nonetheless, this amplitude must be verified whether it complies with standards and manufacturer's specifications.

b. Vibration amplitudes ratio.

$$r_a = \frac{a_2}{a_1} = \frac{Q_2}{Q_1}$$

The amplitude ratio is only a validation of the imposed vibration amplitude in the new spring-damper set requested.

c. Rigidity ratio.

$$r_k = \frac{k_2}{k_1} = \frac{\omega_{n2}^2}{\omega_{n1}^2}$$

The r_k ratio gives an idea of the size and weight of the new springs. As a rule of thumb, the rigidity coefficient is proportional to the spring weight's square root. Therefore, if the rigidity ratio is equal to 4, it is expected that the springs' weight will be duplicated.

d. Friction ratio

This ratio gives an idea of the damping ratio change and the change in the new dampers' weight and size.

The procedure described previously may be extended to other similar cases or modified according to the specific problem at hand. It is recommended to program the above calculations in a spreadsheet to assure calculation accuracy.

10.3.2.4 Example of the General Procedure Applied to the Four Fundamental Vibration Forms

Table 10.2 is created with the general calculation procedure described previously, applied to four identical machines, each having one of the four fundamental vibration forms. They are the same for the four machines, except for the adopted amplitude a_2 of the base-excited case (column 7 of Table 10.2). In this case, an amplitude of 0.07 mm is used because a lower deflection is incompatible with the prefixed values of Q, as was commented before for step 6. In this case, it is physically impossible to reduce the vibration amplitude to 0.045 mm.

Vibration control techniques applied to the four fundamental vibration forms. See Figures 10.2, 10.3, 10.4 and 10.5. Formulas used in each step have the same number as the formulas sequence described previously.

10.3.3 DESCRIPTION OF THE SEVEN BASIC TECHNIQUES

These techniques have been described in previous chapters. However, for didactic reasons, their conceptual philosophy and procedures are briefly described in this section

TABLE 10.2
General Calculation Procedure

Calculation Line number	Physical property	Units	Externally Excited. Lncrease Rigidity and Daamping	Externally Excited. Lncrease Damping Only	Self-excited	Base-excited
Data obtained in the field						
	Weight W	kg	5,000	5,000	5,000	5,000
	Disturbance frequency ω_d	rad/s	100	100	100	100
Input data	Static deflection x_{01}	mm	1.00	1.00	1.00	1.00
	Vibration amplitude a_1	mm	2.00	2.00	2.00	2.00
	Machine velocity	RPM	3,600	3,600	3,600	3,600
	Admissible amplitude (ISO 7919)	mm	0.046	0.046	0.046	0.046
			Requested vibration amplitude and FDTR			
Input data	Adopted amplitude a_2	mm	0.045	0.045	0.045	0.070
	Requested FDTR		0.50	0.50	0.50	5.00
			Point 1 calculation			
1	Machine mass m	kg mass	510	510	510	510
2	Magnification factor Q_1		2.00	2.00	2.00	2.00
3	Rigidity k_1	kg/mm	5,000	5,000	5,000	5,000
4	Natural frequency ω_{n1}	rad/s	99	99	99	99
5	Frequency ratio u_1		1.01	1.01	1.01	1.01
6	Damping ratio ζ_1		0.247	0.247	0.252	0.286
7	Friction coefficient f_1	kg/(m/s)	24,981	24,981	25,466	28,846

(Continued)

TABLE 10.2 (Continued)

Calculation Line number	Physical property	Units	Externally Excited. Lncrease Rigidity and Daamping	Externally Excited. Lncrease Damping Only	Self-excited	Base-excited
			Point 2 Calculation			
8	Frequency ratio u_2		0.50	1.01	0.50	6.00
9	Natural frequency ω_{n2}	rad/s	200	99	200	17
10	Rigidity k_2	kg/mm	20,387	5,000	20,387	142
11	Static deflection x_{02}	mm	0.25	1.00	0.25	35.32
12	Magnification factor Q_2		0.184	0.045	0.184	0.050
13	Damping ratio ζ_2		5.40	11.01	1.14	0.10
14	Friction coefficient f_2	kg/(m/s)	1,100,540	1,111,111	231,906	1,724
			Design Ratios			
15	Frequency ratio u_2/u_1		0.50	1.00	0.50	5.94
16	Amplitudes a_2/a_1		0.02	0.02	0.02	0.02
17	Rigidity ratio k_2/k_1		4.08	1.00	4.08	0.03
18	Friction ratio f_2/f_1 ratio		44.05	44.48	9.11	0.06

The calculation of Point 2 for the base-excited or force transmitted case returns a static deflection (35.32 mm) at Point 2, which may be inadmissible. The problem is the selected admissible vibration amplitude (a_2 = 0.046 mm) that requires a too high FDTR (FDTR = 5, then u = 6). Therefore, the natural frequency is too low (17 rad/s), which finally produces a small value of the spring rigidity coefficient (142 kg/mm). The calculation sequence must be redone, admitting a higher vibration amplitude at Point 2 if the spring-damper set deflection of 35.32 mm is not admissible. Review the comment under Table 10.2 regarding the compatibility between the desired magnification factor at Point 2 and the design frequency ratio.

10.3.3.1 Technique 1. Externally-Excited Machine

Technique 1 displaces the operating point 1 to the left and downward in the Q(u) plot by increasing rigidity and damping. The spring-damper set substitution produces a higher damping ratio, which, in the case of Figure 10.2, is higher than 1. Therefore, at point 2, the motion transients are overdamped, and the time response to external excitation is longer. The amplitude ratio is equal to the magnification factors ratio. In the example of Figure 10.2, this ratio is equal to 0.20. This ratio means that the vibration amplitude is reduced by 80%. Point 2 is moved horizontally to the left; therefore, the assembly's natural frequency is increased.

It is possible to increase only the damping by changing the existing dampers. This damping increase moves the operating point downward at the same frequency as point 1. Therefore, the operating point is displaced to a curve of a higher damping ratio. As rigidity does not change in this technique, the natural frequency remains unchanged. Therefore, the static deflection x_0 and the natural frequency are not changed.

Techniques 1 is used only when the system is running near resonance. If this is not the case, the damping addition may require exaggerated heavy springs and dampers. Figure 1.32 shows that at frequency ratios away from resonance (u = 1), the Q(u) curves are too close to each other; therefore, a small reduction of Q(u) produces an important increase in the damping ratio. Thus, the friction coefficient significantly increases because it is directly proportional to the damping ratio.

Technique 1 increases rigidity and damping; therefore, the operating point displacement to a lower magnification factor curve and higher FDTR has a caveat: spring and dampers are heavier and more expensive. Hence, the weight, size, and cost of springs and dampers limit the rigidity and damping increase of Technique 1, and, consequently, vibration amplitude cannot be reduced in many cases as desired.

The example of Table 10.2 shows that at point 1, the rigidity coefficient is 5,000 kg/mm and at point 2 is four times larger. The exact rigidity ratio between points 1 and 2 is $r_k = 4.08$. This ratio means that springs at point 2 will be $\sqrt[2]{r_k} = 2.02$ times heavier than at point 1. This calculation is a rough estimate of the spring's weight change. The friction ratio is much higher: $r_f = 44.5$, which indicates that dampers are much heavier, and their size may represent a strong challenge to procuring and installation. If the procurement or installation is not feasible, the only solution is to admit a higher vibration amplitude at point 2. For example, if the admissible amplitude is 0.1 mm instead of 0.047 mm, the friction ratio r_f is reduced from 44.5 to 19.1. Even so, this ratio indicates that dampers will be much heavier than dampers at point 1.

A common question in the field would be: why don't we use the same dampers of point 1 and only change the springs? The answer to this question can be found by re-running the general procedure's routine of Table 10.2, which returns an admissible amplitude of 0.31 mm (equivalent to 12.2 mils) and a damping ratio ζ of 0.25. The friction ratio for this case is 2.02. As the resultant amplitude is 0.31 mm, this solution is unacceptable because this vibration amplitude is almost seven times higher than the requested 0.045 mm.

Furthermore, as the damping ratio is below 1, transients that might occur will be oscillating with significant amplitudes. Therefore, a trade-off between vibration amplitude and the weight and size of dampers must be solved. For example, an absorber could be added. Therefore, it is recommended to consult the machine's manufacturer or a specialized company in vibration and discuss different potential solutions.

10.3.3.2 Technique 2. Self-Excited Machine

The philosophy to control self-excited vibration is to increase the rigidity and damping of the spring-damper set, just as is done in Technique 1. See Figure10.3. The comments made about the risks of designing too heavy springs and dampers hold for this fundamental vibration form.

The self-excited vibration is easy to identify as per previous comments because the frequency vibration coincides with a multiple of the machine velocity. The calculation procedure to mitigate the vibration amplitude is the same as that used for the externally excited system. The only difference is in the frequency response and damping ratio formulas.

Table 10.2 of the general procedure shows that the rigidity ratio is 1. The friction ratio is extremely high, 44.48 times the friction of the existing dampers. However, it is important to consult a mechanical designer before procuring the new dampers to assure that the weight and size can be adapted to the existing installation. An unbalanced shaft-rotor may produce this vibration case; therefore, a recommended solution is to send the rotor to a balancing workshop and, after that, decide whether to change the spring damper set or not.

10.3.3.3 Techniques 3 and 4. Base-Excited and Force Transmitted Machine

Consider the case of an unacceptable vibration transmitted by the base to the machine. Section 1.11.3 describes the control philosophy to reduce the transmitted vibration from the base to the machine. In essence, Technique 4 moves the operating point to the right of the Q(u) plot, where the attenuation zone is.

This technique inserts low rigidity springs and dampers between the machine and the foundation. The low rigidity increases the frequency ratio (remember $\omega_n = \omega_{disturbance}/u$) and reduces the system's natural frequency. As this technique aims at putting the operating point in the attenuation zone, the disturbance frequency ratio must be made higher than $\sqrt[2]{2}$.

The magnification factor is diminished by increasing the frequency ratio and reducing the natural frequency ω_n. However, if the damping is too low, vibration transients may last a long time; therefore, it is advisable to add more damping. Another issue of low springs stiffness is the high sensitivity to static forces, making the machine move up and down in an almost unstable motion.[3]

Ratios of Table 10.2 show that the frequency ratio has been increased beyond where the attenuation zone starts. As force transmissibility ratio (Formula 1.83) and vibration amplitude ratios (Formula 1.82) are identical, it is possible to use Figure 1.8 to calculate the springs rigidity and friction coefficient for a base-excited machine. In this figure, the ordinates represent the amplitude ratio and not the transmissibility ratio. Conceptually, they are different properties, but their values are the same; therefore, the procedure used in Section 8.1 holds for the base-excited mode.

10.4 PREDICTING AND PREVENTING HARMFUL VIBRATIONS

Machine vibration is easy to detect in the field by simply putting a hand on the machine. If the environment is not noisy, it is possible to listen to a soft tremor due to vibration. If this situation holds for a long time, some machine parts, like screws, bolts, or nuts, will lose their tightness, and the tremor will turn into a noisy pounding. However, the best way to detect a machine's vibration is not by hand but by using a vibrometer that measures the three kinematic vibration variables: amplitude, frequency, and acceleration[4].

A wide commercial offer of vibrometers exists. Most of them measure the three kinematic vibration variables. It is common to use an instrumentation group formed by a vibrometer, sound level meter and laser aligner to perform vibration controls. This instrumentation allows a fast understanding of vibration problems and helps to correct any anomaly on the spot.

The most common source of turbomachine failure is when the fatigue stress is higher than the fatigue limit. Therefore, it is of utmost importance to organize a monitoring system to predict the most critical machines' vibration and make a fast decision if alarms arise. ISO standards 10816 and 13374 should be used as a guide to organize a monitoring system and assess the results, alarm levels, vibration source identification and other tasks recommended for each specific case.[5] It is also advisable to train one or more persons in vibration monitoring, assessing measurements, and planning mitigation or repair activities.

However, vibration monitoring and interpretation is a sophisticated technique that makes it advisable to contract external support from specialized consultant companies in the case of severe vibration. In some industries, vibration monitoring is an outsourced service that should contact the management plant to discuss vibration symptoms, their causes and create plans and action items.

It should be kept in mind that the cost of avoided failures is usually much higher than the investment in a consultant company, instrumentation, and training people. Some technical publications assure that the return on investment (ROI) of the monitoring system is defined as the ratio between the cost of obviated failure over the total cost of technicians, instrumentation, consultant, etc., is around 36:1. This ratio means that for every dollar invested in the vibration monitoring system, 36 dollars expense in repairs, spares, lack of production and others, are saved.

There is a broad recommended scope of subjects to be mastered by a team of specialists in vibration, namely:

a. Vibration theory. No in-depth math is necessary, although, in complex installations like a power plant, mastering vibration math is a significant plus for decision making.
b. ISO standards, as previously mentioned.

c. Organize training in specialized instrumentation, measuring, acquiring data stored in the instruments, and analyzing results on a computer.
d. Knowledge of the machine history. Identification of parts that may vibrate or have produced vibration in the past.
e. Skills in recording measurements and defining benchmarks. It is essential to determine what vibration indices are acceptable, make a baseline and carry out a periodic verification of indicators performance.
f. Trend analysis of historical series obtained by the monitoring system.

A robust monitoring system and skilled personnel may identify vibration sources by analyzing the frequency spectrum and compare them with typical patterns of different vibration sources.

Modern instrumentation allows measuring position, velocity and acceleration of mechanical parts, such as the shaft, housing, etc. The instrument display shows waves against time, which in general have significant distortion. This distortion happens because different vibration modes may be separated in perfect sinusoidal streams based on the Fourier transform. This graphical information predicts problems but does not show where problems are in the machine. Fortunately, modern instrumentation converts the amplitude against time into an amplitude versus frequency figure, usually known as a frequency spectrum. Figure 10.6 shows the sequences that are produced to identify vibration sources. The Total Time Signal curve is the primary information obtained from instruments in the field. The central figure shows the harmonic components of the previous total time signal. Finally, on the right figure is the Frequency Spectrum. These plots give important information because they allow identifying vibration sources. The complete process is made by instrumentation, based on the mathematics of the Fourier transform.

The total time signal is obtained by a vibration meter in the field, which usually displays an initial confusing wave, as shown in Figure 1.8. This data is collected and then introduced in a specialized software of vibration analysis. This software's output is the central figure of the harmonic components. However, this

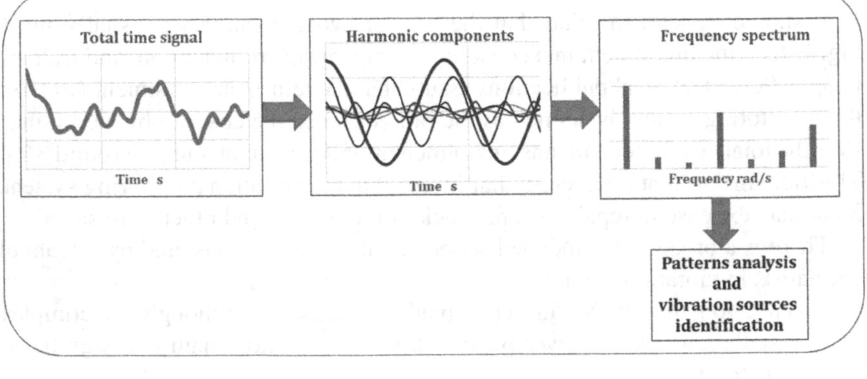

FIGURE 10.6 Frequency spectrum obtained by the Fourier transform.

plot only indicates how many vibration modes exist, along with their amplitude, frequency, and phase. Still, no vibration source identification is obtained from this information, except that a vibration problem exists.

The Frequency Spectrum production is the next step. It shows each component amplitude versus its corresponding frequency. For instance, an unbalanced rotor exhibits only one amplitude at the rotational velocity frequency. In this step, the Fourier transformation is used by the software to produce the frequency spectrum figure. The amplitude and frequency of all vibration modes appear in the spectrum.

The vibration source identification procedure is the fourth step of Figure 10.6 and is graphically represented by the block named Pattern Analysis and Vibration Source Identification. The vibration analyst creates the last step: interpreting oscillograms and preparing a report according to ISO standards, where all vibration sources are identified. The last step is to discuss and decide with the plant technicians' team the vibration causes and mitigating actions and prepare future monitoring and repairs.

About 40 different frequency spectrum patterns exist; each identifies a specific vibration origin, such as bearings in poor shape and unbalanced rotating parts. This comparison with typical patterns is a procedure that requires a skilled vibration analyst to interpret the spectrum plot, make a diagnostic of what happened and recommend the next steps. Guidelines and specifications to assess torsional anomalies are described in standards such as US MIL-STD 167[6], API 610, 617, 618, 619 and RP 684. As these standards are for specific equipment, the designer should carefully choose the most proper case.

10.4.1 Admissible Vibration Amplitude

10.4.1.1 Turbomachine Rotor

Figures and tables to estimate the vibration regime of turbomachine rotors are available in different publications. The application range of some standards is usually described in their text. For example, the ISO 7919 standard expresses that it applies to the following list of equipment.

- steam turbines
- turbo-compressors
- turbogenerators
- turbofans
- electric drives and associated gears, where relevant
- rotodynamic pumps (turbopumps).

API 684 standard recommends that during a balanced rotor's shop test, peak to peak vibration amplitude at maximum velocity must be lower than the values calculated with Formulas 10.1 or 25 μm (1 mil), whichever is less. It is suggested to use the values returned by these formulas to set field alarms and tripping controls.

Formula 10.1 API formula of the maximum admissible peak to peak amplitude after balancing

$$a_{adm} = \sqrt[2]{\frac{12{,}000}{RPM_{max}}} \qquad a_{adm} = 0.0254 \times \sqrt[2]{\frac{12{,}000}{RPM_{max}}}$$

a in mils a in mm

Log-log Figure 10.7 shows Formula 10.1 in mm and mils.

The restriction of admissible amplitude given by Formula 10.1 "or 1 mil, whichever is less" seems to be too strict in practice. Turbomachinery International's article mentioned in the footnote[7] refers to a major oil company that specifies 2 mils tolerance from 0 to 3,000 RPM and recommends using Formula 10.1 for speeds higher than 3,000 RPM. Field engineers give special attention to changes in vibration displacements. They usually say that it is preferable a constant amplitude of 4 mils and not a sudden change from 2.5 to 3.9 mils. This assertion may be true in some cases but cannot be used as a "golden rule," for the absolute value of displacement is always a good indicator of potential risks.

10.4.1.2 Machine Case and Bearings Cap

The vibration amplitude of the machine and bearing is measured on their cases. This amplitude depends on the machine and bearing mechanical status quality and is inversely proportional to the rotor-shaft velocity. Therefore, the formula to calculate the vibration amplitude is:

Formula 10.2 Peak-to-peak vibration amplitude

$$a = \frac{m}{RPM}$$

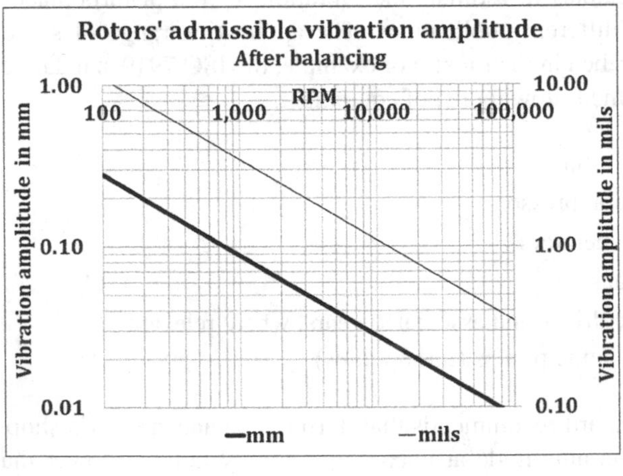

FIGURE 10.7 Admissible vibration amplitude in balanced rotors.

TABLE 10.3
Vibration Regime Classification for Machine Structures and Bearing Caps

Vibration Regime	Metric System: Mechanical Quality Index m | mm×RPM |	Metric System: Velocity mm/s	British System: Mechanical Quality Index m | mils×RPM |	British System: Velocity Inches/s
Very rough	330	15.95	13,000	0.6280
Rough	152	7.98	6,000	0.3140
Slightly rough	76	3.99	3,000	0.1570
Fair	46	1.99	1,800	0.0785
Good	22	1.00	874	0.0392
Very good	10	0.50	380	0.0196
Smooth	5	2.49	192	0.0980
Very smooth	2	1.24	92	0.0490

Coefficient m is the mechanical quality index, given by the |a×RPM| product. With the measured vibration amplitude and RPM, this index is calculated in |mm•RPM| or |mils•RPM| and entered in the second or fourth columns of Table 10.3 to read the vibration regime in the first column.

For example, let be a filtered vibration amplitude measurement of 1.2 mils in a 3,600 RPM turbine's bearing caps. The mechanical quality index m is 4,320 |mil•RPM|. This value is between the slightly rough and rough regimes. Therefore, this regime should be considered an alarming level. In this case, it is recommended to take frequent measurements of the quality index m to track the vibration regime's evolution. As an alternative method, the vibration velocity is measured and entered in the fourth or fifth columns to read the vibration regime's first column.

Amplitude measurements should be filtered; otherwise, the vibration regime could be erroneously determined. See the Engineers Edge site indicated in the footnote[8] for further details.

NOTES

1 Books recommended for Chapter 10: 1. Mechanical Vibrations by Kelly –2. Mechanical Vibrations by Den Hartog – 3. Mechanical Vibration by Benaroya, Nagurka and Han – 4. Marks' Standard Handbook for Mechanical Engineering by Avallone, Baumeister – 5. Fundamentals of Vibration Engineering by Bykhovsky

1. Mechanical Vibrations by S. Graham Kelly. Schaum's Outline Series. McGraw Hill, 1996 – 2. Fundamentals of Vibration Engineering by Isidor Bykhovsky. MIR Publishers, Moscow. 1st published 1972 – 3. Mechanical Vibrations by J.P. Den Hartog. Dover Publications, Inc. New York, 1985, 4th edition – 4. Mechanical Vibration by Haym Benaroya, Mark Nagurka and Seon Han. CRC Press. Taylor and Francis Group, fourth edition 2018. 5. Marks' Standard Handbook for Mechanical

Engineers by Eugene A. Avallone and Theodore Baumeister, Mc Graw Hill, 10th edition - Fundamentals of Vibration Engineering by Isidor Bykhovsky. MIR Publishers, Moscow. 1st published 1972

2 Note: this chapter is strongly grounded on concepts and formulas of Section 1.11; therefore, it is recommended to revisit this section.

3 Consult Fundamentals of Vibration Engineering by Isidor Bykhovsky. MIR Publishers, Moscow. 1st published 1972. Chapter 9, Section 48.

4 Consult the book Gas Turbine Engineering Handbook by Meherwan P. Boyce, 4th edition. 2012 Chapter 16 Spectrum Analysis.

5 See: https://www.iso.org/obp/ui/#iso:std:iso:10816:-3:ed-2:v1:en
https://www.iso.org/obp/ui/#iso:std:iso:13374:-3:ed-1:v1:en

6 Consult MIL-STD-167-1A 2 November 2005 for shipboard equipment MIL-STD-167-1A, 2 November 2005 in the military site http://quicksearch.dla.mil/Transient/3E550B35AEC84A1F87F881DEA909E2EA.pdf

7 a) See: https://www.engineersedge.com/vibration/vibration_severity_figure_13658.htm
b) See: Turbomachinery International magazine, article Shaft Vibrations and Unbalancing Limits in the magazine site: https://www.turbomachinerymag.com/shaft-vibration-and-unbalance-limits-too-stringent-for-some-rotors/.
c) See book Is my machine OK? By Robert Perez, and Andrew Conkey. Industrial Press Inc. New York.

8 See: https://www.engineersedge.com/vibration/vibration_severity_figure_13658.htm

11 Feedback Control Techniques

11.1 INTRODUCTION TO FEEDBACK CONTROL TECHNIQUES

The application of Feedback Control Theory[1] to control vibration is used extensively in modern systems with a significant sophistication degree. For example, the retractable fin used as an anti-roll system in many ships is operated by a feedback control system that detects the roll angle. The system sends a roll signal to a controller that retracts or extends the anti-roll fin[2].

Industrial modern systems use computers where the controller algorithms are programmed. These controllers have a setpoint commanded by the process model, which the real process must follow. However, the loop dynamic is the same as the old control systems without a computer. Therefore, the control systems theory based on the frequency response is valid, though modern control techniques, such as predictive control, have significantly improved process controllability.

Feedback Control Theory is the mathematical basis of the automatic control for many industrial applications, naval, air or land transport, astronautics, military operations, houses, etc. Most human activity fields use automatic controls to assure the efficiency of production processes or human beings' comfort. Furthermore, other knowledge fields also use the feedback control theory for their developments.

An automatic control system keeps the pre-set values of physical quantities of a specific technological process. The control system opposes any disturbance, looking to re-establish the specified conditions. It must then "know" the real physical variables controlled and compare them with the pre-set ones. If these variables do not match, the controller reacts accordingly. The difference between real and pre-set variables is an error signal that the controller must reduce to zero in the shortest possible time and a stable manner.

The Feedback Control Theory is studied using Laplace transform. This mathematical tool allows a differential equation in the time domain to be converted into an algebraic expression in the complex frequency, easy to manipulate to investigate the system's performance[3]. The reader is recommended to review the Laplace transform principles and use the formulas table to obtain the system's time response.

11.1.1 Main Definitions of Feedback Control Theory

Physical systems have dynamic characteristics[4] that are controlled by a special instrument known as the controller. A controller is the watchdog of the system, and its purpose is to send an order to a system to make it act according to

DOI: 10.4324/9781003175230-14

pre-established technical conditions. In the past, and even today, in some scarce cases, the controllers were pneumatic. Today's controllers are electronic. They are independent physical modules or are integrated into a computer or a PLC (Programmed Logic Controller), as was commented before. The controller is programmed with an algorithm designed to produce a desired system's behavior due to setpoint changes or spurious disturbances.

Two system properties are the main goal of the Feedback Control Theory: stability and accuracy. In simple words, it can be said that a system is stable if, after excited, it reacts with a damped transient response. On the other hand, accuracy is a permanent regime's attribute. It is given by the difference between the pre-set (desired) process output and the actual output value. This difference is known as the system's error.

All feedback control systems are prone to be unstable if their controller is improperly tuned with an excessive gain. Therefore, feedback systems' instability is a potential "sickness" that must be prevented in the design stage. Nonetheless, it is common that after a correct design, the system gets unstable during the operation due to its unstable dynamics that tends to produce non-damped oscillations after a setpoint change or disturbance in the process. These post-design problems can be solved by readjusting the controller tuning because controllers are manufactured for a wide spectrum of process dynamics.

The Feedback Control Theory is studied with a unique mathematical body, whose formulas are valid regardless of the specific application. The first step to designing an automatic control system is to prepare a process' model to be controlled, which is the differential equation of motion. These equations are solved in the time domain. However, if the time domain equations are transformed into the frequency domain through the Laplace transform, algebraic formulas are returned. These formulas are easier to manipulate than differential expressions, and their importance is that they can predict the system's stability and accuracy.

Two main types of control systems exist: servomechanism and regulating systems. The first is the system's mass position, velocity, or acceleration control, like a modulating valve that adjusts the fuel flow to a kiln or boiler or an accelerometer to detect acceleration in a vibratory structure. Therefore, their reactions, in general, are in the order of a few seconds.

Regulating systems are slower than servomechanism. They are subject to a fixed setpoint for long periods and must be prepared to overcome external disturbances and return the plant[5] to the setpoint as soon as possible. This control system type is the case of a temperature or combustion oxygen whose values must be constants in time. These processes' reaction time is slow; they may have time constants of minutes to several hours.

11.2 CONTROL SYSTEMS BASICS

A system is defined as a set of physical parts interconnected to act as a unit. Systems are formed by huge or tiny physical processes controlled by the controller's mathematical algorithm. The algorithm's design is one of the most important purposes of the Feedback Control Theory. This design is made with mathematical models that allow analyzing and synthesizing the algorithm according to pre-set stability and accuracy specifications.

The system's output to input ratio, expressed as the Laplace transform, is known as the system's transfer function. This transfer function is the system's physical model that defines its dynamics and predicts its performance when excited by an input signal.

The cascade blocks of Figure 11.1 graphically represent the system's input-output signals. X(s) is the Laplace transform of the input, and Y(s) is the Laplace transform of the output. The transfer function of each block is $G_i(s) = Y_i(s)/X_i(s)$. The block cascade is equivalent to one equivalent block, whose transfer function is the transfer function product of all cascaded blocks from 1 to n. This transfer function is G(s) = Y(s)/X(s). The transfer function "shapes" the output signal Y(s) based on the system's dynamic properties and the input X(s).

In linear systems, the most general expression relating the input and output signals is the equality between the input and the output derivatives' sum with respect to time. If the system has a physical relation between input and output given by integrals, successive derivatives on both sides of the equation make all integrals disappear. Therefore, the general formula of a system's input-output signals is:

Formula 11.1 General input-output formula of a linear system

$$\sum_{i=0}^{m} a_i \cdot \frac{d^i x(t)}{dt^i} = \sum_{j=0}^{n} b_j \cdot \frac{d^j y(t)}{dt^j}$$

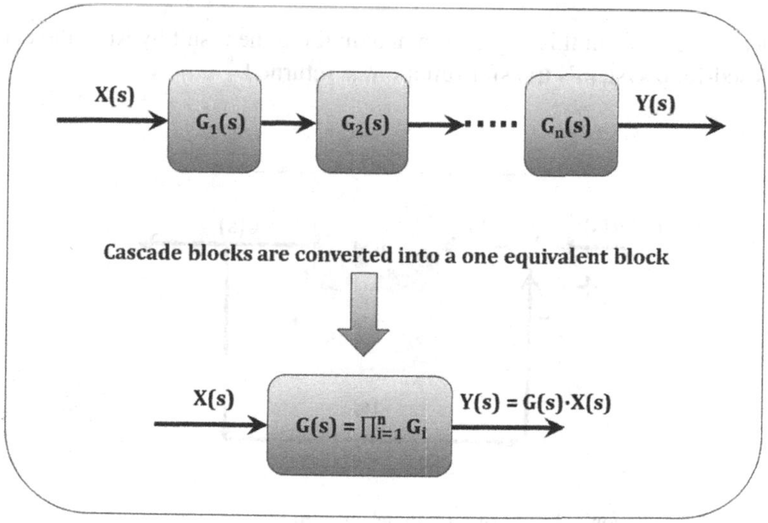

FIGURE 11.1 Block diagram of a cascade system.

The Laplace transform of this equation returns the system's transfer function as an algebraic expression:

Formula 11.2 General formula of a linear system transfer function

$$G(s) = \frac{Y(s)}{X(s)} = \prod_{1}^{n} G_i(s) = \frac{\sum_{j=0}^{n} b_j \cdot s^j}{\sum_{i=0}^{n} a_i \cdot s^i}$$

Controlled system performance does not depend only on the input signal but on its dynamic properties given by the transfer function's mathematical structure. Therefore, the system's transfer function's study predicts the control system's stability or instability and its accuracy for different input signal types before the system is installed.

11.2.1 Closed-Loop Systems

The canonical form (most basic form) of a feedback system is shown in Figure 11.2. The negative feedback signal C(s) is introduced in the signal detector Σ that creates a signal error E(s) = R(s) – C(s), where R(s) is the preset reference or setpoint. The transfer function of the upper forward path tends to cancel the error if the controller inside the forward block G(s) is adequately tuned.

The transfer function H(s) = C(s)/R(s) is derived below:

$$C(s) = G(s) \cdot E(s) = G(s) \cdot \left[R(s) - C(s)\right]$$

Clearing C(s) from this expression and dividing the result by R(s), the canonical closed-loop system's transfer function[6] is returned.

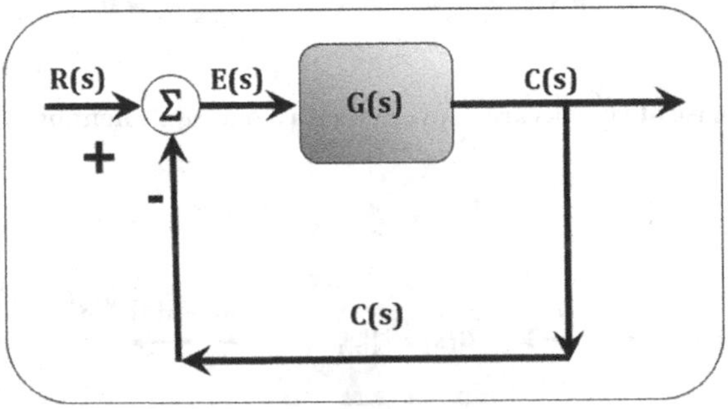

FIGURE 11.2 Canonical form of a feedback system.

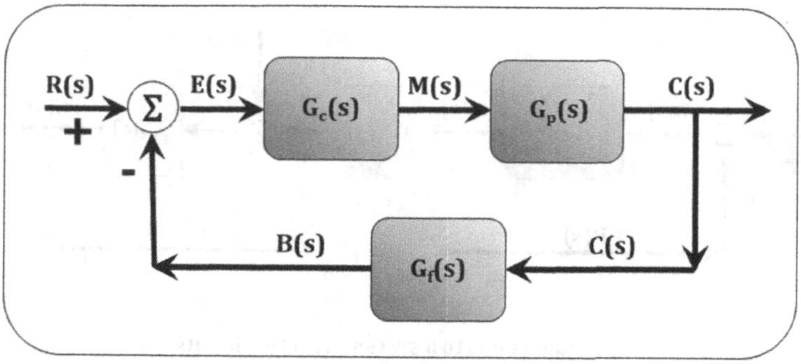

FIGURE 11.3 Block diagram of a feedback control system.

Formula 11.3 Closed-loop transfer function

$$H(s)=\frac{C(s)}{R(s)}=\frac{G(s)}{1+G(s)}$$

Considering the controller and the transducer that measures the controlled variable C(s) in the canonical closed-loop, the result is the automatically controlled system shown in Figure 11.3. This block diagram shows the transfer functions of the controller $G_c(s)$, the controlled system or plant $G_p(s)$ and the feedback transducer $G_f(s)$. This transducer is usually a detector of some physical property that converts it into a weak electrical signal. In industrial installations, it is often used the 4 to 20 mA range.

With the same procedure used to derive the canonical closed-loop system's transfer function, the transfer function of the control system of Figure 11.3 is obtained.

Formula 11.4 Closed-loop transfer function of a feedback control system

$$H(s)=\frac{C(s)}{R(s)}=\frac{G_c(s)\cdot G_p(s)}{1+G_c(s)\cdot G_f(s)\cdot G_c(s)}$$

The closed-loop system performs according to the controller's algorithm. If this is incorrectly designed or not finely tuned at the plant's commissioning, the aftermath could be worse than an open-loop system controlled by an unskilled operator.

Reality has proven that automatic control systems' behavior is not always crystal clear because they are subject to external or internal disturbances. These

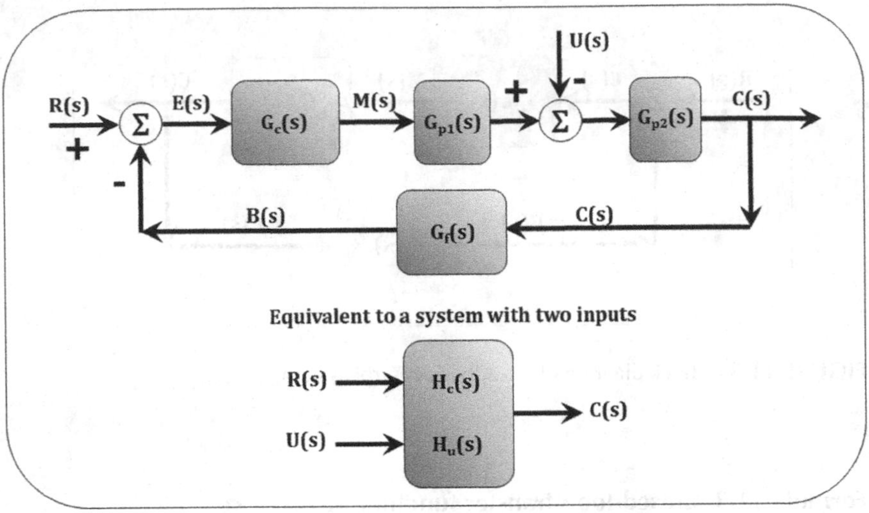

FIGURE 11.4 Feedback control system with an external disturbance.

disturbances are an issue for most types of regulating systems that must be overcome. In general, disturbance U(s) arises somewhere in the plant, splitting the process into two different transfer functions, as shown in the block diagram of Figure 11.4. The result is a two-input system, R and U, where the reference signal R is usually constant, and the disturbance U is mainly random.

In this system, there are two closed-loop transfer functions, defined by the ratios C(s)/R(s) and C(s)/U(s). As the system is linear, the system output C(s) is formed by the sum of two signals, one, named $C_c(s)$, due to the setpoint and the other, designated $C_u(s)$, produced by the disturbance.

Formula 11.5 Formation of the system output C(s)

$$C(s) = C_c(s) + C_u(s) = H_c(s) \cdot R(s) + H_u(s) \cdot U(s)$$

The $H_c(s)$ is the transfer function of the controlled system, and $H_u(s)$ is the disturbed system's transfer function. As the system is linear, to derive $C_u(s)$, the setpoint R(s) is made equal to zero. Conversely, to derive $C_c(s)$, the disturbance signal U(s) is made zero; then, closed-loop transfer functions $H_c(s)$ and $H_u(s)$ are derived with the same procedure used to derive Formula 11.7.

$$C_c(s) = G_{p1}(s) \cdot G_{p2}(s) \cdot G_c(s) \cdot \left[R(s) - G_f(s) \cdot C_c(s)\right]$$

$$C_u(s) = G_{p2}(s) \cdot U(s) - G_{p1}(s) \cdot G_c(s) \cdot G_f(s) \cdot C_u(s)$$

Clearing $C_c(s)$ and $C_u(s)$ of these equations, the closed-loop functions of the setpoint and the disturbance are returned.

$$H_c(s) = \frac{C_c(s)}{R(s)} = \frac{G_c(s) \cdot G_{p1}(s) \cdot G_{p2}(s)}{1 + G_c(s) \cdot G_{p1}(s) \cdot G_{p2}(s) \cdot G_f(s)}$$

Formula 11.6 Closed-loop transfer functions of a regulating system

$$H_u(s) = \frac{C_u(s)}{U(s)} = \frac{G_{p2}(s)}{1 + G_c(s) \cdot G_{p1}(s) \cdot G_{p2}(s) \cdot G_f(s)}$$

As will be discussed next, a system's time response depends on the input signal, R(s) or U(s), and the nature of the transfer function's poles and zeros. The zeros are the root of the numerator, and the poles are the roots of the denominator. It is noted that $H_c(s)$ and $H_u(s)$ have the same denominator; hence they have the same poles, which means that their performances will be similar.

11.3 TIME RESPONSE OF LINEAR SYSTEMS

The term "time response" describes the system's reaction curve of the transient regime as a function of time. The reaction curve depends on the input signal type and the poles nature of the closed-loop transfer function. Figure 11.5 shows the reaction curves for five different types of poles with a unit impulse input. The standard representation of poles and zeros in the s-plane is with crosses for poles and circles for zeros. Stable systems have poles with negative real parts. See cases A and B. Unstable systems have poles with zero or positive real parts. See cases C, D and E.

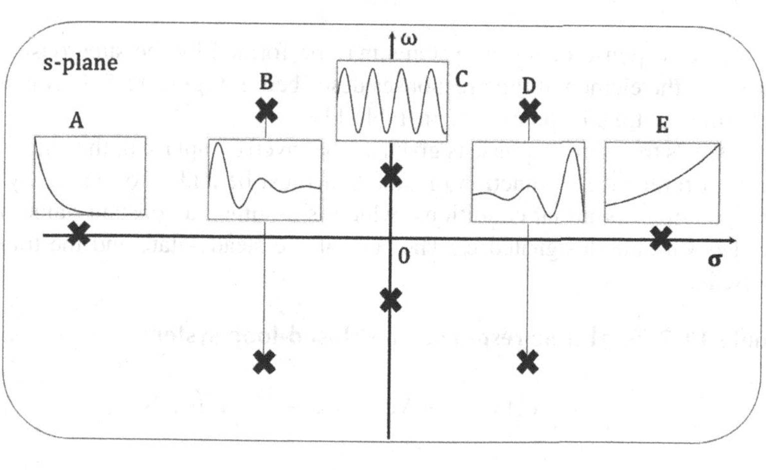

FIGURE 11.5 Elemental time responses in the s plane.

TABLE 11.1
Laplace Transform of Elemental Time Responses to a Unit Impulse Input

Designation	Pole Types	Elemental time Response	Physical Time Reaction
A	Negative real part. Stable system	$e^{-\sigma \cdot t}$	Damped decreasing exponential
B	Complex with negative real part Stable system	$e^{-\sigma \cdot t \pm j \cdot \omega \cdot t}$	Damped decreasing oscillation
C	Imaginary. Unstable system	$e^{\pm j \cdot \omega \cdot t}$	Undamped constant oscillation
D	Positive real part. Unstable system	$e^{\sigma \cdot t}$	Undamped increasing exponential
E	Complex with a positive real part. Unstable system	$e^{\sigma \cdot t \pm j \cdot \omega \cdot t}$	Undamped increasing oscillation

TABLE 11.2
Standardized Setpoint Signals for Accuracy Assessment

Input Signal	Laplace Transform	Time Function
Impulse	$R(s) = \delta(s) = 1$	$\delta(t) = 1$ at $t = 0$ $\delta(t) = 0$ at $t \neq 0$
Step	$R(s) = \frac{1}{s}$	$r(t) = 1$ at $t \geq 0$
Ramp	$R(s) = \frac{1}{s^2}$	$r(t) = t$ at $t \geq 0$
Parabola	$R(s) = \frac{1}{s^3}$	$r(t) = \frac{t^2}{2}$ at $t \geq 0$

The time response of linear systems may be formed by the superposition of some, or all the elemental time responses described in Figure 11.5. Formulas for these elemental time responses are in Table 11.2.

A system's transient response is given by the inverse Laplace of the output. The expression returned is a function of time that must be added to the steady-state of the system at its initial conditions, which is assumed a constant value of the controlled variable designated c_0. The sum of the steady-state and the transient response is:

Formula 11.7 Total time response of a closed-loop system

$$c(t) = c_0 + \Delta c(t) = c_0 + L^{-1}\left[H(s) \cdot R(s)\right]$$

The above formula shows that the Laplace transform calculates the transient conditions and not the permanent regime. For this reason, some control engineers say that the Laplace transform is an "incremental approach" to system dynamics. However, the Laplace transform also provides formulas to calculate the error signal of the permanent regime.

11.4 CONTROL ACTIONS IN CLOSED-LOOP SYSTEMS

There are three basic controller types: proportional P, integral I and derivative D, which means that the controller output of the manipulated variable m(t) is proportional to the error, plus the error's integral, plus the error's derivative with respect to time. Each one has a feature and specific application to optimize the closed-loop system's accuracy, stability, and velocity. The most common combinations are the PI and the PID. Figure 11.13 shows the flow diagram of a PID controller and process, which is self-explicative (Figure 11.6).

The algorithm responds to the following formula.

Formula 11.8 Algorithm of a PID controller

$$m(t) = m_0 + \Delta m(t) = m_0 + K_c \cdot e(t) + \frac{K_c}{T_i} \cdot \int_0^t e(t) \cdot dt + K_c \cdot T_d \cdot \frac{de(t)}{dt}$$

m_0 is the controller's output at the system's steady state. The manipulated variable increment Δm(t) happens when the controller detects an error. The controller's output m(t) returns to the previous or a new steady-state after the error is damped.

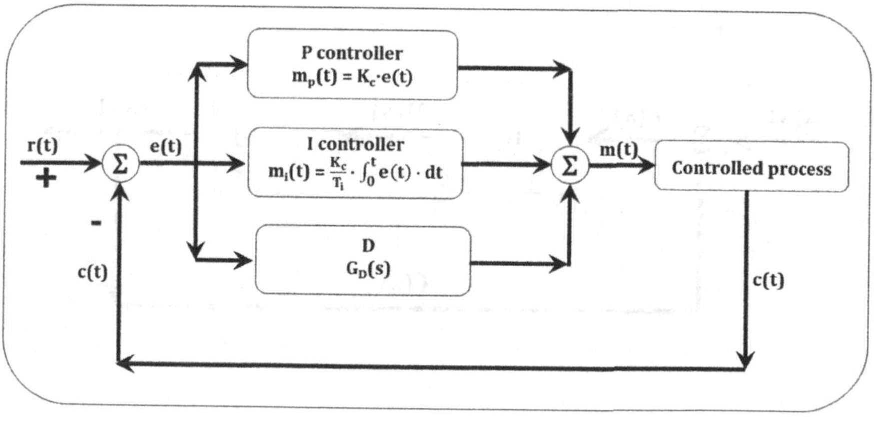

FIGURE 11.6 PID controller algorithm.

11.4.1 Proportional Control Action

The controller's most fundamental function is to send the controlled process a signal proportional to the error. Therefore, if the error signal experiences a sudden change, like a step, the controller's algorithm is:

Formula 11.9 P controller's algorithm

$$m_P(t) = K_c \cdot e(t)$$

Where K_c is the controller's static gain. The transfer function of a P controller is simply its static gain K_c. See Figure 11.7, a closed-loop system with a P controller and unit feedback.

The proportional control has an advantage: it quickly takes a control action as soon as the error is detected. However, it has a disadvantage: the closed-loop system does not achieve zero error condition, as demonstrated below.

11.4.1.1 Error with a P Controller

Friction forces dampen closed-loop transients. After the transient is extinguished, the system output stays in a constant quiet state, which, theoretically, may continue for an infinite time if no new inputs arise. This idea suggests that the limit of a controlled variable for $t \rightarrow \infty$ has a constant value regardless of the passing time. For this reason, it is said that the system is in a permanent regime during this condition. There are two variables of interest evaluated in a closed-loop system in the transient and permanent regimes: the controlled variable c(t) and the error e(t). In the permanent regime, these variables are designated c_∞ and e_∞.

There are four standardized setpoint signals to calculate a closed-loop system error that assess the system's accuracy. They are the impulse, step, ramp, and parabola input signals. The Laplace transform and time function of these signals are in Table 11.4.

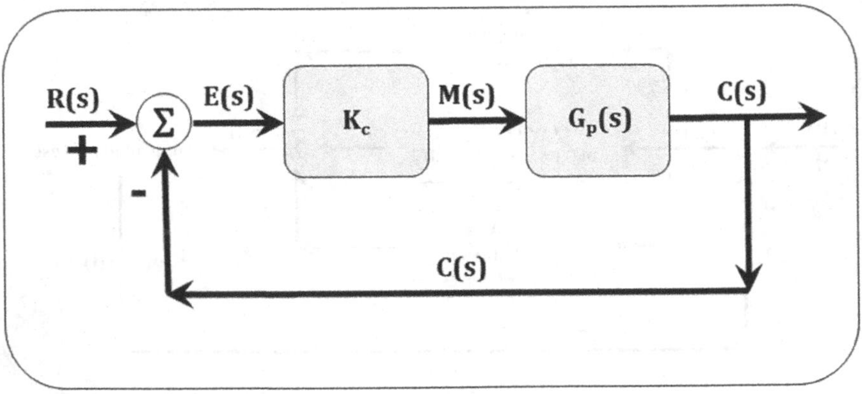

FIGURE 11.7 Closed-loop system with a P controller.

Errors are calculated with any of the above setpoint signals R(s) with the following formula. This formula shows what the controller does for the system's accuracy:

Formula 11.10 Closed-loop transfer function of the error signal

$$E(s) = R(s) - G_c(s) \cdot G_p(s) \cdot E(s)$$

$$E(s) = \frac{R(s)}{1 + G_c(s) \cdot G_p(s)}$$

The error in the permanent regime is given by the final value theorem of the Laplace transform. It says that a function of time value at an infinite time is given by: $\lim_{t \to \infty} f(t) = \lim_{s \to 0} s \cdot F(s)$. Therefore, the application of this theorem returns the permanent regime error.

Formula 11.11 Permanent error of a feedback-controlled system

$$e_\infty = \lim_{t \to \infty} e(t) = \lim_{s \to 0} \frac{s \cdot R(s)}{1 + G_c(s) \cdot G_p(s)}$$

Applying this formula to systems of the form $G_p(s) = K_p \cdot \Pi \prod_{i=1}^{m} (1 + s \cdot T_i) / \prod_{j=0}^{n} (1 + s \cdot T_j)$, where the numerator is not divided by p^n, that is, with no poles at the origin, the above formula with a P controller is:

$$e_\infty = \lim_{t \to \infty} e(t) = \lim_{s \to 0} \frac{s \cdot R(s)}{1 + K_c \cdot K_p}$$

The errors for the standardized inputs are:

Formula 11.12 Errors for standardized setpoint signals with a P controller

$$e_{\text{impulse}} = 0$$

$$e_{\text{step}} = \frac{1}{1 + K_c \cdot K_p} \neq 0$$

$$e_{\text{ramp}} = e_{\text{parabola}} \to \infty$$

The P controller produces zero error only when the system is excited by an impulse signal, not commonly seen in practice. The error is never zero for the often-used step function, and, with ramp and parabola inputs, the error is infinite. Therefore, the P controller, in general, does not satisfy the accuracy requirements of most controlled processes.

The product $K_c \cdot K_p$ is called the position error constant K_{pe}. The error is usually expressed in percentage or in per unit of the setpoint signal, as shown by Formula 11.13.

Formula 11.13 Position error for a step input in per unit and in percentage of the setpoint signal

$$\varepsilon_p = \frac{e_{p\infty}}{R} = \frac{1}{1+K_{pe}} \quad \varepsilon_{p\%} = \frac{100}{1+K_{pe}}$$

In summary, the P controller reduces the error with a step signal input but does not offset it. Only an infinite gain could reduce the error to zero, but infinite gain is impossible for a proportional controller. Furthermore, a controller's too high gain may produce excessive transient overshoots or even instability of the controlled process. Hence the proportional action is a limited control mode because it does not meet the zero-error condition and may produce inadmissible transients or instability whenever its gain is exaggeratedly increased, trying to reduce the position error.

The difference between the step setpoint signal and the controlled variable c in the permanent regime is the error $e_{p\infty}$; thus, the gain K_c may be calculated as a function of this admissible error with the following formula:

Formula 11.14 Static gain of a P controller versus the admissible error and K_p

$$K_c = \frac{1}{K_p} \cdot \left(\frac{1}{\varepsilon_{p\,\text{admissible}}} - 1 \right)$$

As will be seen in the next section, PI controllers offset the error with a step signal. Their cost is not much different from a P controller; hence, P controllers are not usually applied in regulating processes, where accuracy is essential.

11.4.1.2 Time Response With a P Controller

In this example, a second-order system process with transfer function $G_p(s)$ is controlled by a P controller. This case demonstrates the influence of the P controller's static gain on the closed-loop time response. See Figure 11.8. The open-loop transfer function G(s) and the closed-loop transfer function H(s) are:

Formula 11.15 Transfer function of the controlled system (left) and closed-loop system (right) with a P controller

$$G(s) = K_c \cdot G_p(s) = K_c \cdot \frac{K_p \cdot p_1 \cdot p_2}{(s+p_1)\cdot(s+p_2)} \quad H(s) = \frac{N(s)}{D(s)} = \frac{K_c \cdot K_p \cdot p_1 \cdot p_2}{(s+p_1)\cdot(s+p_2) + K_c \cdot K_p}$$

Where p_1 and p_2 are the open-loop poles, K_p is the static process gain, and K_c is the P controller's static gain. The poles of H(s) are the roots of the denominator of H(s) or characteristic equation:

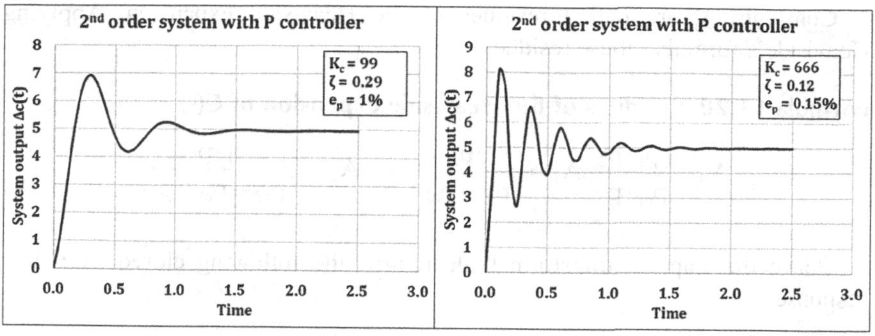

FIGURE 11.8 Second-order system with $\zeta<1$ with P controller at $K_c = 99$ and $K_c = 666$.

Formula 11.16 Characteristic equation of H(s)

$$D(s) = (s + p_1) \cdot (s + p_2) + K_c \cdot K_p = 0$$

The roots of the characteristic equation are the closed-loop poles. Their values depend on the open-loop poles p_1 and p_2 and the gains K_c and K_p

Formula 11.17 Closed-loop system poles

$$p_{c1,c2} = -\frac{p_1 + p_2}{2} + \frac{1}{2} \times \sqrt[2]{(p_1 - p_2)^2 - 4 \cdot K_c \cdot K_p}$$

The Laplace transform of the output is the transfer function H(s) times the input R(s). In this example, this input is a step of magnitude R, whose Laplace transform is R/s. See Table 11.2.

Formula 11.18 Laplace transform of the closed-loop system output

$$C(s) = H(s) \cdot R(s) = \frac{K_c \cdot K_p \cdot p_1 \cdot p_2}{(s + p_{c1}) \cdot (s + p_{c2})} \cdot \frac{R}{s}$$

Then, applying the Heaviside expansion, the output C(s) is converted into the following sum of terms. This conversion makes it easy to anti-transform each of them.

Formula 11.19 Expansion of C(s)

$$C(s) = K_c \cdot K_p \cdot R \cdot \left(\frac{A_0}{s} + \frac{A_1}{s + p_{c1}} + \frac{A_2}{s + p_{c2}} \right)$$

Constants A_i are called residues of the Heaviside expansion. Applying Heaviside's formula[7], these residues are:

Formula 11.20 Residues of the Heaviside expansion of C(s)

$$A_0 = \frac{p_1 \cdot p_2}{p_{c1} \cdot p_{c2}} \quad A_1 = \frac{p_1 \cdot p_2}{(p_{c1} - p_{c2}) \cdot p_{c1}} \quad A_2 = \frac{p_1 \cdot p_2}{(p_{c2} - p_{c1}) \cdot p_{c2}}$$

Hence the Laplace transform table returns the following closed-loop time response:

Formula 11.21 Closed-loop time response

$$c(t) = K_c \cdot K_p \cdot R \cdot \left(A_0 + A_1 \cdot e^{-p_{c1} \cdot t} + A_2 \cdot e^{-p_{c2} \cdot t}\right)$$

If the closed-loop poles p_{c1} and p_{c2} are real numbers, they are the inverse of time constants T_1 and T_2; then, the closed-loop system is a second-order system with a damping ratio higher than 1, whose time response is an S-shaped curve. No oscillation happens in this case. The condition of this exponential response is: $(p_1 - p_2)^2 - 4 \cdot K_c \cdot K_p > 0$

If the closed-loop poles are complex conjugate, they are of the form: $\sigma \pm j \cdot \omega$. The time response is obtained by replacing this expression in Formula 11.21 to see the damped oscillatory response formula:

Formula 11.22 Damped oscillatory time response. Exponential terms

$$c(t) = K_c \cdot K_p \cdot R \cdot \left[A_0 + e^{-\sigma \cdot t}\left(A_1 \cdot e^{j \cdot \omega \cdot t} + A_2 \cdot e^{-j \cdot \omega \cdot t}\right)\right]$$

According to Euler's formula, imaginary exponents indicate that the response is damped oscillatory:

Formula 11.23 Damped oscillatory time response. Oscillatory terms

$$c(t) = K_c \cdot K_p \cdot R \cdot A_0 \cdot \left[1 + e^{-\sigma \cdot t} \cdot \left(\frac{A_1 + A_2}{A_0} \cdot \cos \omega \cdot t + j \cdot \frac{A_1 - A_2}{A_0} \cdot \sin \omega \cdot t\right)\right]$$

Thus, introducing Formulas 11.20 of A_0, A_1 and A_2, and the ratio $\sigma/\varphi = \zeta/(1-\zeta^2)$ the time response is:

$$c(t) = K_c \cdot K_p \cdot R \cdot A_0 \cdot \left[1 + e^{-\sigma \cdot t} \cdot \left(\frac{A_1 + A_2}{A_0} \cdot \cos \omega \cdot t + j \cdot \frac{A_1 - A_2}{A_0} \cdot \sin \omega \cdot t\right)\right]$$

This formula represents the transient of many machines because their damping ratio is usually lower than 0.12.

What Changes Does the Adjustable Gain of the P cController Produce?

According to the final value theorem, the closed-loop transfer function at an infinite time is the limit of H(s) for $s \to 0$. Applying this theorem to the H(s) expression of Formula 11.15, the result is $H_\infty = K_p \cdot K_c/(1 + K_p \cdot K_c)$, hence the permanent value of $c_\infty = H_\infty \cdot R$. The closed-loop dynamic gain H_∞ is always lower than 1. Thus, the error is never zero if the controller is only proportional. The error may be reduced by increasing K_c; however, the K_c increase is limited because the K_c gain regulates the transient overshoot and frequency. Both properties increase when K_c is increased. See in Figure 11.8 the transient curve for two different values of K_c; 99 and 666. This adjustment decreased the damping ratio from 0.29 to 0.12, and the peak value increased from 7 to 8. The frequency escalated from 9.9 rad/s to 25.6 rad/s.

In summary, the higher the controller's gain, the lower the permanent error and the higher the transient overshoots. As the controller has only proportional action, the system's error never falls to zero. However, in some P controllers, it is possible to introduce a signal manually added to the controller's output to offset the error. Of course, this is valid until the manipulated variable needs to be changed to another operating point. In that case, the offset signal must be reintroduced.

11.4.2 Proportional Plus Integral Control Action

The residual error produced by a P controller may be offset if active control action is applied until the error is zero. Active means that the manipulated variable keeps increasing or decreasing while the error exists. This control action is obtained by integrating the error signal; therefore, the controller type is an integral controller, designated with the letter I. The combination of a proportional and an integral controller is designated PI (Proportional and Integral), and its time algorithm is:

Formula 11.24 PI controller algorithm

$$m_{PI}(t) = K_c \cdot e(t) + \frac{K_c}{T_i} \cdot \int_0^t e(t) \cdot dt$$

The I controller output rate is K_c/T_i ratio; then, this action is stronger as the integration time grows shorter. When a step error of magnitude e appears, the proportional action produces an output equal to $K_c \cdot e$. The P output is the same one minute later, but the I output is $K_c/T_i \cdot e$. This signal output means that the ratio of the I output to the P output is $1/T_i$. Then it is said that the I output has repeated the P output $1/T_i$ times in one minute. For these reasons, the I controller's knobs for adjusting the integral action are marked in repeats of the proportional action per minute (or per second) or simply in repeats per minute (or per second).

The Laplace transform of the PI controller's algorithm is the anti-transform of Formula 11.25:

Formula 11.25 Transfer function of a PI controller

$$G_{PI}(s) = K_c \cdot \left(1 + \frac{1}{s \cdot T_i}\right) = K_c \cdot \frac{s + \frac{1}{T_i}}{s}$$

Hence, PI controllers have a pole at s = 0 and a zero at $-1/T_i$. In standard commercial controllers or in PLCs or computers where this algorithm is implemented, the static gain K_c and the integration time T_i have a broad range of adjustments to optimize many system types.

11.4.2.1 Error With a PI Controller

The Laplace transform of a closed-loop system's error with a PI controller is:

Formula 11.26 Transfer function of a process controlled by a PI controller

$$E(s) = \frac{s \cdot R(s)}{s + K_c \cdot \left(s + \frac{1}{T_i}\right) \cdot G_p(s)}$$

Taking the final value theorem from the Laplace table, the error for a step input is

Formula 11.27 Permanent error for a step input with PI controller

$$e_\infty = \lim_{t \to \infty} e(t) = \lim_{s \to 0} s \cdot E(s) = 0$$

If R(s) is a ramp, its Laplace transform is R/s^2; therefore, the error is constant and equal to:

Formula 11.28 Error with ramp input and PI controller

$$e_\infty = \frac{T_i \cdot R}{K_c \cdot K_p}$$

Therefore, the PI controller offsets the error with step input but not with a ramp input. For a parabola input, the error is infinite. In summary, P controllers take quick actions proportional to the current error value. Instead, I controllers take a steady growing action as long as the error persists. This action offsets errors produced by impulse or step inputs but not by a ramp or parabola setpoints.

11.4.3 Proportional Plus Derivative Action

The PI controller reduces the system's errors but does not predict the error tendency because its output is only proportional to the current error magnitude and the error history since it arises. Instead, derivative action is based on the error velocity, which means that derivative action detects the "urgency" of quick control action. This action is made by the controller regardless of the error sign.

The derivative action is combined with proportional action, as shown by the following PD algorithm.

Formula 11.29 PD controller algorithm

$$m_{PD}(t) = K_c \cdot e(t) + K_c \cdot T_d \cdot \frac{de(t)}{dt}$$

By measuring the error velocity, the PD controller generates a correcting action before the proportional controller action. This anticipation improves the system stability and transient response reducing the overshoots. The PD transfer function is formed by the sum of two terms: the proportional plus the derivative:

Formula 11.30 Transfer function of a PD controller

$$G_{PD}(s) = K_c + K_c \cdot T_d \cdot s = K_c \cdot T_d \cdot (z_d + s)$$

Where $z_d = 1/T_d$.

An example of the derivative action in a second-order system that has a damping ratio lower than 1. The closed-loop transfer function with a PD controller is:

$$G(s) = \frac{N(s)}{D(s)} = K_c \cdot (1 + T_d \cdot s) \cdot \frac{\omega_{nG}^2}{s^2 + 2 \cdot \zeta_G \cdot \omega_{nG} \cdot s + \omega_{nG}^2}$$

Formula 11.31 Open and closed-loop transfer functions with PD controller

$$H(s) = \frac{K_c \cdot (1 + T_d \cdot s) \cdot \omega_n^2}{s^2 + 2 \cdot \zeta_G \cdot \omega_{nG} \cdot s + \omega_{nG}^2 + K_c \cdot (1 + T_d \cdot s) \cdot \omega_{nG}^2}$$

Subscript G stands for variables of the open-loop system. After grouping the coefficients of s in the characteristic equation, formulas of the closed-loop damping ratio and natural frequency are derived as follows:

Formula 11.32 Closed-loop damping ratio and natural frequency

$$\begin{aligned} D(s) &= s^2 + \left(2 \cdot \zeta_G \cdot \omega_{nG} + K_c \cdot T_d \cdot \omega_{nG}^2\right) \cdot s + \omega_{nG}^2 + K_c \cdot \omega_{nG}^2 \\ &= s^2 + 2 \cdot \zeta_H \cdot \omega_{nH} \cdot s + \omega_{nH}^2 \end{aligned}$$

$$\zeta_H = \frac{2 \cdot \zeta_G + K_c \cdot T_d \cdot \omega_{nG}}{2 \cdot \sqrt[2]{1+K_c}} \qquad \omega_{nH} = \omega_{nG} \cdot \sqrt[2]{1+K_c}$$

In these formulas, subscript H stands for variables of the closed-loop system. The closed-loop damping ratio is set at the desired value by adjusting the T_d's derivative time and the static gain K_c. In general, K_c and T_d's ideal adjustment is to make the closed-loop damping ratio equal to 0.707, where both the overshoot and settling time are minimum. The closed-loop is also a second-order system but with a different damping ratio and natural frequency. Same as happens with a P controller, the closed-loop natural frequency is higher with a PD controller than the open-loop frequency. The designer must choose K_c and T_d to achieve a closed-loop system with the desired damping ratio and natural frequency. Figure 11.9 shows an example of damping ratio curves $\zeta_H = f(K_c, T_d)$ for a second-order system with a PD controller. The plot is valid for an open-loop system with a damping ratio equal to 0.2 and natural frequency equal to 50 rad/s.

Figure 11.10 shows a second-order system's time response with a damping ratio lower than 1, with two types of controllers: P and PD. It is notorious that the derivative action has stabilized the transient response of the P controller. The smallest damping ratio occurs when the derivative time is zero; see Figure 11.9. At $T_d = 0$, the damping ratio is equal to $\zeta H = \zeta G / \sqrt[2]{1+K_c}$. This formula returns the damping ratio with a P controller only. See the ζ_H Formula 11.32.

As the derivative action is with respect to time, noisy signals produce excessive control actions that may strike the control system. These types of signals are typically produced by electromagnetic induction originated in power systems apparatus or cables next to control instrumentation. Hence, the error signal received by the derivative controller must be previously passed through a low pass filter. In general, this filter is a first or second-order system with a damping ratio higher

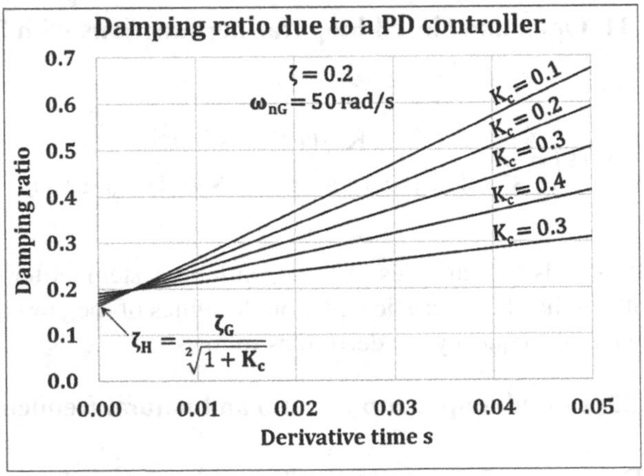

FIGURE 11.9 Example of closed-loop damping ratio with PD controller.

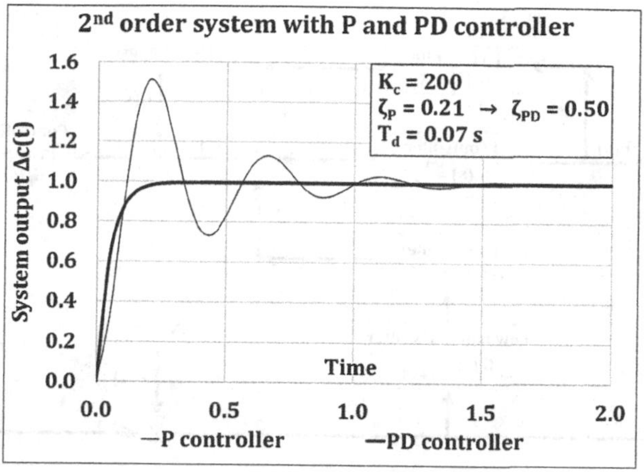

FIGURE 11.10 Time responses with P and PD controllers.

than 1. In general, noisy signals are damped by the process because longer time constants than the noise usually form the last. It means that the same process acts as a low passband filter.

Another issue created by the derivative action is a step disturbance in the error signal or a sudden setpoint change. If the derivative module receives a step input, its output has, theoretically, an infinite amplitude. This change is usually called "derivative kick." The derivative kick instantaneously hits the system with a large abrupt manipulated variable transmitted to the controlled system. This problem is prevented by removing the error from the detector's error signal, as shown in Figure 11.11. This type of controller is designated P-D. If the controller has integral action, it is designated PI-D and not PID as controllers where the error is introduced in the derivative module. There are other controllers' configurations that restrict undesired actions from the derivative controller's part.

Therefore, in practice, the derivative controller is restricted by a low bandpass filter; thus, pure derivative action does not exist in standard controllers. Therefore, the transfer function of restricted derivative controllers includes a low bandpass filter, as shown in Figure 11.11.

Formula 11.33 Transfer function of a PD restricted controller

$$G_{PD}(s) = K_c + \frac{K_c \cdot T_d \cdot s}{1 + T_r \cdot s} = K_c \cdot \left(1 + \frac{T_d}{T_r}\right) \cdot \frac{(s + z_d)}{(s + p_d)}$$

The zero and pole of the derivative action are $z_d = 1/(T_d+T_r)$ and $p_d = 1/T_r$. Typically, the time constant T_r of the low bandpass filter is 10% of the T_d's derivative time.

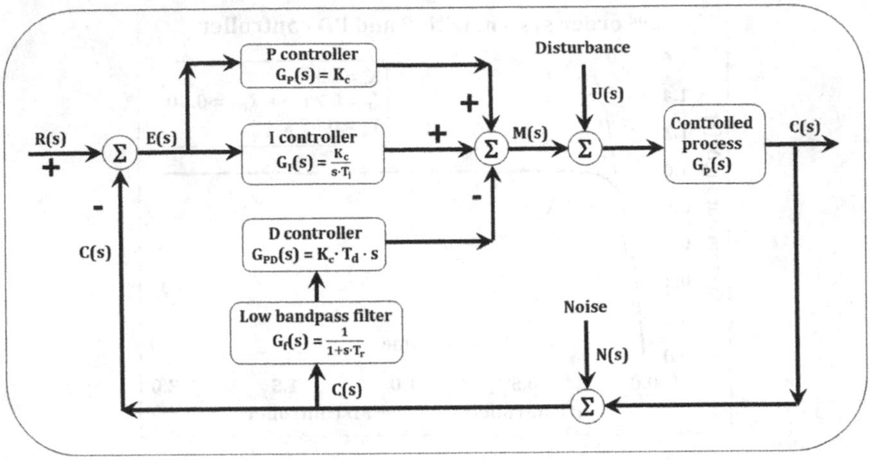

FIGURE 11.11 PI-D controller with no setpoint input R(s) to the D controller.

11.4.4 PID Control Action

PID controllers reunite the features of P, I and D controllers in only one algorithm. Figure 11.11 shows the flow diagram of a process driven by a PI-D controller. Each controller's module produces an output, finally merged in only one manipulated variable to command the process. A time algorithm with a step setpoint input is used below to understand a PI-D controller's output.

The Laplace transform of the PID control algorithm and its frequency response (reminder: substitute s by j • ω) is the sum of the three control actions discussed in the last two sections.

Formula 11.34 Laplace transform and frequency response of the PID algorithm

$$G_{PID}(s) = \frac{M(s)}{E(s)} = K_c \cdot \left(1 + \frac{1}{s \cdot T_i} + \frac{T_d \cdot s}{1 + T_r \cdot s}\right)$$

$$G_{PID}(j \cdot \omega) = \frac{M(j \cdot \omega)}{E(j \cdot \omega)} = K_c \cdot \left(1 + \frac{1}{j \cdot \omega \cdot T_i} + \frac{T_d \cdot j \cdot \omega}{1 + T_r \cdot j \cdot \omega}\right)$$

The PID controller's time response with a step error E is formed by the sum of three response times, each corresponding to one controller type output. These signals are independent one from the other, except the proportional action, whose gain K_c affects the three outputs. This concept is readily apprehended, looking at the following formula:

Formula 11.35 PID output with a step error E

$$m(t) = L^{-1}\left[G_{PID}(s)\cdot\frac{E}{s}\right] = K_c \cdot E \cdot \left(1 + \frac{t}{T_i} + \frac{T_d}{T_r}\cdot e^{-\frac{t}{T_r}}\right)$$

Refer to the chart in Figure 11.12: the curves stand for a PID controller's control actions when a step error E appears. The proportional and derivative controllers immediately react with a step pulse of amplitude m(0) = K_c • (1 + T_d/T_r) • E, approximately equal to 11 times the proportional output K_c•E. This high pulse is that the T_d/T_r ratio is typically equal to 10, as commented before. After the zero instant, the derivative output exponentially decays with the time constant T_r; therefore, this pulse may be considered extinguished in a 3×T_r period. After this quick response, the output decays; then, the integral action drives the output by linearly increasing the manipulated variable until the error is offset.

Controllers' settings used in chart 11.12 and error input		
K_c	1.0	
T_i	0.3	seconds
T_d	1.0	seconds
T_r	0.1	seconds
Error E	1.0	step

The derivative action result is that the controlled process rapidly receives a strong command proportional to the error velocity. The chart in Figure 11.12 shows that the integral output has been advanced by a time T_a, meaning that the error will be offset before a PI controller's action.

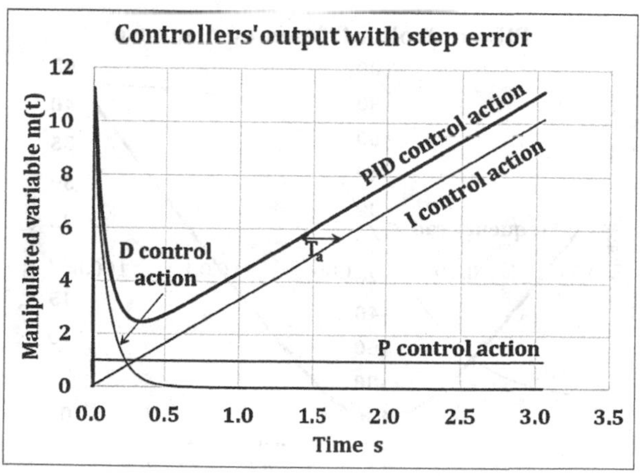

FIGURE 11.12 Controllers' output m(t) and setting with a unit step error.

The PID frequency response formula may be written as a function of zeros and poles instead of the previously algebraic expression. For example, a PID controller with a pure derivative action ($T_r = 0$) has the following frequency response formulas.

Formula 11.36 Unrestricted PID controller's frequency response

$$G_{PID}(j\cdot\omega)=\frac{M(j\cdot\omega)}{E(j\cdot\omega)}=K_c\cdot\left(1+\frac{1}{j\cdot\omega\cdot T_i}+T_d\cdot j\cdot\omega\right)$$
$$=K_c\cdot T_d\cdot\frac{(j\cdot\omega+z_1)\cdot(j\cdot\omega+z_2)}{j\cdot\omega}$$

$$\varphi_{PID}(j\cdot\omega)=\tan^{-1}\frac{\omega}{z_1}+\tan^{-1}\frac{\omega}{z_2}-\frac{\pi}{2}$$

The zeros are the numerator's roots of the module $G_{PID}(j\bullet\omega)$ of Formula 11.36. The formulas' zeros are:

Formula 11.37 Zeros of the PID controller

$$z_1=-\frac{1}{2\cdot T_d}+\frac{1}{2}\cdot\sqrt[2]{\frac{1}{T_d^2}-\frac{4}{T_i\cdot T_d}}\qquad z_2=-\frac{1}{2\cdot T_d}-\frac{1}{2}\cdot\sqrt[2]{\frac{1}{T_d^2}-\frac{4}{T_i\cdot T_d}}$$

Figure 11.13 shows the frequency response of a PID controller in a Bode plot. Two features are remarkable: at high frequencies (see phase curve at frequencies higher than ω = 0.40 rad/s), the phase shift is positive, which predicts a good influence on stability and transient overshoot. This good property is due to the

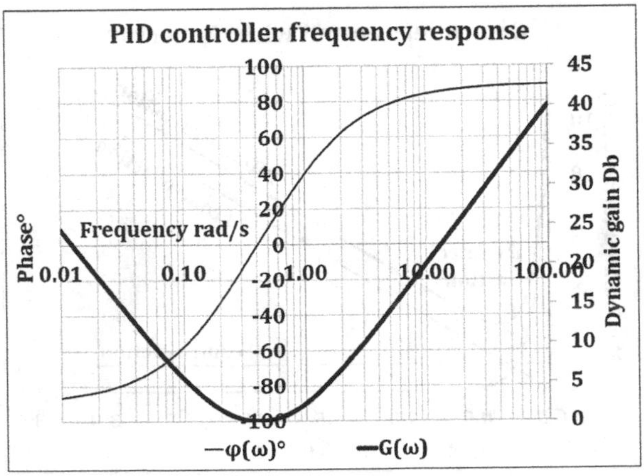

FIGURE 11.13 Bode chart of the PID controller frequency response.

derivative action. The other important feature is that the dynamic gain increases at 20 dB/dec rate as frequencies decrease below 0.1 rad/s. At zero frequency, the gain is infinite; therefore, the position error will be zero.

11.5 CLOSED-LOOP STABILITY

It is well known that feedback control systems may suffer a kind of "sickness" usually known as a lack of stability. Under this condition, closed-loop systems oscillate with constant or increasing amplitudes. Controlled systems do not tolerate this oscillation. They have been designed for one of two purposes: to follow a changing setpoint under strict conditions of accuracy and velocity (servomechanism) or to keep constant a process output regardless of random disturbances affecting the process (regulating system). Of course, disturbances are unpredictable and cannot be prevented, but a correct controller tuning must damp them quickly.

The controller's tuning aims at suppressing any type of instability. Methods to achieve closed-loop stability have been studied since the 1930s. An extensive bibliography about the four most important Automatic Control methods of representing the systems' frequency eases the designing and controller settings of feedback systems. They are:

a. The root-locus method by Evans
b. Nyquist plot
c. Bode plot
d. Black plot

All of them are named after their inventors. Nyquist, Bode and Black plots are methods based on open-loop frequency response. A suggested list of books is at the end of this chapter, though it is also recommended to consult the many specialized internet sites in automatic control systems. In this chapter, the Bode plot is used to explain the system's stability and avoid its instability.

Once a closed-loop system starts to oscillate, it stays oscillatory if the output signal c(t) is shifted exactly 360° when it travels through the error detector and the controlled process. 360° shift angle means that the oscillatory wave returns to the process output with the same phase it had before, which produces a constant or enlarged oscillation amplitude. The phase shift is produced in two places: the error detector and the controller-process set. As the reference signal r(t) and the feedback signal b(t) have opposite signs, the error signal always shifts −180° in the error detector. But there is an added phase shift when the error propagates through the controller and the process. If this added phase shift equals −180°, the system may be unstable because the total shift is 360°. The phase shift produced in the error detector cannot be changed; therefore, it is always −180°. Therefore, the control system design or the controller adjustment in the field purports to avoid a phase shift of −180° at a frequency where the dynamic gain $|G_c(j\omega)\bullet G_p(j\omega)|$ is equal to 1 or higher in the controller-process assembly.

The phase shift in the process depends on its dynamic components, such as first and second-order systems combination. Can this shift be equal to −180°? The answer is yes. A first-order system may shift the phase up to -90° and a second-order system up to −180° at an infinite frequency. If a third system is added, then the total shift may be lower than −180° at intermediate frequencies. This discussion concludes that for a system to be unstable, the controller-process must be, at least, a third-order system.

Is the −180° phase shift sufficient condition to make a system unstable? The answer is no because closed-loop signals are attenuated or amplified as they propagate through the controller and process. A signal is attenuated if the dynamic gain of the controller and the process is lower than 1. In this case, the system will tend to be stable.

These conceptual considerations conclude that a closed-loop system is unstable if its dynamic gain is 1 or higher at the frequency where the phase shift is −180°. Conversely, a closed-loop system is stable if the dynamic gain is lower than 1 at the frequency of the −180° phase. If so, it is said that the system has absolute stability. However, this condition does not say "how" stable the system is. The relative stability answers this question as explained ahead.

11.5.1 Absolute Stability Determination

In this section, a third-order system's absolute stability controlled by a P controller is analyzed. The dynamic gain G is obtained from the system's open-loop transfer function, replacing the complex frequency s by the imaginary frequency j•ω. Therefore, to obtain a system's frequency response, the real part of the complex frequency is set to zero.

Formula 11.38 Open-loop dynamic gain

$$|G(j\cdot\omega)| = \frac{p_1\cdot p_2\cdot p_3\cdot K_c\cdot K_p}{|(j\cdot\omega - p_1)|\cdot|(j\cdot\omega - p_2)|\cdot|(j\cdot\omega - p_3)|}$$

If p_1, p_2 and p_3 are real numbers, their inverses are the time constants T_1, T_2, and T_3 of each first-order system that composes the process. In this case, the process is controlled by a P controller, whose static gain is K_c. The process' static gain (for s = 0) is K_p. The phase angles of each process' section are summed to obtain the controller-process total shift angle.

Formula 11.39 Open-loop phase shift

$$\varphi(\omega) = -\tan^{-1}(\omega\cdot T_1) - \tan^{-1}(\omega\cdot T_2) - \tan^{-1}(\omega\cdot T_3)$$

Charts of Figures 11.14–11.16 graphically assess the same system's absolute stability under four different conditions, achieved by adjusting the controller's

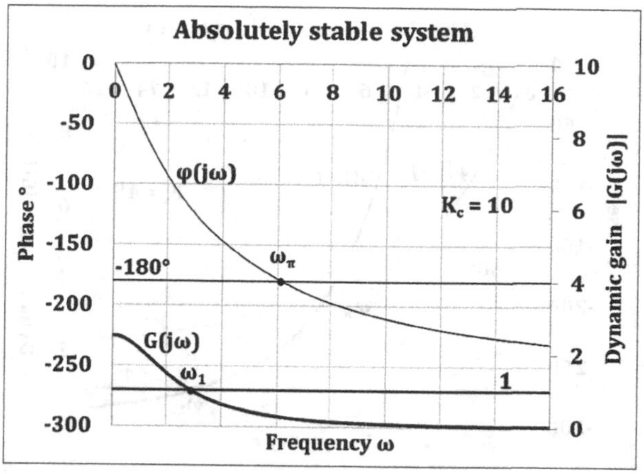

FIGURE 11.14 Dynamic gain and phase of an absolutely stable system.

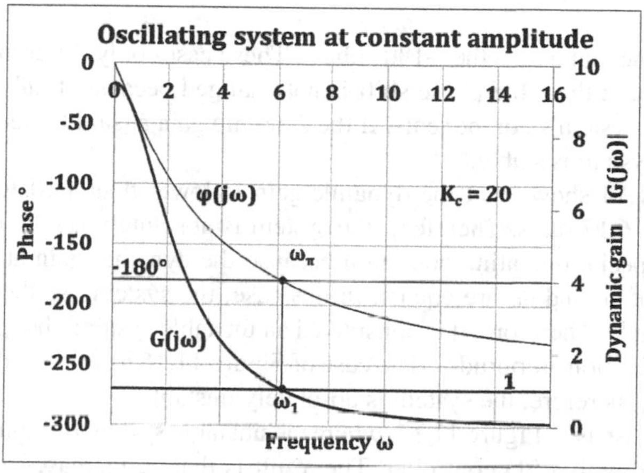

FIGURE 11.15 Dynamic gain and phase of an oscillating system at constant amplitude.

gain K_c. These cases are a) stable with $K_c = 10$, b) oscillating at constant amplitude with $K_c = 20$, c) unstable with $K_c = 45$, and d) unstable compensated with a PD controller of $K_c = 45$. The process constants are: $T_1 = 0.5$ s, $T_2 = 0.25$ s, and $T_3 = 0.2$ s and the process static gain is $K_p = 20$. The four figures show the dynamic gain curve with a thicker line than the phase curve. Two singular points have been marked in each figure: the frequency ω_π at which the phase shift is $-180°$ and the frequency ω_1 of dynamic gain equal to 1. The relative position of these points determines if the system has absolute stability. Dynamic gain values are read on the right vertical axis and the phase shift angles on the left axis.

The dynamic gain curve intersects a horizontal line corresponding to a gain equal to 1 at the frequency ω_1. Similarly, the phase curve intersects a thin

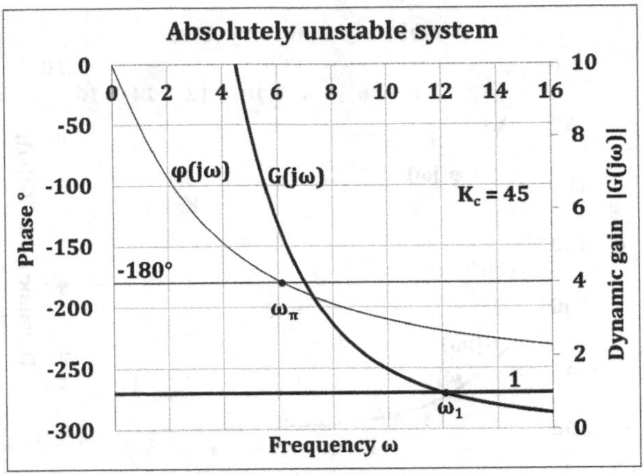

FIGURE 11.16 Dynamic gain and phase of an absolutely unstable system.

horizontal line that marks the −180° phase. These cases only differ in the controller's static gain; then, the phase shift is not changed because it only depends on the dynamic system's components. If the dynamic gain at said frequency is lower than 1, the system is stable.

Figure 11.14 shows that the dynamic gain is lower than 1 (0.30) at the frequency $\omega_\pi = 6.17$ rad/s. Therefore, the system is absolutely stable. Figure 11.15 displays a special operating condition because the dynamic gain at ω_π equals 1. It means that ω_1 and ω_π are equals. In this case, the system oscillates with constant amplitude. Therefore, it is considered an unstable system, though it does not amplify oscillation amplitudes. The case of Figure 11.16 has at ω_π a dynamic gain equal to 5.6; therefore, the system is absolutely unstable.

The last case is in Figure 11.17, where the unstable system of Figure 11.16 was compensated with a PD controller. The result is that ω_π increases to 16.4. rad/s. At this frequency, the dynamic gain is 0.38; then, the system is absolutely stable. The derivative action has compensated the instability and increased the natural frequency. The controller's dynamic gain and phase shift are calculated with the following formulas.

Formula 11.40 Dynamic gain and phase shift of a PD controller

$$G_c(\omega) = \frac{T_d}{T_r} \cdot \frac{\omega}{\sqrt[2]{(1/T_r)^2 + \omega^2}} \quad \varphi_c(\omega) = \tan^{-1}(\omega \cdot T_d) - \tan^{-1}(\omega \cdot T_r)$$

By adding a PD controller, the system's open-loop dynamic gain and phase shift are changed; hence, Formulas 11.40 must be included in Formulas 11.38 and 11.39, as follows:

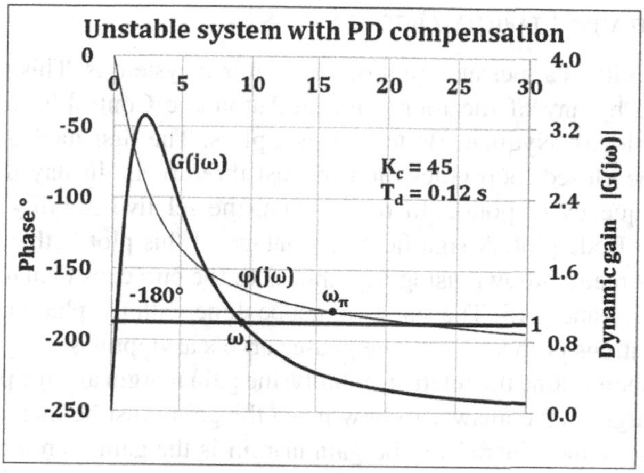

FIGURE 11.17 Unstable system converted to a stable system with a PD controller.

Formula 11.41 Third-order system compensated by a PD controller

$$\left|G(j\cdot\omega)\right| = K_c\cdot K_p\cdot\frac{T_d}{T_r}\cdot\frac{\omega}{\left|(j\cdot\omega-p_1)\right|\cdot\left|(j\cdot\omega-p_2)\right|\cdot\left|(j\cdot\omega-p_2)\right|\cdot\left|\left(j\cdot\omega-\frac{1}{T_r}\right)\right|}$$

$$\begin{aligned}\varphi(j\cdot\omega) = &-\tan^{-1}(\omega\cdot T_1)-\tan^{-1}(\omega\cdot T_2)\\ &-\tan^{-1}(\omega\cdot T_3)+\tan^{-1}(\omega\cdot T_d)-\tan^{-1}(\omega\cdot T_r)\end{aligned}$$

Adding a PD controller has increased the frequency ω_π, which means that the bandwidth is wider; therefore, the system operates faster than a single P controller system. The other important change is that transient overshoots are lower, as seen in Figure 11.10. How was this achieved? Because the derivative action has moved the ω_π frequency to a chart's zone where the dynamic gains are lower than 1. It happened because the derivative term's contribution to the total phase at ω_π is a positive angle equal to $\tan^{-1}(\omega \bullet T_d)$. In summary, a derivative action is a valuable tool in servos controlling mechanisms' position driven by a changing setpoint, where speed and short transients are important, for example, the control of airplane ailerons or computer numeric control tools used in many industries.

11.5.1.1 Numerical Determination of the Absolute Stability

Another important method to determine if a system is absolutely stable is calculating the closed-loop transfer function's poles. As discussed in Section 11.4, complex poles with positive real parts produce an oscillatory response with exponentially increasing amplitudes. Theoretically, these amplitudes reach an infinite value at an infinite time. Therefore, if the poles have a zero or positive real part, the system is unstable.

11.5.2 Relative Stability Determination

Relative stability is a measurement of how stable a system is. This measurement is performed by any of the four classical Automatic Control Theory methods, namely root-locus, Nyquist, Bode or Black plots. The first method graphically calculates the closed-loop poles, and the last three plots display the open-loop system's frequency response. In this section, the relative stability is explained based on the Bode plot. A significant advantage of this plot is that the dynamic gain curve is readily drawn using asymptotes to the curve, as explained in charts of Figures 8.14 and 8.15. The same cannot be done with the phase curve, though some publications propose using the phase curve's asymptotes.

Two numbers define the relative stability: the gain margin and the phase margin. The gain margin is the answer to how much the gain must be increased to make the system unstable. Therefore, the gain margin is the gain increment needed to get unit gain at the ω_π frequency. Its formula is $G_m = 1 - |G(\omega_\pi)|$.

See the gain margin indicated in Figure 11.18. If the gain margin is in dB, its formula is $G_{mDb} = -|G(\omega_\pi)|_{dB}$. If the gain margin in dB is positive, the system is stable; otherwise, it is unstable. Figure 11.18 shows that at $\omega_\pi = 6.3$ rad/s, the dynamic gain is negative: -11.7 dB. Hence, the gain margin is 11.7 dB, and the system is stable. However, if the dynamic gain at 6.3 rad/s is increased by 11.7 dB, the system will start to oscillate with constant amplitude at that frequency. If the dynamic gain increase is higher than 11.7 dB, the system will oscillate with increasing amplitudes.

The system of Figure 11.19 is unstable because the dynamic gain is positive at ω_π, which means that transients are undamped and crescent oscillations. It is also important not to confuse negative dynamic gain with negative gain margin. At frequency ω_π, a negative dynamic gain has a positive gain margin and vice versa.

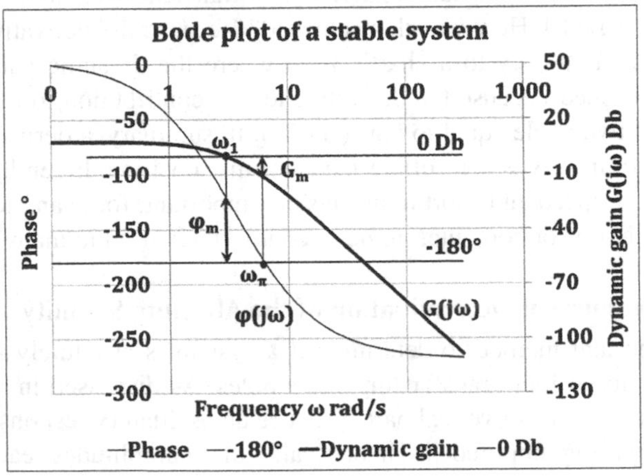

FIGURE 11.18 Gain and phase margins of a stable system.

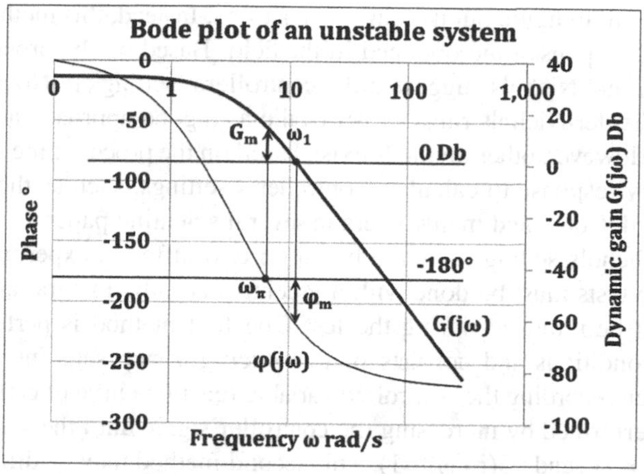

FIGURE 11.19 Gain and phase margins of an unstable system.

The phase margin is conceptually defined by how much the phase must be delayed to unstable the system. The phase margin is the phase change to obtain −18O° phase shift at ω_1 frequency. Therefore, according to this definition, its formula is $\varphi_m = 180 + \varphi(j\omega_1)$. If the phase margin is negative, the system is unstable. See the phase margins indicated in Figures 11.18 and 11.19.

The relative stability is given by the values of the gain and phase margins. Typically, the recommended phase margin is in the 30° to 60° range, and the gain margin is in the 2 to 10 dB range. If the system has low gain and low phase margin values, a derivative action may overcome the problem. As the derivative action has a positive phase shift, it is possible to design a PD controller so that the PD's positive phase reduces the phase shift in the zone of frequencies ω_π and ω_1.

In summary, the Bode plot is a powerful tool to foresee relative stability problems in the design stage or assess the field's optimum controller settings.

11.6 CONTROLLER SETTINGS CALCULATION

There are several procedures to design a control system of an industrial process or device; all aim to determine the best controller type and calculate its settings. As controller settings are an extensive topic in Automatic Control, only one of today's many methods is described in this text. The chosen method is the Ziegler-Nichols, after their inventors. Nonetheless, control engineers must master the many compensation procedures described in the Feedback Control Theory bibliography, where frequency plots and root-locus are intensively used.

11.6.1 Ziegler-Nichols Tuning Methods

The Ziegler-Nichols rules[8] are a heuristic method specially designed for calculating controllers' settings when the mathematical process model is complicated, and it is difficult to figure out its transfer functions. Instead, this method resorts to experimental response time obtained in the field. Based on the process response time, Ziegler and Nichols suggested the controllers' settings in November 1942. Today, the Ziegler-Nichols rules are still valid as a good approach to the controllers tuning. However, other methods exist, based on the process time response and the frequency response to calculate controller's settings, such as those proposed by Rutherford, Coon, and many others in several scientific papers[9].

Ziegler-Nichols setting rules are based on two different experimental procedures. These tests must be done with a P controller only. Integral and derivative actions must be removed during the test. The first method is performed under open-loop conditions and consists of introducing a step setpoint signal to the plant and then recording the controlled variable reaction curve of c(t). The second method is performed by increasing the controller's gain until the system starts to oscillate ($\omega_0 = \omega_\pi$, and $|G(j \bullet \omega)| = 1$). This second method may be difficult to do in some plants because oscillations may damage equipment or products in progress. Process engineers, responsible for quantity and quality of production, are usually reluctant to allow the Ziegler-Nichols second method.

11.6.1.1 Ziegler-Nichols Method Based on the S Reaction Curve

This method applies to processes that react exponentially to the unit step input with an S-shaped reaction curve. This reaction curve is typical of systems with two or more first-degree systems and no second-order systems with a damping ratio lower than 1. There are two parameters measured on the output curve, c(t), used to approximate the process transfer function to a time constant, T_p, and a delay time, T_L (also called dead time or effective lag). During the delay time, the process reaction is zero. See Figure 11.20. These two constants are measured by drawing a tangent line to the c(t) curve in the curve's straight part, where the

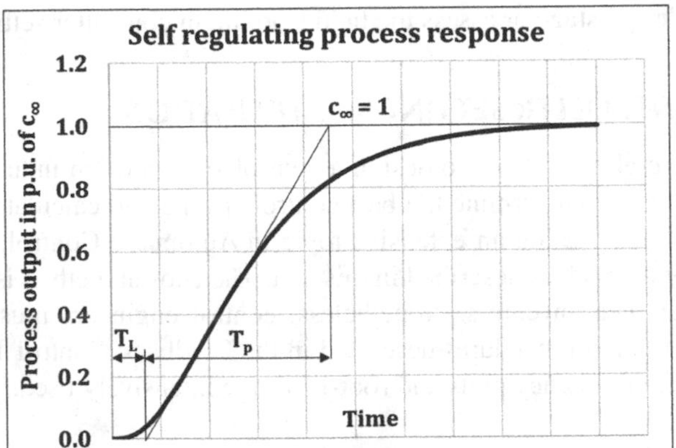

FIGURE 11.20 Process reaction curve T_p and T_L calculation.

inflection point is positioned. Extending this tangent line from the c_∞ ordinate to the time axis, the process's time constant T_p is returned on the chart.

The transfer function of a delay time is $e^{-s \cdot T_L}$. Its frequency response is $e^{-j \cdot \omega \cdot T_L}$. The dynamic gain and phase shift of this expression is derived with Euler's formula, which proves that the dynamic gain is 1 (or 0 dB) and the phase shift is $-\omega \bullet T_L$. Hence, the delay time does not change the system's dynamic gain but contributes to the total phase shift with negative angles, impairing the system's stability.

The process reaction curve of Figure 11.20 suggests that the process is equivalent to a delay time T_L and a first-degree system with time constant T_p, whose transfer function is Formula 11.42.

Formula 11.42 Open-loop transfer function returned by a step input test

$$G_p(s) = \frac{C(s)}{M(s)} = \frac{K_p \cdot e^{-s \cdot T_L}}{1 + s \cdot T_p}$$

This transfer function is easy to obtain in practice by putting the controller in manual condition and introducing a step setpoint with the corresponding controller's knob. Nothing prevents this same operation if the controller is programmed in a PLC (Process Logic Controller) or computer.

Once T_L and T_p are known, Ziegler and Nichols, in their paper of November 1942, suggest the settings shown in Table 11.3, which have proved to be a good approach to the controller settings. If using these settings, the system does not perform satisfactorily; it may be improved by adjusting the controller's static gain K_c. This procedure is heuristic, meaning that it may be enhanced in the field by introducing slight changes to the Ziegler-Nichols settings until the time response is satisfactory.

The PID settings of Table 11.3 are derived for a controller with no restriction in the derivative action. Replacing PID settings in the PID transfer function, the result is:

TABLE 11.3
Ziegler-Nichols Settings Based on Time Reaction Curve

Controller Type	Kc	Ti	Td
P	$\frac{T_p}{T_L}$	∞	0
PI	$0.9 \times \frac{T_p}{T_L}$	$\frac{T_L}{0.3}$	0
PID	$1.2 \text{ to } 2.0 \times \frac{T_p}{T_L}$	$2 \times T_L$	$0.5 \times T_L$

Formula 11.43 PID controller transfer as a function of zeros and poles with Z-N settings

$$G_c(s) = K_c \cdot \left(1 + \frac{1}{s \cdot T_i} + s \cdot T_d\right) = 0.6 \times T_p \cdot \frac{\left(s + \frac{1}{T_L}\right)^2}{s}$$

Therefore, PID controllers adjusted with Ziegler-Nichols settings, with no derivative restriction, place two double zeros on the negative real axis and one pole in the coordinate's origin. The zeros advance the phase, which means moving up the phase curve, shifting it away from −180°. This action stabilizes and speeds up the system. The pole s is the integral action, which produces an infinite module of $G_c(s)$ at zero frequency; therefore, the position error is zero.

Example of Controller's Setting Calculation Based on the S Reaction Curve

In this section, a third-order system controlled by a PID controller, adjusted with Ziegler-Nichols settings, is analyzed. The system transfer function is given by Formula 11.42. Hence the closed-loop performance of a first-degree system with a delay time is assessed below, based on the Bode plot of $|G_c(j\omega)\bullet G_p(j\omega)|$. The open-loop transfer function is:

Formula 11.44 Open-loop transfer function of the controller and process

$$G(s) = G_c(s) \cdot G_p(s) = \left[K_c \cdot \frac{\left(s + \frac{1}{T_L}\right)^2}{s}\right] \times \left[K_p \cdot \frac{e^{-s \cdot T_L}}{s + \frac{1}{T_p}}\right]$$

The brackets in Formula 11.44 only differentiate the controller from the process; hence, they are not mathematically necessary. The Bode plot shown in Figure 11.21 corresponds to the following case:

- Process time constant and delay time measured in the S reaction curve:

$K_p = 1$
$T_p = 10.00$ minutes
$T_L = 0.10$ minutes

- Controller settings calculation:

$T_i = 2 \times 0.10 = 0.2$ minutes
$T_d = 0.5 \times 0.10 = 0.05$ minutes

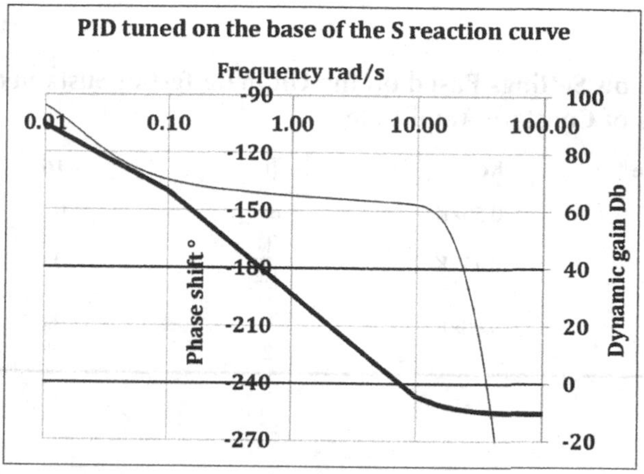

FIGURE 11.21 Bode plot of the system with PID tuned on the base of the S reaction curve.

- The controller's static gain K_c was adjusted, in a spreadsheet model, until the phase margin was 30°. The resultant gain margin of this operation is 9 dB. Both values are within the recommended range.

The phase curve's abrupt lag in the 10-100 rad/s decade is mainly due to the T_L delay time's harmful influence, whose phase shift is directly proportional to the frequency and the dead time. Were it not for this dead time, a process with one zero and two poles has a low phase shift at high frequencies, tending to -90° at infinite frequency. These figures prove that the delay time is a difficult element to control. Some control systems, more sophisticated than PID controllers, have been designed to cope with a delay time process.

11.6.1.2 Ziegler-Nichols Method Based on Ultimate Dynamic Gain and Frequency

In this method, the closed-loop system is forced to oscillate with constant amplitude to measure the static gain and frequency under such conditions. This oscillation is the condition indicated as case C in Figure 11.5 of elemental time responses in the s plane. Both static gain and frequency are called the ultimate gain K_u and the ultimate period of oscillation T_u, respectively. This condition is achieved by controlling the system with a P controller and adjusting its static gain until the system has sustained oscillations. If the controller has integral and derivative actions, they must be disabled before performing the test. The static gain K_u must produce a dynamic gain of 1 at the frequency ω_π, so the system becomes oscillatory with constant amplitudes.

TABLE 11.4
Ziegler-Nichols Settings Based on the Ultimate Test of Sustained Oscillations of Constant Amplitude

Controller Type	Kc	Ti	Td
P	$0.50 \times K_u$	∞	0
PI	$0.45 \times K_u$	$\frac{T_u}{1.2}$	0
PID	$0.60 \times K_u$	$\frac{T_u}{2}$	$\frac{T_u}{8}$

Based on the ultimate test results, Ziegler and Nichols proposed the setting formulas shown in Table 11.4 as a function of K_u and T_u. Therefore, by replacing in the controller's transfer function the values of the said table, the controller adjusted with Ziegler-Nichols settings is returned.

For example, for a PID controller, the replacement of Ziegler-Nichols settings in the transfer function 11.36 returns the following expression:

$$G_{PID}(s) = K_c \cdot T_d \cdot \frac{(s+z_1)\cdot(s+z_2)}{s} = 0.075 \times K_u \cdot T_u \cdot \frac{\left(s+\frac{4}{T_u}\right)^2}{s}$$

Where:

Formula 11.45 PID controller with Ziegler-Nichols ultimate settings

$$K_c \cdot T_d = 0.075 \times K_u \cdot T_u \; z_1 = z_2 = -\frac{4}{T_u}$$

The expected time response curve of systems tuned with any of the Ziegler-Nichols methods is an oscillatory transient with a ¼ decay ratio when the system is excited by a disturbance. If the system has a second-order dominant mode, the damping ratio corresponding to the ¼ decay ratio is 0.23. However, it is now preferred to achieve a higher damping ratio close to 0.707 because it produces the minimum settling time with a 4% peak.

Both K_u and T_u result from the ultimate test; however, if the process transfer function is known, the following formulas may be used instead of the test.

Formula 11.46 Ultimate static gain and period

$$K_u = \frac{1}{\left|G_p(j\cdot\omega_\pi)\right|} \; T_u = \frac{2\cdot\pi}{\omega_\pi}$$

The transfer function $G_p(j \bullet \omega)$ can be known by the S reaction test described in the last section, which is usually easier to perform than the ultimate test.

Example of Controller's Setting Calculation Based on Ultimate Gain and Period

Consider an example of a system with a PID controller. The system's frequency response with a PID controller, adjusted with the ultimate gain and oscillation period, is given by Formula 11.47.

Formula 11.47 Open-loop transfer function of controller and process

$$G(j\cdot\omega) = \left[0.075\times K_u \cdot T_u \cdot \frac{\left(j\cdot\omega + \frac{4}{T_u}\right)^2}{j\cdot\omega}\right] \times \left[K_p \cdot \frac{e^{-j\cdot\omega\cdot T_L}}{j\cdot\omega + \frac{1}{T_p}}\right]$$

The process static gain is $K_p = 1$. The first-order system's time constant is $T_p = 10$ minutes, and the dead time is $T_L = 0.1$ minutes. The Ziegler-Nichols oscillation test returned the Bode chart in Figure 11.22, showing the system's ultimate conditions.

This plot is easily identified because $\omega_\pi = \omega_1$. The K_u and T_u calculation is based on the ultimate frequency test results.

$$\omega_\pi = \omega_1 = 15.77\,\text{rad/min}$$

$$T_u = \frac{2\cdot\pi}{\omega_\pi} = 0.4\,\text{min}$$

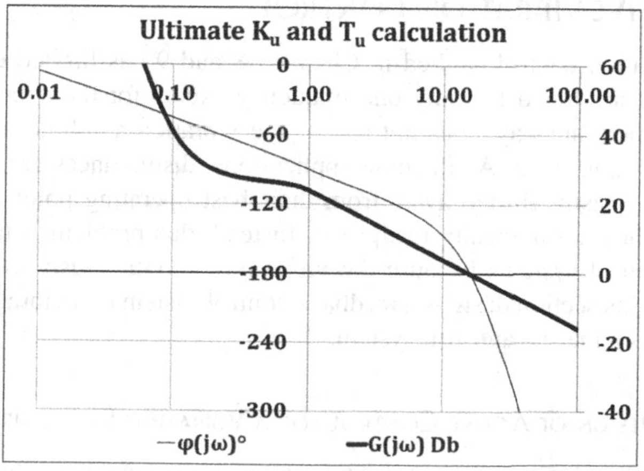

FIGURE 11.22 Ultimate test to measure K_u and T_u.

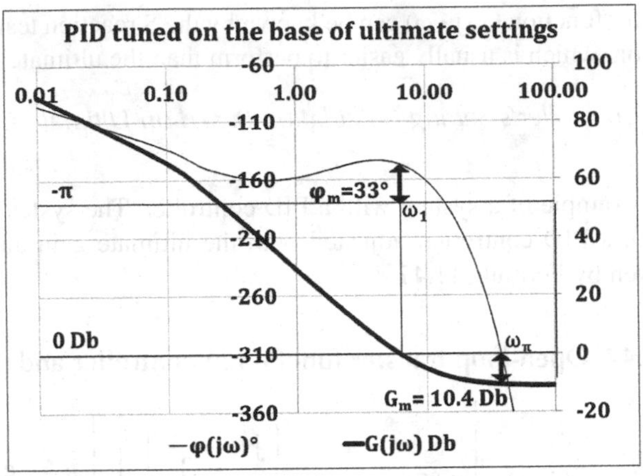

FIGURE 11.23 Bode plot of a system with a Ziegler-Nichols PID compensation.

$$\left|G(j\cdot\omega_\pi)\right| = 0.063$$

$$Ku = \frac{1}{0.063} = 15.85$$

After the PID settings are calculated, the system's gain and phase margins that they produce must be measured in a Bode plot to determine if they are acceptable. See chart in Figure 11.23.

The PID's results tuned with Ziegler-Nichols settings returned a gain margin of 10.4 dB and a phase margin of 33°, which are considered satisfactory. Some specialists recommend 8 db[10] minimum gain margin and others 6 dB.

11.7 ACTIVE VIBRATION CONTROL

Passive vibration control studied in Chapters 8 and 9 has limited effectiveness because they are tuned for only one frequency except for the damped Frahm's absorber, which, anyway, does not reduce the vibration amplitudes to zero discussed in Section 9.3.5. As in some applications, disturbances are random; the absorbers are easily shifted away from their best operating point and are useless unless they are manually readjusted. Instead, this problem is tackled using feedback control systems because the vibration is continuously monitored and controlled. This section designs a feedback control system of a vibrating structure and describes a marine anti-roll system.

11.7.1 Design of Active Control for a Vibrating Structure

The example used in this section is a two stores base-excited structure, displayed in Figure 11.24. It corresponds to the fundamental vibration form described in Section 1.11.3.

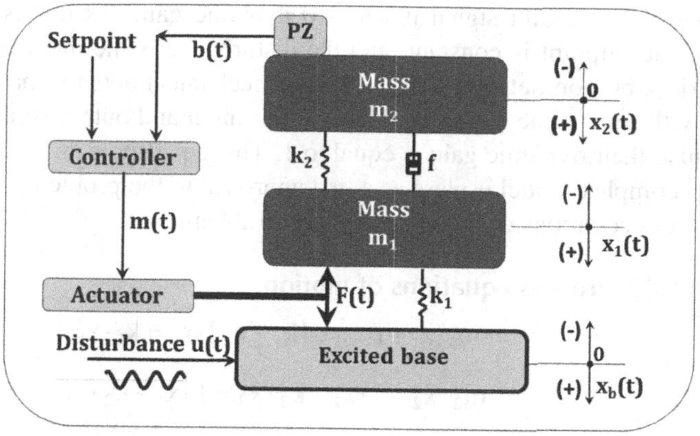

FIGURE 11.24 Vibration active control of the base-excited structure.

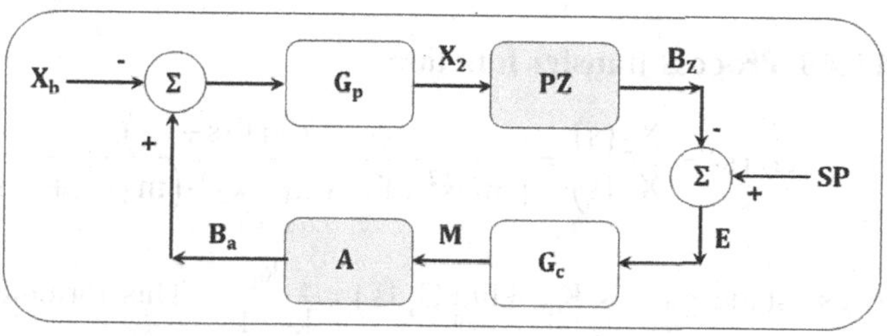

FIGURE 11.25 Block diagram of the structure and control system.

The block diagram of the structure and the vibration control system is in Figure 11.25. It is a system with two degrees of freedom: $x_1(t)$ and $x_2(t)$. The model to be developed was conceptually described in Figure 11.4. The transfer functions of this system are two, one for the setpoint input and the other for the disturbance input. See Formulas 11.6.

The excitation is harmonic and produced by a close machine in the 0.2 to 0.8 rad/s frequency range. This vibration is not admissible for the structure; hence it is requested to reduce its amplitude at least by 90% in the frequency range mentioned before. The total mass of the machine and the second-floor structure has a mass m_2. The first floor supports the load of mass m_1. There is a spring-damper set k_2-f between the first and second floor. The first floor is linked to the base by a spring k_1. The base is non-rigid and is excited by a harmonic disturbance force u(t); therefore, the foundation vibrates with amplitude x_b and transmits vibration to the upper floors. The variable to control is the vibration amplitude of the second floor. A piezoelectric device PZ measures this vibration amplitude. The outgoing signal from PZ is sent to the controller. The controller's manipulated variable m(t) is introduced into the actuator, creating an alternate force F(t), opposed to the disturbance force u(t). The actuator is an electromagnetic device, and its force is equivalent to a spring of adjustable stiffness in parallel with the spring k_1.

The controller setpoint signal is set at 0 dynamic gain. As this is a regulator system, the setpoint is constant, and the disturbance is the input signal. The piezoelectric vibration detector and the electromechanical actuator have transfer functions with small-time constants. Hence, their input and output signals are the same because their dynamic gain is equal to 1. This equality is an assumption to avoid a too complex model useless as a first approach to the problem.

The process equations of motion of Figure 11.24 are:

Formula 11.48 Process equations of motion

$$m_1 \cdot \ddot{x}_1 + f \cdot \dot{x}_1 + (k_1 + k_2) \cdot x_1 = k_1 \cdot x_b$$

$$m_2 \cdot \ddot{x}_2 + f \cdot \dot{x}_2 + k_2 \cdot x_2 = f \cdot \dot{x}_1 + k_2 \cdot x_1$$

Taking the Laplace transform of the above equations and assuming no initial conditions, the structure open-loop transfer function is returned and shown in Formula 11.49.

Formula 11.49 Process transfer function

$$G_p(s) = \frac{X_2(s)}{X_b(s)} = \frac{k_1 \cdot (f \cdot s + k_2)}{(m_1 \cdot s^2 + f \cdot s + k_1 + k_2) \cdot (m_2 \cdot s^2 + f \cdot s + k_2)}$$

The process static gain is $K_p = \lim_{s \to 0} G_p(s) = \frac{k_1}{k_1 + k_2}$. This formula indicates that in a permanent regime, the vibration amplitude x_2 is lower than the base exciting amplitude x_b, which, anyway, is not admissible.

11.7.1.1 Converting the Transfer Function to Obtain the Frequency Response

The denominator of the transfer function 11.49 can be converted into a polynomial of fourth-order:

Formula 11.50 Denominator of $G_p(s)$

$$D_p(s) = s^4 + a_3 \cdot s^3 + a_2 \cdot s^2 + a_1 \cdot s + a_0$$

Coefficients of this polynomial are:

Formula 11.51 Coefficients of the polynomial $D_p(s)$

$$a_3 = \frac{m_1 + m_2}{m_1 \cdot m_2} \cdot f \quad a_2 = \frac{f^2 + m_1 \cdot k_2 + (k_2 + k_2) \cdot m_2 + f \cdot (k_1 + 2 \cdot k_2)}{m_1 \cdot m_2}$$

$$a_1 = \frac{f \cdot (k_1 + 2 \cdot k_2)}{m_1 \cdot m_2} \quad a_0 = \frac{k_2 \cdot (k_1 + k_2)}{m_1 \cdot m_2}$$

The process has little friction forces, except for the damper under the second floor. As the structure harmonically vibrates, it is inferred that the polynomial

$D_p(s)$ is formed by two second-order systems of $\zeta<1$. Therefore, the structure's transfer function is:

Formula 11.52 Process transfer function formed by two second-order systems

$$G_p(s)=\frac{\frac{k_1}{k_1+k_2}\cdot\frac{\omega_{n1}^2\cdot\omega_{n2}^2}{z_p}\cdot(s+z_p)}{\left(s^2+2\cdot\zeta_1\cdot\omega_{n1}\cdot s+\omega_{n1}^2\right)\cdot\left(s^2+2\cdot\zeta_2\cdot\omega_{n2}\cdot s+\omega_{n2}^2\right)}$$

This expression meets the condition seen before that $K_p=\lim_{s\to 0}G_p(s)=\frac{k_1}{k_1+k_2}$.

The zero $z_p = k_2/f$ is the inverse of the spring-damper set time constant, installed between m_1 and m_2. This set is a first-order system. The zero z_p is the only zero that the transfer function has; as was seen before, this zero improves the system stability because its positive phase from $\omega = 0$ to ∞ is opposed to the denominator's negative phases.

Coefficients of Formula 11.50 are given by Formulas 11.51 as a function of the structure's physical parameters: inertia, stiffness, and damping. The two-term product of the denominator of Formula 11.52 produces the following set of equations, where coefficients a_0 to a_3 are known. Therefore, ζ_1, ω_{n1}, ζ_2 and ω_{n2} are cleared from this set of equations:

Formula 11.53 Equations of ζ_1, ω_{n1}, ζ_2 and ω_{n2}

$$a_0=\omega_{n1}^2\cdot\omega_{n2}^2 \qquad a_1=2\cdot\omega_{n1}\cdot\omega_{n2}\cdot\left(\zeta_1\cdot\omega_{n2}+\zeta_2\cdot\omega_{n1}\right)$$

$$a_2=\omega_{n1}^2+4\cdot\zeta_1\cdot\omega_{n1}\cdot\zeta_2\cdot\omega_{n2}+\omega_{n2}^2 \qquad a_3=2\cdot\left(\zeta_1\cdot\omega_{n1}+\zeta_2\cdot\omega_{n2}\right)$$

After these four properties are cleared from the set of Formulas 11.53, they are used in the transfer function 11.54 and frequency response 11.55.

11.7.1.2 Frequency Response

As theory dictates, the frequency response is obtained by replacing the complex frequency s by $j\cdot\omega$ in transfer function 11.52. This non-dimensional frequency response formula must be plotted in the Bode diagram to verify the system's characteristics needed to design the controller algorithm.

Formula 11.54 Process frequency response

$$G_p(j\cdot\omega)=\frac{X_2(j\cdot\omega)}{X_b(j\cdot\omega)}$$

$$=\frac{\frac{k_1}{k_1+k_2}\cdot\frac{\omega_{n1}^2\cdot\omega_{n2}^2}{z_p}\cdot(j\cdot\omega+z_p)}{\left(-\omega^2+j\cdot 2\cdot\zeta_1\cdot\omega_{n1}\cdot\omega+\omega_{n1}^2\right)\cdot\left(-\omega^2+j\cdot 2\cdot\zeta_2\cdot\omega_{n2}\cdot\omega+\omega_{n2}^2\right)}$$

The terms between the denominator's parenthesis are designated $D_1(j\cdot\omega)$ and $D_2(j\cdot\omega)$ to simplify the following expressions, so, the process' module in dB and phase of the frequency response are:

Formula 11.55 Module and phase of the process frequency response

$$\left|G_p(j\cdot\omega)\right|_{dB} = 20\times\left[\log\left(\frac{k_1}{k_1+k_2}\cdot\frac{\omega_{n1}^2\cdot\omega_{n2}^2}{z_p}\right) + \log\left|(j\cdot\omega+z_p)\right| - \log\left|D_1(j\cdot\omega)\right| - \log\left|D_2(j\cdot\omega)\right|\right]$$

$$\varphi(\omega) = \tan^{-1}\left(\frac{\omega}{z_p}\right) - \tan^{-1}\left[\frac{2\cdot\zeta_1\cdot\frac{\omega}{\omega_{n1}}}{1-\left(\frac{\omega}{\omega_{n1}}\right)^2}\right] - \tan^{-1}\left[\frac{2\cdot\zeta_1\cdot\frac{\omega}{\omega_{n2}}}{1-\left(\frac{\omega}{\omega_{n2}}\right)^2}\right]$$

Having derived the structure frequency, it is possible now to design the controller and verify if the second floor's vibration amplitude is reduced to at least 90% as requested.

11.7.1.3 Controller's Design

The chart in Figure 11.26 shows the Bode plot of the base-excited structure of fourth-order. The parameters used in this example are:

$$z_p = 0.1\,rad/s \quad \zeta_1 = 0.01 \quad \omega_{n1} = 0.5\,rad/s \quad \zeta_2 = 5$$
$$\omega_{n2} = 1.0\,rad/s \quad k_1 = 3.0\,t/mm \quad k_2 = 1.0\,t/mm$$

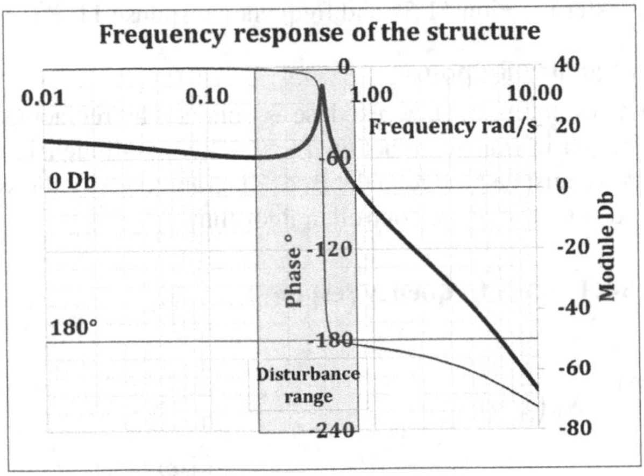

FIGURE 11.26 Bode plot of the structure. Open-loop.

The spring-damper set between the first and second floor is a first-order system with a time constant of 10 s. Therefore, as the zero z_p is the inverse of this time constant it is $z_p = 0.1$ rad/s.

The structure base is excited by an external source in the 0.2 to 0.8 rad/s range. The structure's frequency response is shown in chart 11.26. The disturbance range is shown with a shadowed area in gray color. In this range, the amplitude curve is initially flat at 11 dB. It means that in the flat zone, the excitation input amplitude X_b is 3.6 times amplified on the second floor. Inside the disturbance range, there is a resonant peak of 33.8 dB, equivalent to an amplification factor of 50 times on the second floor. Thus, 1 mm amplitude of X_b in the base produces 50 mm amplitude on the second floor. Hence, installing an active system to reduce these amplitudes by at least 90% is requested. The structure experiences a sudden delay of the phase angle at the frequency of the amplitude peak produced by the structure's inertia and rigidity and the damper on the second floor. After this frequency, the phase curve falls at a much lower rate, below the −180°, suggesting that the PID must advance the phase to prevent resonance problems at frequencies beyond 1 rad/s.

It is proposed to install a PID controller to overcome the amplification in the disturbance zone and assure stability at frequencies higher than 1 rad/s.

After the frequency response curves were simulated for different settings, the following parameters were chosen for the PID controller.

$$K_c = 12.0$$
$$T_i = 25\,\text{s}$$
$$T_d = 6.25\,\text{s}$$

With these settings, the controller's frequency response is depicted in the chart in Figure 11.27.

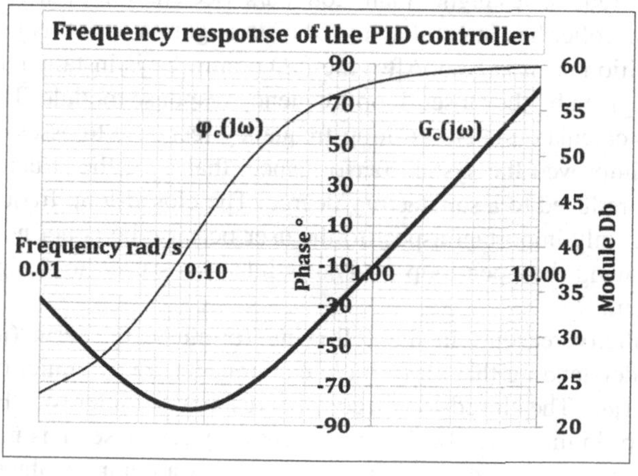

FIGURE 11.27 Frequency response of the PID controller.

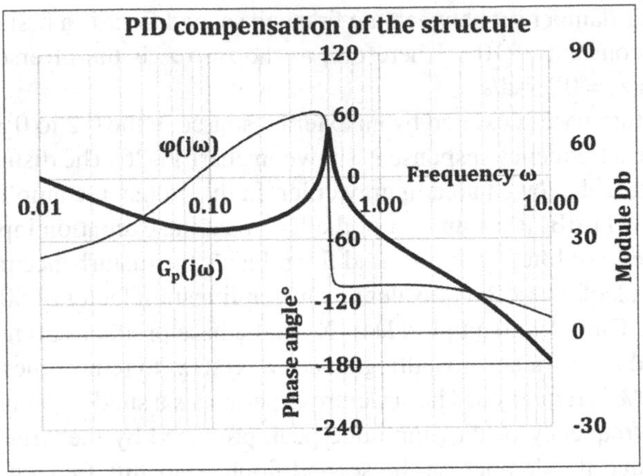

FIGURE 11.28 Open-loop frequency response of the structure with a PID controller.

The result of these settings produces the open-loop frequency response of Figure 11.28. The controller's phase angle prevents instability because this angle does not reach −180°. Thus, it is impossible to measure the gain margin because the phase curve does not cross the −180° line. Conversely, the phase margin can be measured at a frequency of 7 rad/s, where the amplitude is 0 dB. The result is a 58° phase margin. Therefore, the system stability is assured. At low frequencies, the module grows until it is infinite at zero frequency; therefore, the accuracy for a step input is also assured.

If chart 11.28 is compared with chart 11.26, it is noted that the phase curve has no values under the −180° line, which means that the system cannot be unstable. This stabilizing action is due to the phase advance of the PID, whose phase curve is positive at frequencies higher than 0.08 rad/s. See chart in Figure 11.27.

As the controller has integral action, the static gain at $\omega = 0$ is infinite; therefore, the position error is zero. After the PID controller is installed, the vibration amplitude is given by the closed-loop frequency response module. Therefore, it is necessary to calculate the closed-loop frequency response to assess if the active control has improved the system performance, that is, if the structure vibration amplitude is reduced to a satisfactory degree. The closed-loop frequency performance is the only important aspect to the user because he is not normally interested in knowing the open-loop analysis's intricacies used to design the active control system.

For simplicity reasons, in the following formulas, the term (jω) has been dropped, understanding that all formula components are dynamic; thus, they are functions of (jω). The closed-loop frequency response of interest is $H_u = X_2/X_b$ of Formula 11.6. In this case, the feedback frequency response G_f is not considered because it has negligible dynamic properties. H_u is a complex algebraic expression; therefore, it is a vector formula obtained from the open-loop response. The

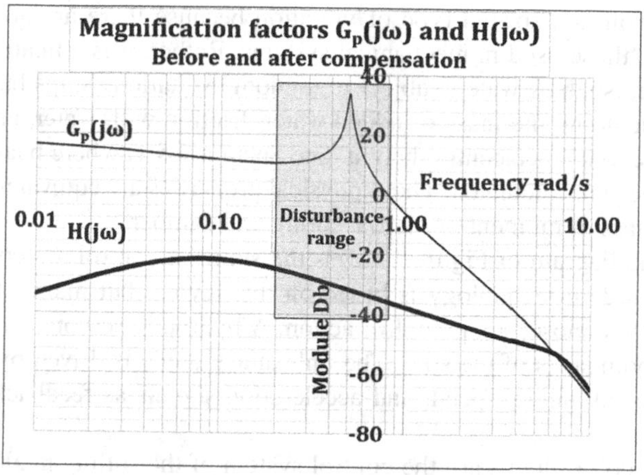

FIGURE 11.29 Closed-loop frequency response with a PID controller.

vector $\overline{H}_u$ is the ratio of two vectors: $\overline{G}_p$ and $(\overline{1}+\overline{G}_c\cdot\overline{G}_p)$. The formula to calculate the H_u modulus is derived from simple trigonometric considerations on these vectors diagram, which returns Formula 11.56.

Formula 11.56 Closed-loop modulus based on the open-loop modulus and phase

$$|H(j\omega)| = \frac{|G(j\omega)|}{1+|G(j\omega)|^2+2\cdot|G(j\omega)|\cdot\cos\varphi((j\omega))}$$

The chart in Figure 11.29 shows the closed-loop frequency response $|H_u|$ along with the structure's frequency response $|G_p|$ to visualize the benefits of the active system with respect to the freely vibrating structure. It shows that in the excitation frequency range (0.2 to 0.8 rad/s), the magnification factor has an average value of -30 dB equivalent to an attenuation factor of 0.03. It signifies that for 1 mm amplitude at the base, the second-floor vibration amplitude is 97% lower. Thus, vibration amplitudes are reduced as requested by the structure's owner.

11.7.2 Absorption of Ship's Roll

The ship's movements are excited by water waves. Since they are alternating movements, rolling, pitching, and others, they can be studied with some concepts and formulas of the vibration theory as demonstrated in Chapter 7. Several systems have been developed to mitigate the ship's oscillation, discussed in the same chapter. In this section, the application of automatic control to retractable fins is described.

A ship's roll is a special type of vibration because there are no springs and dampers like those used in mechanical systems. Rather, it is a floating body in a fluid medium, such as water, subject to random excitation. Ships have moments of inertia, including the mass of added water, friction with water, and the ship's reactions due to its metacentric height. See Section 7.5.1, where beam-ship natural frequencies are calculated. These physical properties are equivalent to systems formed by mass or moment of inertia, spring and damper.

The block diagram of Figure 11.30 is the simplest control system of the rolling angle. Modern technology is based on this layout, but many other features complement the closed-loop control action. A hydraulic servomechanism drives the fins on both sides of the ship. The fin's attack angle is driven by a controller that uses the roll angle, speed, and acceleration signals as feedback to the fin's controller.

The grey color blocks are the control system of the rolling angle. Inside this system, another feedback system, shown in black blocks, assures the fin is at the correct attack angle inside the water (see Section 7.3.4). This secondary loop is a hydraulic servomechanism that follows the command signal given by the roll controller. It means that the roll controller's output is the setpoint for the hydraulic servomechanism. The roll controller is usually a PDD. It has two derivative actions in cascade. Therefore, the controller has an action that depends not only on the error velocity but also on its acceleration. The feedback signal is taken from gyroscopes and accelerometers that detect rolling conditions for the controller.

The simplest model of ship rolling is the equation of motion of a rotating second-order system. Of course, CFD science and its associated software have developed more sophisticated and exact models to design the optimum hull shape for each case.

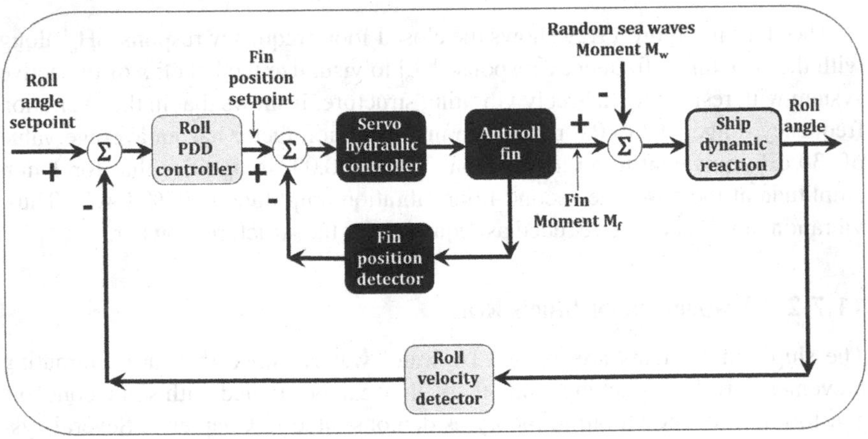

FIGURE 11.30 Control system of ship roll.

Formula 11.57 Simple equation of motion of a rolling ship

$$(J_s + J_a)\cdot\ddot{\vartheta} + f\cdot\dot{\vartheta} + k\cdot\vartheta = M_w(t) + M_f(t)$$

J_s is the ship's moment of inertia, and J_w is the added water moment of inertia. See Section 7.5.1.

f is the friction coefficient of the water on the hull plus the air with the superstructure. In general, this friction is small, and in some calculations, it is negligible.

k is the product Δ·GM, equivalent to a rotating system's rigidity. See Section 7.3.3.

M_w is the random moment produced by the water.

M_f is the equilibrating moment created by the fin.

The ship's oscillation absorption is produced by the equilibrating moment of two fins: one on the ship's port side and the other on the starboard side. The following formula gives this moment:

Formula 11.58 Equilibrating moment produced by the fin[11]

$$M_f = (\rho_w \cdot V^2 \cdot c_l)\cdot A_f \cdot d_f$$

The term between the parenthesis of Formula 11.58 is the water pressure under the fin. A_f is the fins' area, and d_f is the fins' moment arm. The water pressure depends on the product of the water density ρ_w times the square of the fin's relative velocity V with respect to the flow velocity and the lifting coefficient c_l that depends on the fin angle α between flow direction fin and the fin's geometric axis.

Formula 11.59 Lifting coefficient formula

$$c_l(\alpha) = \left|\frac{\partial c_l(\alpha)}{\partial \alpha}\right|_{\alpha=0} \cdot \alpha$$

The active fin stabilizer is one of the most efficient stabilizer systems. Its percentage roll absorption is 90%, a good response compared with passive anti-roll tanks, whose efficiency is 60% to 70%. However, it is not effective at speeds slower than 6 knots. Of course, this is not a problem because the cruiser velocity of modern ships is much higher than 6 knots. It has the advantage of needing low power, but the disadvantage is that it may be damaged during a rough sea operation. Due to its mechanical size and the sophisticated control systems, its initial cost is high[12]. In summary: active control is the most effective roll stabilizer; however, its modelling is difficult because it deals with random disturbance and nonlinearities.

BIBLIOGRAPHY ABOUT FEEDBACK CONTROL SYSTEMS

Books

1. Analysis and Design of Feedback Control Systems. Thaler and Brown.
2. Control Systems. Analysis, Design and Simulation. John W. Brewer.
3. Feedback and Control Systems. Distefano III, Stubberud, Williams.
4. Théorie et Calcul des Asservissements Linéaires. J. Gille, P. Decaulne, M. Pelegrin.
5. Linear Control Systems. John J. D'Azzo and Constantine H. Houpis.
6. Automatic Control Systems. Benjamin Kuo.
7. Feedback Control Systems. Alex Abramovici and Jake Chapsky.
8. Feedback Control of Dynamic Systems. Gene Franklin, J. Powell, et al.
9. Theory and Problems of Laplace Transforms. Murray Spiegel.
10. Model Predictive Control. Eduardo Camacho and Carlos Bordons.
11. Process Control Systems. F.G. Shinskey.
12. State Space and Linear Systems. D. Wiber.
13. Process Dynamics and Control. D. Seborg, T. Edgar and D. Mellichamp.

Classical Papers Based on the Frequency Response

1. Optimum Settings for Automatic Controllers. J. Ziegler and N. Nichols.
2. The Calculation of Process Control Settings from Frequency Characteristics. N. Ream.
3. How to Find Controller Settings from Process Characteristics. G. Coon.
4. How to Reckon Basic Process Dynamics. Leslie Zoss.
5. Non-overshooting PD and PID controllers Design. M. Tabatalbael and R. Barati-Boldaji.
6. The Regeneration Theory. H. Nyquist.
7. The Frequency Response Method. R.H. Macmillan.
8. Frequency Response Analysis for Industrial Automatic Control Systems. D. St Clair, W. Coombs and W. Owen.
9. The Practical Application of Frequency Response Analysis to Automatic Process Control. C. Rutherford.
10. Frequency Response Analysis and Controllability of a Chemical Plant. A. Aikman.
11. Frequency Response Analysis of Continuous Flow Systems. H. Kramers and G. Alberda.
12. A Method of Estimating Dynamic Characteristics of Physical Systems. Sidney Lee.
13. Control System Behavior Expressed as a Deviation Ratio. J. Janssen.
14. The use of zeros and Poles for Frequency Response or Transient Response. W. Evans.
15. Determination of Transient Response from Frequency Response. A. Leonhard.

16. A Uniform Approach for the Optimum Adjustment of Control Loops.
17. Frequency Response Data Presentation, Standards and Design Criteria. R. Oldenburger.
18. Optimum Settings for Automatic Controllers. J. Ziegler and N. Nichols.
19. The Calculation of Process Control Settings from Frequency Characteristics. N. Ream.
20. How to Find Controller Settings from Process Characteristics. G. Coon.
21. El Control Óptimo Frente al Control Convencional en Sistemas Típicos de Industria de Procesos. V. Koppel.

NOTES

1 Recommended books for Chapter 11: 1. Automatic Control Systems by Benjamin Kuo. Wiley & Sons, Incorporated, 1999 – 2. Modern Control Engineering by Katsuhiko Ogata. Pearson Education Inc, 5th edition 2015 – 3. Feedback and Control Systems by Joseph J. Distefano III, Allen R. Stubberud, Ivan J. Williams. McGraw Hill. Schaum's Outline Series. 1967. – 4. Digital Control of Dynamic Systems by Gene F. Franklin, J. David Powell and Michael L. Worman. Addison-Wesly Publishing Company, 1990. – 5. Analysis and Design of Feedback Control Systems by Thaler and Brown. McGraw Hill Book Co. Inc. 2nd edition. – 6. Feedback Control Systems by John J, D'Azzo and Constantine H. Houpis McGraw Hill Inc. 1965. – 7. Control Systems: Analysis, Design and SimulationBy John W. Brewer. Prentice-Hall; 1st Edition (January 1, 1974) – 8. Théorie et Calcul des Asservissements Linéaires , by Jean Charles. Gille, Paul Decaulne, and Marc Pelegrin. Dunod GF, 1967

2 See the paper An Overview of Roll Stabilizers and Systems for Their Control K.S. Kula Gdynia Maritime University, Poland in https://www.researchgate.net/publication/ of TransNav Journal.

3 There exists an extensive bibliography about Laplace transform in Calculus and Feedback Control Theory books. The reader is referred to them for consulting purposes during the study of this chapter. This text assumes that the reader knows how to apply the Laplace transform to solve differential equations.

4 The term dynamic refers to a system's attribute that depends on the frequency.

5 In this text the terms plant and process are used indistinctively.

6 Closed-loop system and feedback system must be considered synonymous.

7 Heaviside's formula of residues: $A_i = \lim_{s \to -p_{ci}} \left[(s + p_{ci}) \cdot C(s) \right]$. C(s) must be under the form of Formula 11.18.

For A_0 the pole is $p_{ci} = 0$.

8 It is recommended to consult:

1. The Handbook of PI and PID Controller Tuning Rules, 2nd Edition, Aidan O'Dwyer Imperial College Press.
   K.J. Åström and T. Hägglund – 2. PID Controllers: Theory, Design, and Tuning, 2nd Edition, Instrument Society of America, Research Triangle Park, North Carolina, 1995. [2] C.-C. Yu, - 3. Autotuning of PID Controllers, Springer-Verlag, New York, 1999.

9 Some influential classic papers about Automatic Control studies

1. The Regeneration Theory. By Harry Nyquist. Bell Systems Technical Journal, volume 11, pp 126-147, 1937

2. The Practical Application of Frequency Response Analysis to Automatic Process Control by Rutherford. IEEE Convention May 1949
3. The Use of Zeros and Poles for Frequency Response or Transient Response. By Walter R. Evans. ASME Headquarters, May 11 1953, Paper No, 53 -A-24
4. How to Find Controller Settings from Process Characteristics. By Geraldine A. Coon from Taylor Instrument Co. Magazine Control Engineering.
5. Optimum Settings for Automatic Controllers. By J. G. Ziegler and N. B. Nichols. Presented ASME Annual Meeting Dec 1 to 5, 1941. New York.

10 Consult the paper Frequency Response Data Presentation, Standards and Design Criteria. By Rufus Oldenburger. Presented at the ASME Annual Meeting New York, 4th of December 1953.

11 See the following papers:

1) Application of particle swarm optimized PDD2 control for ship roll motion with active fins by Melek Ertogan, Seniz Ertugrul and Metin Taylan. Article in IEEE/ASME Transactions on Mechatronics · January 2015. https://www.researchgate.net/publication/283188230
2) An Overview of Roll Stabilizers and Systems for Their Control K.S. Kula Gdynia Maritime University, Poland.
3) Fin based active control for ship roll motion stabilization Neha Sunil Patil, Awanish Chandra Dubey, and V. Anantha Subramanian Department of Ocean Engineering, Indian Institute of Technology Madras, Chennai-600036, India. https://doi.org/10.1051/matecconf/201927201026

12 Consult the comparison table of several roll stabilizers in: https://www.marineinsight.com/naval-architecture/roll-stabilization-systems/

Index

W

Z

Printed in the United States
by Baker & Taylor Publisher Services

Printed in the United States
by Baker & Taylor Publisher Services